Ecological Economic and Socio Ecological Strategies for Forest Conservation

Felix Fuders • Pablo J. Donoso
Editors

Ecological Economic and Socio Ecological Strategies for Forest Conservation

A Transdisciplinary Approach Focused on Chile and Brazil

Springer

Editors
Felix Fuders
Economics Institute, Faculty of Economics
and Administration
Universidad Austral de Chile
Valdivia, Chile

Pablo J. Donoso
Institute of Forests and Society
Universidad Austral de Chile
Valdivia, Chile

ISBN 978-3-030-35378-0 ISBN 978-3-030-35379-7 (eBook)
https://doi.org/10.1007/978-3-030-35379-7

This Springer imprint is published by the registered company Springer Nature Switzerland AG.
The registered company address is: Gewerbestrasse 11, 6330 Cham, Switzerland

Foreword

This Foreword was to have been written by Manfred Max-Neef, who, as a pioneer in ecological economics and as rector of the Universidad Austral de Chile, made large intellectual and institutional contributions to the works collected in this volume. His untimely death prevented this, but the book nevertheless is one of his legacies to us and bears the marks of his DNA in the writings of those whom he has influenced.

Along with the authors in this collection, I count myself as one who has benefited from Manfred's intellect and generosity. Just one example, after the Pinochet dictatorship, there was a conference in Santiago that Manfred organized to consider the full range of various economic systems that Chile might adopt for the future. At one end of the spectrum was ecological economics, and Manfred invited me to speak on that subject. At the time, I was employed by the World Bank which felt it had to approve all speeches by employees and declined to clear mine. I told Manfred, expressing regrets. Manfred then wrote to the World Bank Chile Department stating that the World Bank's economic views would be presented by another participant, not by me. I was invited to speak in my independent professional capacity on the subject of ecological economics, and it was just a coincidence that I was currently employed by the World Bank. After 17 years of dictatorship, Chile was eager for free speech on economics, and he doubted that the World Bank really intended to restrict such speech and was confident that they would upon reconsideration grant permission. However, if for some reason that should not be possible, then he kindly requested a letter explaining the reasons, so that he might publicly read it at the conference to explain my absence. I was granted permission. Manfred was persuasive!

This ability to persuade carries over to the subject of ecological and social economics and is continued in the convincing studies of his younger colleagues in this collection. General principles of ecological and social economics, while persuasive on their own, are much more convincing when applied to concrete specific applications, as done in the articles that follow. Specific issues of forest conservation in southern Chile and Brazil are studied from transdisciplinary viewpoints, and the incoherence between neoclassical economics and the biophysical sciences is investigated. Ecosystem services of forests, such as habitats for biodiversity, carbon sinks, and rainfall regulators, both at the local and global levels are studied. The role of forests as critical natural capital is explained, both in their stock-flow aspect as producers of goods and in their fund-service aspect as producers of services. Also investigated are means of paying for such services which are often non-rival and non-excludable, and therefore not suitable for private ownership and markets, but increasingly scarce and in need of efficient allocation. Valuation of ecosystem services is considered, as well as the problem of minimal scale of forest coverage necessary to maintain such services. The balance between extractivism and conservation is studied. Likewise explained is the threat of high interest rates and present value maximization to exploited species that must grow at slower rates. More generally, environmental opportunity costs of economic growth are identified and explained. The role of indigenous wisdom needed as a brake on the haste of anthropocentric culture is recognized. And the importance of preservation and beneficial use of urban forests is not overlooked.

This volume is a big contribution to understanding the proper use of forests in general, the specific use and preservation of the particular forests of southern South America, and how general ecological economic principles can be concretely applied to using and conserving this magnificent dowry of living wealth. I am sure Manfred would have been very proud to endorse the book! I think it also justifies his prescience many years ago in including ecological economics among the visions for the future of the Chilean economy considered at the conference he organized. That vision is still far from realized, but this book brings it closer.

School of Public Policy, University of Maryland, College Park, MD, USA
August 2019

Herman Daly

Preface

Transdisciplinarity goes beyond the concept of interdisciplinarity, since it is conceived as the capacity of researchers from several disciplines and stakeholders to jointly define a problem and seek for solutions, which requires open minds not only for collaboration but also an effort to understand the disciplines and perspectives of others. The structure of the great majority of universities in terms of faculties and departments reinforces the unidisciplinary formation that nowadays is offered, especially at the undergraduate levels as our dear colleague Manfred Max-Neef (†) once stated in his "Foundations of Transdisciplinarity" (*Ecological Economics*, 2005). According to his point of view, a first step toward a necessary transformation of science and the form we educate our students should occur at the level of postgraduate programs oriented, whenever possible, around thematic areas instead of specific disciplines. As an example, a postgraduate program in *forest conservation* could call together not only forest engineers and agronomists but also economists, lawyers, chemists, biologists, etc. and achieve transdisciplinarity in each of them, considering that the result would not be the study of forests as seen from the perspective of the forest engineer, or of the agronomist, or of the biologist, but as seen in an integrated manner. The foundation of the "Transdisciplinary Research Center for Socio-ecological Strategies for Forest Conservation" (TESES) at the Universidad Austral de Chile (UACh) is such an attempt.

This book was conceived by the members of TESES, who are mostly professors at the UACh. It contains contributions by some researchers of TESES plus distinguished researchers from around the world, combining the fields of economics, ecology, biology, anthropology, sociology, and statistics. It is not, however, simply a collection of works written by the authors from different disciplines, but rather, each chapter attempts to be in itself transdisciplinary. We acknowledge that a transdisciplinary effort to progress toward sustainable management of natural resources is long-standing, but we hope that this book will be a significant contribution to enhancing social, managerial, and political approaches to forestry management, helping to protect forest ecosystem functions and services. This, in turn, should benefit local communities and society as a whole, as it reduces the negative externalities of forestry management and enhances future opportunities.

This book attempts to give a transdisciplinary approach to find ecological-economic and socio-ecological strategies for forest conservation. It combines economic, ecological, and social aspects related to forest conservation strategies to provide a holistic view of this complex topic. Overall, a nice outcome of transdisciplinarity would be achieving governance systems that are able to sustain natural and social communities. In this way, we wish to contribute to the design of a resilient *human-forest model* that takes into account the multiculturalism of local communities, including aspects of ecological economics, development economics, and land use planning. The book mixes theoretical concepts – some might even be categorized as philosophical – and practical approaches in Brazil and Chile, providing concrete lessons based on real experiences in the region. Although it focuses on cases in Brazil and Chile, the results might be applied to other regions too, i.e., we hope that the case studies and proposals provided will be useful for a broad audience of readers concerned with natural resource sustainability.

Valdivia, Chile
August 2019

Felix Fuders
Pablo J. Donoso

Contents

Notes on Editors and Contributors

Editors

Felix Fuders completed his M.A. in International Business Administration and Ph.D. in Economics and Social Sciences both at the University of Erlangen-Nuremberg, Germany. He is professor and researcher at the Universidad Austral de Chile. Currently, he serves as director of the Economics Institute as well as director of SPRING Latin America, a Master of Science program in development planning for growing economies jointly offered in the universities in the Philippines, Ghana, Tanzania and Germany, and of the Economic Policy Chapter of the Transdisciplinary Research Center for Socio-ecological Strategies for Forest Conservation (TESES). He has been a visiting professor at the Münster University of Applied Sciences, Germany, and visiting researcher at the RLC Campus, University of Bonn, Germany.

He is author and coauthor of publications regarding regional economic integration (EU and MERCOSUR), regulatory economics, as well as ecological economics and monetary policy. His main research focuses on studying economic as well as moral-ethical problems inherent in our financial system that he believes to be the most relevant but least recognized reason for market failure. He is convinced that the effect that the money interest rate exerts on money supply and aggregate demand is not fully understood and that many social, economic and especially ecological problems find their roots in an inadequate monetary policy. In this context, he currently works on explaining why the privatization of natural resources is neither a sustainable nor an allocative efficient solution to what Hardin once called the "Tragedy of the Commons" and why we should distinguish between man-made products and pure natural resources concerning the assignation of property rights. He further advocates a model of an economy in which the driver to achieve allocative efficiency is neighborly love instead of competition.

He is the chairman of the Foundation Natural Economic Order (www.inwo.de), Frankfurt, and member of the Association for Sustainability, Berlin and the Network Sustainable Economics, Berlin. He has published in internationally renowned publishing houses such as Palgrave MacMillan, Springer, HART Publishing, and Duncker & Humblot and is a regular speaker at congresses and scientific meetings at national and international levels.

Pablo J. Donoso is forestry engineer of the Universidad Austral de Chile. He earned his M.Sc. and Ph.D. in Forest Resources Management both at the State University of New York, College of Environmental Science and Forestry. He is professor and researcher in the Department of Forests and Society, School of Forest Sciences and Natural Resources, Universidad Austral de Chile (UACh), Valdivia, Chile, and served as vice-dean of the Faculty of Forest Sciences (2007–2010) and as director of the Department of Silviculture (2011–2012).

Currently, he is the director of the Transdisciplinary Research Center for Socio-ecological Strategies for Forest Conservation (TESES) at the Universidad Austral de Chile. His main research areas are forest dynamics and silviculture, especially of native forests, but he has also dealt with issues related to forest policies in Chile. His publication record includes 65 articles in WOS as well as 4 edited books and 26 book chapters, many of which are the result of numerous competitive funded research projects as well as cooperation with peers and graduate students. He has served as the major professor of nearly ten graduate students. The main goal of his research has been to establish first the ecological foundations for management of mostly mixed forests and then to evaluate results of the implementation of silviculture in diverse types of native forests, including plantations of native species. He is a strong believer that mostly long-term silvicultural experiments will provide strong information to support sustainable forest management (SFM), and in that direction in most of his research project, he has established permanent plots, with more than 100 of these throughout south central Chile, many of them in the Llancahue Reserve (1300 ha) near the city of Valdivia. However, he is aware that SFM will eventually result from good governance of social-ecological systems, and for that reason, TESES and some NGOs are considered by him important to advance in that direction.

Contributors

José Aylwin is a lawyer specialized in human rights and indigenous peoples and professor at the School of Law at the Universidad Austral de Chile, Valdivia. At the beginning of the 1990s, he participated in the Special Commission of Indigenous Peoples (Comisión Especial de Pueblos Indígenas (CEPI)) working on the draft bill of the current Indigenous Act. He was the director of the Institute of Indigenous Studies at Universidad de La Frontera (UFRO) between 1994 and 1997. He was also the coordinator of the Program on Indigenous Rights of the Institute (2002–2004). He has conducted studies and researches on the rights of indigenous peoples in North America (Master of the British Columbia University, Canada) and Latin

America for ECLAC (UN), University of Montana, Inter-American Institute of Human Rights, and IWGIA of Denmark. He has participated in several international conferences, connecting with national and international organizations in the fields of indigenous and environmental rights. He is author of several publications regarding human rights and the rights of indigenous peoples in Chile as well as abroad.

Jan Börner is full professor for Economics of Sustainable Land Use and Bioeconomy at the University of Bonn, Germany, with applied research experience in Latin America, Africa, and Europe. Between 2012 and 2017, he was Robert Bosch Junior Professor for Economics of Sustainable Natural Resource Use at the Center for Development Research (ZEF) in Bonn, where his work focused on the economic analysis and evaluation of tropical forest conservation policies. Before, he worked as a research associate at the Center for International Forestry Research (CIFOR) and the International Center for Tropical Agriculture (CIAT) based in Brazil, where he was involved in several global research projects on REDD+ and human environment interactions. His current research agenda expands from national-level environmental policy analysis toward the role of global trade and consumption patterns in affecting ecosystem services provision from ecologically sensitive landscapes.

Liviam Elizabeth Cordeiro-Beduschi is forestry engineer. She completed her M.Sc. in Ecology of Agroecosystems in the "Luiz de Queiroz" College of Agriculture (ESALQ) and her Ph.D. in the Graduate Program in Environmental Science (PROCAM/IEE) of the Institute of Energy and Environment both at the University of São Paulo, Brazil. She served as international consultant for the Forestry Department of the United Nations Food and Agriculture Organization (FAO). She carried out agroforestry research with the Brazilian Agricultural Research Company (EMBRAPA) and with the National Environment Fund of the Ministry of the Environment of Brazil (FNMA/MMA). Currently, she is researcher at the Forest Governance Research Group at the USP and partner of the Forestry Engineers Association for Native Forests of Chile (AIFBN). She develops research on public policies and governance of native forests and sustainable rural development with focus on community forest management in South America.

Jiří Dušek completed his M.Sc. in Geobotany and Applied Ecology from the Biological Faculty of the South Bohemia University České Budějovice, Czech Republic, and his Ph.D. in Plant Anatomy and Physiology from the Masaryk University in Brno, Czech Republic. Currently, he is researcher at the Global Change Research Institute (CzechGlobe) of the Czech Academy of Sciences. He is a principal researcher of wetland ecosystem station belonging to the national research infrastructure (CzeCOS) and the European research infrastructure of Integrated Carbon Observation System (ICOS). He has published about 30 scientific publications and is coauthor of 3 utility models related to gas measurements by chamber method in wetlands.

Alfredo Erlwein completed his M.Sc. in Holistic Science (systems and complexity in ecology) from Schumacher College, University of Plymouth, England, and his Ph.D. in "Sustainable Use of Bioenergy," from the Interdisciplinary Center for Sustainable Development of the Georg-August-Universität, Göttingen, Germany. He is agronomist from Pontifical Catholic University, Chile, and currently is professor at the Institute of Agricultural Engineering and Soils, Faculty of Agricultural Sciences, and at the Transdisciplinary Center for Environmental Studies and Sustainable Human Development (CEAM) both at Austral University of Chile. His research lines include sustainable use of energy (biofuels technology, energy balance and ecological footprint of energy use, organic waste management, bioenergy, and climate change) and the territorial planning of rural ecosystem (power and territory, cognition and territory, landscape ecology, ecological design).

Joshua Farley is professor in Community Development and Applied Economics and Fellow at the Gund Institute for Ecological Economics at the UVM and is president elect of the International Society for Ecological Economics. He holds a Bachelor of Science in Biology (Grinnell), a master's degree in International Affairs (Columbia), and a Ph.D. in Agricultural and Resource Economics (Cornell). Though trained in neoclassical economics, he has never accepted many of its core axioms or ethical underpinnings. His broad research interests focus on the design of economic institutions capable of balancing what is biophysically possible with what is socially, psychologically, and ethically desirable. His previous positions include program director at the School for Field Studies' Centre for Rainforest Studies, executive director of the UMD Institute for Ecological Economics, and Fulbright Fellow and visiting researcher at the Universidade Federal de Santa Catarina. His recent research focuses on agroecology, farmer livelihoods, and ecosystem services in Brazil's Atlantic Forest, the economics of essential resources, redesigning financial and monetary systems for a just and sustainable economy, and harnessing humanity's capacity for cooperation to address prisoner's dilemmas. He is coauthor with Herman Daly of *Ecological Economics: Principles and Applications*.

Isabel Jurema Grimm is researcher at the Ecossocioeconomics Research Center (NECOS) and the Research Group for Alternative Development, Innovation, and Sustainability (GPADIS) at the Federal University of Mato Grosso and coordinator of the Postgraduate Master's Program in Governance and Sustainability (PPGS) of the Higher Institute of Management and Economics (ISAE). She received her Ph.D. in Environment and Development from the Federal University of Paraná (UFPR). She also works with research lines in governance, urban management, "ecossocioeconomics", climate change, and low carbon tourism.

Christian Henríquez Zúñiga completed his B.A. in Business Tourism Administration from the Universidad Austral de Chile and his M.A. in Regional Development from the University of Blumenau, Brazil. He is a Ph.D. candidate in Human Sciences at the Universidad Austral de Chile with stays in the University Federal de Parana (UFPR) and University of São Paulo, Brazil; researcher at the Transdisciplinary Center for Environmental Studies (CEAM) and coordinator of the Right Livelihood College Austral Campus, both in the Universidad Austral de Chile; and lecturer at the Catholic University of Maule, Chile.

Teodoro Kausel received his Civil Engineering degree from the Universidad de Chile and his Ph.D. in Economics from the University of Münster, Germany. He has worked in Chile, Germany, and Botswana (Africa), where he acted as a government consultant in energy and natural resources matters. Since 1992, he is professor at the Universidad Austral de Chile, Valdivia, dedicated to research and teaching in regional economy, economic regulation, climate change, energy, and university policy.

Jozef Kiseľák received his Ph.D. in Applied Mathematics from Comenius University, Slovakia. He worked as a scientific project assistant at Johannes Kepler University in Linz. Currently, he is an assistant professor at P.J. Šafárik University in Košice, Slovakia. His research interests are in the field of dynamical systems with applications in biology, medicine, ecology, and finance. He also deals with optimal designs of experiments and information theory. He has published research articles in high-impacted international journals of mathematical and statistical sciences and is a referee of several international journals in the frame of pure and applied mathematics.

Daniela Manuschevich is professor at the Geography School in the Universidad Academia de Humanismo Cristiano in Santiago, Chile, since 2017. She is an early-career transdisciplinary researcher, seeking to connect the economic, political, cultural, and environmental changes fostered by tree farm expansions in Chile. She has published works on historical tendencies of land use change, policy-based scenarios of land use change, ecosystem service modeling, as well as discourse analysis of the forest policy in Chile. The chapters presented in this book are the first results of a 2-year ethnography with peasants living in the fringes of tree farm expansion and historical forest degradation. She is a Fulbright scholar and member of the International Society of Ecological Economics.

Alejandro Mora-Motta holds a bachelor's degree in Economics and a master's degree in Development and Environment. He is currently writing his Ph.D. thesis in the area of Development Studies at the University of Bonn, Germany. He is a Junior Researcher at the Center for Development Research and a member of the Right Livelihood College (RLC) Campus Bonn since August 2015. Between 2016 and 2017, he pursued his field work in Chile in collaboration with the RLC Campus Universidad Austral de Chile, Valdivia. His doctoral project addresses the question how the territorial transformation caused by tree plantations has affected local well-being in rural areas in the Los Rios region in southern Chile. He has research experience in the intersection of the fields of political ecology, ecological economics, and development studies, with applied work in Colombia and Chile.

He is currently interested in studying how development processes, particularly in forestry, affect local communities and in which way sustainable alternatives may emerge.

Nicolas Nazal holds a bachelor's degree in Business from the Universidad Diego Portales, Chile, and a master's degree in Economics from Waikato University in Hamilton, New Zealand. He is currently working on his Ph.D. in Forest Ecosystems and Natural Resources at the Universidad Austral de Chile, Valdivia. His research interest focuses on anthropogenic impacts on forest ecosystems and policies aimed at recovering native forest and sustainable socioeconomic and ecological restoration. Currently, he is a lecturer of Ecological Economics at the Natural Resources and Forest Sciences Faculty, Universidad Austral de Chile, and director of the Small Business Development Center, a government project operated jointly with the Universidad Austral de Chile. He formerly taught a course on Sustainable Economics and International Trade at the Faculty of Economics an Administration. He has over 10 years of work experience in private firms in the shipyard industry and in the salmon farming industry as CFO and project manager.

Iván Oliva completed his B.Sc. in Biology from Pontifical Catholic University of Valparaíso and his Ph.D. in Educational Sciences from Pontifical Catholic University of Chile. Currently, he is a lecturer at the Pontifical Catholic University of Valparaíso and at the Institute of Education Sciences, Faculty of Philosophy and Humanities, Universidad Austral de Chile.

Guillermo Pacheco Habert is a licentiate in Tourism Management from the Universidad Austral de Chile (UACh) and holds an M.A. in Development of Regional Societies from CEDER, ULA. He currently is writing his Ph.D. thesis in Territorial Studies at the "Center of Regional Development Studies and Public Policies" (CEDER), Universidad de Los Lagos, Chile. He is a junior scientist at the Right Livelihood College (RLC), Center for Development Research (ZEF), University of Bonn, Germany, and researcher at "Transdisciplinary Environmental Studies Center" (CEAM) and at the Tourism Institute, Universidad Austral de Chile. He has 10 years of experience in research and the development of projects related to local economies focused on an ecological and solidarity economics, territorial dynamics, landscape production, community-based tourism, and participation of rural and indigenous communities in decision-making processes. He is cofounder of the "Trawün" Cooperative formed in 2017 by peasants with indigenous background, dedicated to family agriculture and community-based tourism in the Panguipulli Municipality located in the southern Chilean Los Andes Mountains.

Roberto Pastén received his M.Sc. and Ph.D. in Economics from the Universidad de Chile and University of Alabama, USA, respectively. He is associate professor of Economics at the Universidad Austral de Chile and former Justice at the Third Environment Court in Chile. He possesses a vast experience in environmental and natural resource economics, risk and uncertainty, economic analysis of law, and econometrics of time series and dynamic panels and published numerous papers in high-impacted journals. He has been an associate professor of Economics at the Universidad de Talca, Chile, and visiting professor in the master and doctoral program at the Universidad de Chile and of Environmental Economics at Groningen University, Netherlands. In 2017, he became inducted in the Hall of Fame of the Faculty of Business and Economics, Universidad de Chile, as the most remarkable postgraduate economist.

João Henrique Tomaselli Piva received his B.A. in Business Administration from the Universidade Positivo, Curitiba, where he is technician in Environmental Protection. Currently, he is studying a master's program in Environment and Development at the Federal University of Paraná in the research line of Ecossocioeconomics of Organizations.

Jennifer E. Romero is a forestry engineer from the University of Chile. She received her M.Sc. in Resources Management and Environmental Studies from the University of British Columbia. She has focused her training and professional development on project management and on environmental policy, with emphasis on Chilean and Argentinean forests. She is currently the executive director of the Chilean NGO Foresters for Native Forests (AIFBN). Communication is an important part of her studies and practice. During the last 3 years, she has been the chief editor of the journal *Bosque Nativo* that is published by AIFBN and has written and edited publications related to forest policy in book chapters and also in massive media for the general public, generally with the aim of raising awareness for the conservation of native forests.

Hugo Rosa da Conceição completed his B.A. in International Relations from the University of Brasília, Brazil; his M.Sc. in Environmental Governance from the University of Freiburg, Germany; and his Ph.D. from the University of Bonn, Germany. He has worked as a program assistant at the United Nations Environment Program and as a junior professional associate at the World Bank Offices in Brazil, working on conservation programs in the Brazilian Amazon. His research focuses on forest protection policies, especially incentive-based policies to reduce deforestation in the Amazon Region, such as REDD+ and Payments for Environmental Services.

Carlos Alberto Cioce Sampaio completed his master's and Ph.D. degrees in Planning and Organizational Management for Sustainable Development from the UFSC with internship in Social Economy (EHESS, France). He has postdoctoral fellowships in Ecossocioeconomics (Universidad Austral de Chile), Corporate Cooperativism (Universidad Mondragon), and Environmental Sciences (WSU). He is an administrator at PUCSP, researcher at CNPq, professor of the Graduate Program in Regional Development at FURB, environmental manager at UP, and environment and development manager at UFPR. Moreover, he was Fulbright Foundation scholar (USA) in 2015 and visiting professor at the Brazilian Center for Contemporary Studies (CRBC) at the School of Higher Studies in Social Sciences

(EHESS) in Paris and the Center for Environmental Studies at the Universidad Austral de Chile (UACh) in 1996 and 2005, respectively. He currently coordinates the Center of Ecossocioeconomics and, in partnership, the Center for Public Policies. He can be seen as a pioneer in theoretical and empirical research on the topic Ecossocioeconomics, comprising planning and organizational management for sustainable territorial development, socio-productive and constitutional arrangement of community-based solidarity, and sustainable tourism in Latin America. He published 117 papers in journals and 183 studies in national and international scientific congresses, as well as 13 books and 59 book chapters.

Abdon Schmitt Filho is professor at the University of Santa Catarina (UFSC), Brazil. His teaching and research efforts focus on designing agroecological systems that synergically interconnect ecological restoration, rural livelihood, and renewable agriculture in southern Brazil. He is the coordinator of the Silvopastoral Systems and Ecological Restoration Lab (LASSre), a participatory action research initiative that worked with more than 622 farmers to make the transition from conventional farming to agroecological systems. Today, some of these farmers are partners in a project of ecological restoration of Atlantic Forest with focus on High Biodiversity Silvopastoral Systems and multifunctional riparian forests. He was a visiting professor at Gund Institute for Environment at the University of Vermont, USA, when he became a Gund IEE affiliate. His research addresses agroecology, silvopastoral systems, sustainable livelihoods, and ecological restoration of Atlantic Forest.

Milan Stehlík completed his Ph.D. in Statistics from Comenius University, Slovakia, and Formal Habilitation at Johannes Kepler University, Austria. Currently, he is full professor at the University of Valparaiso, Chile, and associate professor at Johannes Kepler University, Austria. He is member of the Linz Institute of Technology, member of the Institute of Mathematical Statistics, and invited member of Gwalior Academy of Mathematical Sciences. He published more than 170 scientific papers and gave more than 180 scientific talks. His research interests include experimental design, extremes, exact testing, life data modeling, medical and ecological statistics, economic applications, and reliability theory.

Till Stellmacher holds a master's degree in Development Geography and a Ph.D. in Agricultural Sciences. He is program coordinator and senior researcher at the Center for Development Research (ZEF), University of Bonn, Germany. He coordinates the Right Livelihood College (RLC) Campus Bonn since 2011. His main field of scientific expertise is governance, forest management, and environmental change in the rural Global South. His Ph.D. thesis "Governing the Ethiopian Coffee Forests" shows of local insights of forest use, management, and conservation in contexts of legal pluralism and agricultural transformation. He has several years of work experience in empirical development research in Ethiopia, Bangladesh, India, Tanzania and Burkina Faso. He is currently focusing on questions related to the future of smallholder farming and forest governance.

Manuel von der Mühlen completed his Master of Science (M.Sc.) degree in Regional Development Planning and Management from TU Dortmund and the Universidad Austral de Chile. He is also a candidate for the Master of Advanced Studies (MAS) in Development and Cooperation at ETH Zürich. He is a development planner with a passion for participatory planning processes. He has worked on participatory land use planning (PLUP) together with indigenous peoples and local communities in South America and Asia. While working for the OneMap Myanmar project from the Centre for Development and Environment (CDE), he led the research activities to produce customary land use maps and other knowledge products. Currently, he works for Plan International Switzerland and is responsible for the coordination of different projects. His main project involves flood resilience building in flood-prone communities in El Salvador and Nicaragua.

Part I
General Thoughts on Transdisciplinarity, Economics and Ecology

Chapter 1
Towards a Transdisciplinary Ecological Economics: A Cognitive Approach

Alfredo Erlwein, Iván Oliva, Felix Fuders, and Pablo J. Donoso

an extractive economy is a terminal economy.
Thomas Berry

1.1 Introduction

It is clear that humankind and especially western civilization is in the middle of a multidimensional global crisis that involves all aspects of human living, including its own long-term survival (Max-Neef 2010; IPCC 2011). It is also clear that through the advance of environmental sciences in the last two centuries,[1] we have been able to clearly identify the main scientific causes of such crisis, and the technical solutions to escape from it. Moreover, the environmental sciences were able to predict the arrival of such global crisis with deep and accurate level of details since

[1]Malthus stated one of the first scientific critics to growth as a way of progress, with a simple mathematical demonstration of the demographic limits. After 220 years of its publication, his critic remains valid.

A. Erlwein (✉)
Institute of Agricultural Engineering and Soils, Faculty of Agrarian Sciences. Transdisciplinary Center of Environmental Studies (CEAM), Universidad Austral de Chile, Valdivia, Los Ríos, Chile
e-mail: aerlwein@uach.cl

I. Oliva
Facultad de Filosofía y Humanidades. Instituto de Cs. De la Educación, Universidad Austral de Chile, Valdivia, Región de Los Ríos, Chile

F. Fuders
Economics Institute, Faculty of Economics and Administration, Universidad Austral de Chile, Valdivia, Chile

P. J. Donoso
Institute of Forests and Society, Universidad Austral de Chile, Valdivia, Chile

F. Fuders and P. J. Donoso (eds.), *Ecological Economic and Socio Ecological Strategies for Forest Conservation*, https://doi.org/10.1007/978-3-030-35379-7_1

at least 50 years ago.[2] In other words, the science developed until half a century ago was capable to:

- Predict the current global environmental crisis
- Understand the main causes of it
- Propose what kind of human behaviour (development) was necessary to resolve it

Therefore, we can claim that the global environmental crisis, and our capacity to manage it, is not due to a lack of scientific knowledge or unknown facts. Although the complexity of the natural world implies unpredictability in many processes and phenomena, the drivers and causes of such global change are known since long ago. In fact, in the last decades, the environmental sciences have not neglected the causes originally identified, but rather reinforce its validity through an ever-increasing stock of evidences. It is somehow frustrating to realize that most of the scientific work on environmental sciences in the last decades simply sum more evidences to what was already identified as the main causes of the environmental crisis. In other words, science has focussed on *knowing* more, but not necessary *understanding* more (Max-Neef 2005), perhaps because the main cause of the global crisis is not a scientific one and goes beyond the scope of natural sciences or sciences as a whole. This suggests that it would be incorrect to expect a solution for the crisis from the field of natural science and technology, as under our view the duty of science is already accomplished. That is, setting specific questions regarding the global crisis and proving with evidences the hypothesis that answer them.

As societies have not yet seriously tried to abolish those main drivers of the current crisis, identified long ago by natural sciences, the relevance of technology in proposing a concrete solution (e.g., renewable energies) is limited by the existence of discrete natural boundaries. In other words, technological solutions may move the frontiers of the feasibility, and therefore resolve the problem temporarily, but as long as, for example, an unlimited economic growth imperative exists, the system is determined to collapse. In other words, the civilization and its economy cannot run through "business as usual" if the main biophysical causes of the crisis have not been addressed.

Consequently, we identify a dramatic gap between what is *known* and what it is *done*. Therefore, the core of the environmental crisis lays not in a crisis of knowing (science), but in a *crisis of consciousness*, regarding our relationship with nature and ourselves, and our ability to detach from old patterns of behavior and to change according to what evidences, for so long and so clear, are telling us.

[2]Although there are many previous scientific works giving the same framework, "The Limits to Growth" (Meadows et al. 1972) of the Club of Rome addressed many problems that we are facing today.

1.2 Cognition

In this chapter, a background of cognitive science is used to deal with the problem of transdiscipline regarding the study of forests and nature in general, and its consequences in the way the rationality of neo-classical economics works. However, cognitive science is part of a broader set of disciplines and sciences that share a holistic, systemic, transdisciplinary and postmodern approach to knowledge. In these terms, many of the postulates in this article can be also expressed in other terms by those disciplines, that include phenomenology, systemic epistemology, second order cybernetics, quantum physics, to name a few. Also, this chapter is a very brief summary of referential concepts of cognitive science that is requested in the development of the next chapters.

Cognition is the science of the process of knowing. Since it has inputs from different disciplines (philosophy, neuroscience, semiotics), it is an "interdisciplinary discipline". Cognition can be considered "a promising starting point towards an appropriate and unifying paradigm" in relation to the needs of interdisciplinary approaches (Röling 2000: 5). Among different trends within the science of Cognition, the Santiago Theory of Cognition developed by Humberto Maturana and Francisco Varela provides for the first time a scientific theory that overcomes the Cartesian division of mind and matter (Capra 1997). One of the big insights of the theory is to resolve the crossroad about the relation between the brain and mind. For Maturana and Varela (1980) the relationship between mind and brain is simple and clear: the mind is not a thing, but a process, and the brain is a structure through which the mind is performed. In the same theory, life and the process of knowing are united. In other words, the central insight of the Santiago theory is the identification of cognition with the process of life (Capra 1997). The mental world is not separate from the physical process of living, but an intrinsic characteristic of this process, as living always entails knowing. In words of Maturana and Varela (1980) "Living systems are cognitive systems, and living as a process is a process of cognition. This statement is valid for all organisms, with or without a nervous system".

As cognition is a research field about knowledge, it is well related to the philosophical study of knowledge. *Epistemology* is that branch of philosophy that deals with the nature of knowledge and creation of new knowledge (Novak and Cañas 2008). This gives special relevance to the discussion of the role of language under a cognitive prospect, as philosophy and sciences are realized through language.

1.3 The Emergence of the Environment

Curiously, what we understand as the environment is ever changing, and the valorization of the environment also changes accordingly. This is so, because we as observers actually do not depend on our environment, but on our ecological niche.

The first is what is distinguished by an observer, and the second is where the domain of existence of any living being actually occurs (Maturana and Mpodozis 2000). For example, if we observe a tree, we can distinguish its environment simply by looking at the tree's surrounding. However, what we can see of the tree's surrounding is only what our senses allows us to see. Actually, the tree interacts and depends on many more features than those that we can distinguish for its process of living. We cannot see the underground interactions of millions of root radicals exchanging elements with the soil, nor can we see (but roughly infer) the gas exchange of its leaves with the atmosphere. The sum of all features of the environment is what sets the niche as the actual domain of living of the tree.

The same phenomenon happens to us in relation to our environment. Our environment is not an objective set of discrete resources, but an always changing perception as long as we can distinguish new elements of our niche. Here, a great paradox takes place: as long as we human species have an ever-increasing impact on the biosphere, new elements of the niche emerge from our alteration of it, setting new perceptions of our environment. For example, 200 years ago, we gave no attention to the electromagnetic environment, or to the air and water quality in most of the cities. The attention started when the original features of our niche started to be altered. Just as the French surgeon René Leriche (1879–1955) stated, that "*Health is life lived in the silence of the organs*" (Fantuzzi 2014), as we feel the organs just when they become ill, the new features of the environment frequently appear as a result of our modification of the natural features of our niche. Certainly, there is a coherence between this view and the classical economical view of demand and supply: as scarcer a resource becomes, it gets more expensive (ceteris paribus). So, why so many features of our environment have not been valued even until today by neoclassical economics, considering that some of them are essential for our survival? We will come back to this topic in the discussion of the quantitative focus of classical economics.

1.4 Distinctions and Objectivity

The operation of *distinction* is the basic operation that an observer performs in the praxis of living, through the specification of an entity by operationally cleaving it from a background (Maturana 1988). In the operation of distinction an observer brings forth an entity (a unity, a whole) as well as the medium in which the entity is distinguished and entails in this latter all the operational coherences that make the distinction of the unity possible in his or her praxis of living.

As the perception is determined by the structure of the nervous system, the observer can distinguish only what triggers in him or her a cognitive process that takes place and is defined by the characteristics of the nervous system as such. In this way, the perception is limited by the structure of the observer. As we cannot perceive what is out of our cognitive domain, the process of cognition is driven under a *Structural Determinism.*

The nervous system has an organization closed to the outside, and only operates realizing *senso-effector* internal correlations (Maturana 1989). This way, through our senses the environment triggers in us a cognitive process that takes place and is determined by the nervous system, in which such environment does not participate. Because of that, we cannot refer to an external reality independent of us (Maturana 1988).

In life, in the cosmos, in the existence in general everything is, directly or indirectly, related to everything else. As everything is moving, under certain scales of space and time every phenomenon has a connection with any other, and with the wholeness. Many of those relations are happening at every moment, although not fast or slow enough to be perceived by us, or simply not sensorially (materialistically) felt, just as gravity, synergy or empathy. We can measure some variables related to them, but actually not see them. We do not see the history of a person, although his/her history allows him/her to be there. We do not see the mining field from which a knife has come, although the mining place and the metal of the knife have been mutually transformed, so that there is a relation between the knife and how the mine looks today. Knife and mine are parts of one whole process, related to the history of culture (mining, cooking, etc.), to the process of tectonics that had made the iron emerged, to the history of the supernova in which the iron was made, to the origin of the universe. But in our experiences, those processes or the components of them seem to us as being different realities. Those realities appear from a physiological process that allows us to set boundaries of a phenomenon, so as to specify it as a unity and therefore allows us to see it as an individual experience.

Most probably, this fact played a role in the adaptation of organisms to environments, developing the capacity of differentiating events and developing skills accordingly, but also to become speaking organisms, as distinctions seem to be the base for the development of language. In fact, what we "name", that is, what we identify with a word (concept, idea), is commonly not a "thing" but a distinction, so that there is a strong relation between what we distinguish and what we specify linguistically.

First distinctions were probably sensorial encounters with nature, which explains why there is a correlation between ancient times and languages that are simple and resemble nature. As distinctions are recursive, that is, they can be configured from within previous distinctions, language and human life (culture) has gotten more and more complex (and less natural) building with time an "artificialized" society that has lost contact with the natural world.

The act of distinguishing happens to us as a cognitive process realized by our nervous system. If we cannot distinguish, we look, but we will not see; the process of distinction is at the core of our adaptation as living organisms. It takes place in ourselves, in our own nervous system, and not in the phenomena we are distinguishing. It is a dynamic internal experience that allows us to take contact with the world we live in. But as an internal process, it is not the world itself what we perceive, but only our human way to apprehend it. What we see is not a world independent of us, as the objectivity concept implies, but the result of the encounter of us with the world as perception is a process that results from this encounter.

Therefore, we are involved in what we perceive and distinguish. As perception is realized through the body, it is a personal experience, not sharable with others. We can *speak* to others about the experience, but not transfer our experience to others. Moreover, as different human groups (that is, different networks of conversations) follow different paths of interactions among them and with their environment, those different human groups frequently (if not ever) perform different distinctions, which explain why different cultures see the world so differently.

Deep in the origin of the western culture was the believe in the existence of a natural law that can be rationalized and comprehended. With arise of Enlightenment, the believe in the immanency of the physical reality, governed by hard incorruptible laws independent of us humans, set the core for believing in an objective reality. In fact, the very essence of the objectivity is the domain of a physical reality, as its root is the "object",[3] that is, what can be touched, felt by the senses. In this way, the objective world is essentially a materialistic world, as only counts what can be seen, accounted, quantified. However, objects are also not realities independent from who sees them. The very act of distinguishing an object entails a linguistic act and, therefore, it is dependent on a specific culture that sets the object as such.

In everyday life, objects appear to us as immanent realities. Somehow, they have become the symbol of an immanent reality independent of us, and likewise gave the name for that modern paradigm: the objectivity. As a difference observed by our visual perception is the basic process of distinction (i.e. setting different colors), entities that have visible boundaries are more universally distinguished, that is, the consensus among observers will be almost absolute (borders are visible to everybody). This particularity of "things" gave the study of objects (like in physics) such a robust degree of unquestionability. Additionally, objects are also *discrete unities*, which allow them to be *counted or numbered*. That opens the objects to be taken into "account" by the field of mathematics, which are then welcomed to modern natural sciences, that base most of its methods on a *quantitative* approach.

"Things", that we have called objects, are indeed specific distinctions configured by us as observers. Consequently, as those distinctions are useful, they began to pervade the linguistic domain (the domain of collective coordination) so that we name them and, in this way, objects start to become realities for us, ever more "solid" as more people and time are involved.

Hence, what western society has stated as objective knowledge is in fact knowledge that is consensual, consolidated as "truths" among human groups from a specific background in an specific age. That is why objective truths are ever changing through the history, as new knowledge changes the mainstream believes through different cultural processes (von Bertalanfy 1955; Popper 1959; Kuhn 1970). This way, according to Varela et al. (1991) what we named as objective is

[3]Etymology (https://www.etymonline.com/): from Medieval Latin objectum "thing put before" (literally "thrown against"). Late 14c., "tangible thing, something perceived or presented to the senses".

in fact an "intersubjective" phenomenon, as the subjective individual knowledge is consolidated through the collective consensus.

1.5 Distinctions and Language

Language made us become humans. Language is not simply a form of communication but is what has defined the human species as a social species that self organizes through linguistic interactions. As Maturana (1988) states, "whatever takes place in the praxis of living of the observer takes place as distinctions in language through languaging, and this is all that he or she can do as such". He also proposed (Maturana 1989):

(a) That language is a manner of living in recurrent consensual coordination of actions.
(b) That the human manner of living entails among other things, a braiding of languaging and emotioning that he calls conversation.
(c) That human beings arise in the history with the origin of language, and the constitution of a lineage defined by the conservation of an ontogenetic phenotype that includes conversations as part of it.
(d) That the magnitude of the involvement of the brain and anatomy of the larynx and face in speech as our main manner of languaging indicates that language cannot have arisen later than two to three million years ago.
(e) That rationality pertains to the operational coherences of languaging and that different rational domains are constituted by different basic notions that are accepted a priori, that is, on preference.
(f) That responsibility and freedom are a function of our awareness of the participation of our emotions (preferences) in the constitution of the rational domains in which we operate.

We do not perceive an objective reality independent of us but, instead, we perceive what reality triggers on us, mediated by the process of distinguishing that is performed and determined by our own nervous structure. Let us say again that our "reality" is built upon a set of distinctions instead of objects or discrete unities of information from the "outside world". Therefore, there is a deep relation between distinctions and language as a coordination process among our human groups. As relational processes, distinctions can be considered direct stimuli (like colors or shapes), or relations between those stimuli (like distinguishing a shape or color as an object), or relations that have no stimuli or sensual dimension (like any kind of non-material relation: observations of order, danger, similarity, justice, categories, etc.).

Language shares that relational nature. Words are far more correlated with distinctions than with physical "things", as words can have abstract or relational meanings, just as distinctions do. In other words, we propose that words are originated by distinctions that become common and then are shared within a human group through a process of coordination of actions. That happens in the course of the collective use of such words that become consensual. Words are then not objective realities but refer to specific distinctions. However, although we become coordinated through language, experience is always personal. According to Maturana (1988):

> ...objectivity in parenthesis[4] entails the multiversa, entails that existence is constitutively dependent on the observer, and that there are as many domains of truths as domains of existence, she or he brings forth in her or his distinctions.

1.6 Language and Ideas

Ideas occur in language and, therefore, are strictly related to the nature of linguistic distinctions. That is why ideas can be communicated, have certain logic (coherence), and can be understood. In fact, any distinction in the linguistic domain is simultaneously an idea, including the concept of "reality" that takes place in language.[5] Mathematics is also a language that has its own validity codes and logic. As words, also numbers are distinctions, not objective realities.[6]

Commonly language operates within an internal rationality that gives language its coherence. In western societies, such structure is given by logic: a reality that is structured and through this logic structure behavior is able to be inducted, deducted and predicted, giving us the feeling of sense and the power of control of nature. In the words of Maturana and Varela (1980):

> Language does not transmit information and its functional role is the creation of a cooperative domain of interactions between speakers through the development of a common frame of reference, although each speaker acts exclusively within his cognitive domain where all ultimate truth is contingent to personal experience.

Language entails a rational mind, but there are other mental realms. These realms belong to the pre-linguistic or non-linguistic mind. If we use the term "non-linguistic" (or "non-rational"), we jump into a cognitive domain to which we cannot speak about directly, because obviously we can only speak through a linguistic interaction. With that we refer to a level of cognition that is not intermediate through the rational mind, as found in animals, pre-linguistic children, in a contemplative or life-in-danger experiences, that can also be achieved with meditation techniques. It is the state of mind in which reason is not operating. An experience can be non-linguistic, but it becomes linguistic as soon as we think or talk about it.

---

[4]With parenthesis is meant: Objectivity under question.

[5]*Reality* is a word, and therefore it is a linguistic construct and not an objective reality.

[6]In the logic and validity codes created by the language "mathematics" $1 + 1 = 2$ may be defined as a "reality".

1.7 Ideas and Culture

As words and ideas emerge in the collective interplay of social communication, they change with the evolution of such communication as another kind of biological process. In the process of being educated, we are taught to perform certain configurations of distinctions, so that such configurations end up being the way we perceive the world. The dominant ideas of every era configure the cultural background that sets the identity of every human group, community or country. In this way, the culture can be considered as a network of conversations (Maturana 1997). Any network of conversations entails a way to see and approach the world. As every network develops under its own environmental and social background, everyone develops a different view of reality.

1.8 Culture, Disciplines and Science

Defined as a closed network of conversations conserved through generations (Maturana 1997), cultures are any community of members related by a discrete (with boundaries) network of conversations, not only determined by countries or land-based groups. In this sense, any discipline, as it entails a community that shares a common tradition and specific knowledge, is a smaller network with operational closure, being simultaneously part of a culture. The etymology comes directly from the Latin "*disciplina*", (instruction given, teaching, learning, knowledge) which in turn comes from the word "*disciple*" (pupil, student, follower),[7] as followers of a master or a tradition regarding a specific knowledge. Just as any discipline, different sciences have their own different masters (or science founders) and are also defined by a common cultural background of history and ideas. In general, scientific disciplines roughly share the bases of scientific knowledge,[8] including the scientific method, the rationality and the prevalence of evidences over beliefs. However, every scientific discipline has its own way to describe reality which are commonly not transferable to another. This is precisely why every discipline is fairly closed in linguistic terms, that is, each person uses different logics, driven by different sets of distinctions (that is why they are focused on different problematics of reality).

There is only a thin line between disciplines and sciences, acknowledging science as a group of disciplines. In this way, a scientific truth is also a contextual consensus, situated in space and time and enclosed within the so called "scientific community".[9] Science is also a network of conversations that shares a common

[7]https://www.etymonline.com/.

[8]It is therefore very difficult to find a universal definition of science, and there is no universal consensus about a definition of science among the scientific world.

[9]In this context it is noteworthy that the former German health minister Ulla Schmidt in an official statement wrote that it was an "international consensus" that HIV caused AIDS (Schmidt 2004).

rationality. Contrary to the common belief that natural sciences were "hard" sciences in terms of their supposed objectivity, and social sciences were "soft" –as they would be subjective and impossible to prove and in which postulates would depend always on whom state them– natural sciences share the subjectivity with social sciences.

Science as just another set of human knowledge, is primarily a collective process. That is, it works through the exchange of ideas, the recognition of consolidated ideas that are taken as common base, and the exercise of demonstration or, in other words, a process in which a statement must be proven to the (scientific) community. The approval of a certain statement depends more on the time in history and place, than on an (supposed) inherent value of truth. In fact, there is no such thing as an imminent scientific truth, but an ever-changing set of assumptions that is evolving, built upon the preexisting knowledge and sometimes with radical changes (what then often is called "revolution"). This is what is at the core of the works of Popper (1959) and Kuhn (1970) with the context of paradigm, and that is coherent with the cognitive model that is being developed in this chapter. The sets and beliefs change through time, but they are not something without frame or just relative; at each time they seem to be solid, as every age has its own consolidated truths, just as many other sets of knowledge. For science the earth was flat 500 years ago, and now it is spherical. In both periods there were consensuses that defined the scientific truth of each one, just as the intersubjective nature of language as the main source of knowledge. As we have not arrived yet to the ultimate truth, we will always experience changes in the way we see and think the world in the future.

So, paradoxically natural science as a collective, linguistic and historical phenomenon is also a social process that follows what has been described by social sciences and not natural sciences. That is why social science abandoned the concept of objectivity much before natural sciences did. Seen from this cognitive perspective as a cultural phenomenon, the line that separates natural and social sciences –soft and hard sciences–, seems to be purposeless.

1.9 Science and Values

Values are like a bridge between rationality and emotions, as they integrate ideas, symbols and deep feelings of the community that affiliates to them. They are brought up by every individual community, inspired in the self-image of the group, in terms of how they perceive themselves, what is deeply proper of them, or what they would like to be. In this sense, values are rooted in emotions that drive those images, like expectancy, affection, yearnings, trusts, hope, or the believe in a positive nature of life and the community. Values determine what is important for each culture, and express the common preferences, configuring the bases for the political, legal and economic decision processes. From that prospect, values as guiding concepts are embedded in any culture, including science as a cultural phenomenon.

As any cultural process, science has its own set of basic beliefs, that have been obscured by the idea of objectivity or the belief that the scientific knowledge is a universal, immanent truth (so that there is nothing to *believe*, only to *know*). The *leitmotivs* of sciences are the ideas of progress, and development, common good, well-being or contribution to society. Concepts like truth, knowledge (in contrast to ignorance) or so-called "natural laws"[10] can also be seen as values of science. Another value implicit in modern science, but not necessarily previous to modern era, is the foundational place of rationality in science. In other words, the belief that the world can be described, understood and predicted through reason. Much related to this is the search for prediction, that requires accuracy, revealing a natural affinity of science to certainty. This is a very important topic, as the search for certainty and the ability to predict have naturally pushed modern sciences towards the realm of quantitative approach and specialization. Yes, through this two facts science has indeed become more accurate in predictions but has left unattended a huge dimension of the living. In choosing quantities, current reductionist science has left the world of qualities. In choosing specialization, as a tool to get more details of the phenomena studied, science has gotten atomized and increasingly unable to resolve real-life problems.

Modern natural science, and coherently neoclassical economics, have chosen numbers as a practical solution to the problem of demonstration, but with that a huge portion of the human life has been cut off and set aside. This could be the cause of the lack of understanding of the current ecological and social crisis and, the lack of effectiveness in its solution (by applied sciences). Quantities are discrete distinctions that do not allow to deal with complex phenomena, which commonly imply the interplay of different disciplines that cannot interact under a paradigm of objectivity. Happiness, development, health, politics or the environment are matters that are not located in a specific discipline, and because of that no discipline has been able to entirely resolve such problems. Water, life or beauty do not belong to a scientific specialty, as their nature cannot be reduced to a discrete set of knowledge. Science has been dramatically successful in discovering new insights of the world but is far away to understand our existence. It is the other way around: the sense of life that every community sets for itself determines what sciences and disciplines it will study.

1.10 Neoclassical Economics, Science and Values

In its expectation to resembling a natural science (Smith and Max-Neef 2011), neoclassical economics took the way of a quantitative approach to study economic matters. In doing so, it has chosen to work with the quantifiable phenomena, primarily materialistic, as objects are easily accountable, like we have explained above. That approach has set apart most of the qualitative dimensions of human

[10]It has not been proven yet that the Universe behaves always in the same manner, an assumption that is implicit in the concept of natural law.

living. However, as stated above, many relevant bases of human culture are non-material and those bases have been, intentionally or not, neglected by neo-classical economics, evidencing its incapacity to deal with the real world.

As stated before, any discipline is set under a certain set of believes and preferences that are chosen a priori. In the case of neo-classical economics, there is a big remanence of the classical view, in which founders of the Enlightenment also developed a set of values. For example, for Bacon "the real and legitimate goal of sciences is the endowment of human life with new inventions and riches", as he believed that intangible, philosophical or spiritual approaches do not matter to human happiness (Freudenthal and McLaughlin 2009). Somehow, the relation of such mindset and the neoclassical approach, focused on materialism and consumerism, go hand in hand: as stated above, there is no such thing as value-free science and the same argument goes also for neoclassical economics. According to our perception, some values, assumptions or beliefs of neo-classical economics are:

- Human beings are predictable and can be studied as a physical phenomenon.
- Human beings are selfish and therefore competition as the driving force of the economy is natural.[11]
- Monetary wealth is something positive, and it is an indicator of personal success.
- Humanity has the right to appropriate any useful natural resource.
- Human development is about material things.
- Human happiness is about having money (or things), and the more the better.
- Regarding the long-dated critics to economic growth: natural sciences are not so relevant to be taken seriously.
- Regarding the use of GDP (Gross Domestic Product) as welfare indicator: past and future generations are irrelevant for the economy.
- Regarding enormous impacts on life and people from economic development initiatives: economists know the value of things better than other visions, and therefore economic arguments are enough (first priority) to justify them.[12]

Considering the latter, there are many facts that indicate that in the last decades, neoclassical economics has behaved as an isolated discipline that cannot be considered scientific at all. As stated in previous sections in this chapter, the main criteria of science are:

- To consider the scientific knowledge previously existing
- To openly discuss its ideas and being able to argument against eventual critics
- To support statements with logical reasoning coherent with the rationality of science

[11] A model of an economy in which neighbourly love instead of competition is the driving force of the economy has been described (Fuders 2017; Fuders and Nowak 2019).

[12] This assumption is especially relevant, as it implies that economists understand what is at stake with such environmental impacts.

- In the case of sciences based on mathematics, to be able to predict future scenarios
- To be able to demonstrate its hypotheses

Those criteria are definitively not fulfilled by neoclassical economics. Just the first criterion would be enough to let this discipline fall out of science. When the concept of (perpetual) economic growth is analyzed, a big contradiction with the first law of thermodynamics[13] is found. This is not just a banality; we are taking about what is considered one of the most robust and fundamental scientific laws of the traditionally considered the most robust science (physics). It is simply unimaginable to pretend that such law would not apply to economics. In simple terms, the economic growth as defined by neoclassical economics does not have a base on natural science. Even worse, this contradiction is known since at least four decades, but the economists of this approach have not taken such argument into account even until today. After decades of evidences against the expected function of the parameter, neoclassical economists and politicians still use and support indicators like GDP growth. To neglect critics is an option. As any idea, the GDP is a cultural construct that is chosen by will, and not by its objective essence, so that it is an option instead of a truth. As any option, there must be a reason to have chosen this option, such as ideology, convenience, etc.

In the face of a multidimensional global crisis, the society searches for a way to escape this crisis, debating solutions and assessing all the available methods to address the abundant set of problematics linked to this crisis. As we stated at the beginning of this chapter, we believe that science has the technical solutions for most of the current environmental problems since many decades. However, for a solution to become reality it must surpass a big gate: the economic feasibility.

We can clear most of the polluted waters on earth, replace most of the polluter energy sources by clean ones, reduce the discarded residual materials, reduce the extraction/exploitation of diminishing natural resources or endangered species, decrease substantially the waste of materials and energies, increase the amount and size of alternatives to the mainstream market products. Yes, there is enough knowledge and existing technological solutions to address these main global problematics. But society does not implement them. Why? Because it is economically not viable (see also Chap. 2 in this book).

Natural sciences have gained status as many of their achievements seem to be simply undebatable. That kind of power let people to think that if something is stated by science, it must be true. As most people assume that economics is a science, there is a tendency to accept the argument of "economical unviability" as something that must be true. Therefore, society accepts such argument as valid and considers it

---

[13]In brief, it states that energy (and matter) cannot be created nor destroyed, but only transformed. Related to the production of goods and services this means that they cannot be produced out of nothing. If real GDP increases, so has to increase the use of energy and material input, unless a purely qualitative transformation takes place. Obviously, that is not the case of a quantitatively driven science, like neo-classical economics.

superior to other arguments that can involve even the surviving of animals, people or entire ecosystems. But the evidence shows that the classical economic arguments are not objective realities nor scientific facts rooted in natural sciences. On the contrary, many of the economic rationale is in direct contradiction with the core of natural sciences, here especially the economic growth imperative, since nothing in nature grows forever.

1.11 Towards a Transdisciplinary Ecological Economics

Economics (from Greek "oikonomia") is the science of managing the "oikos", the scarce recourses of a household, in an efficient manner (Aristotle 1995). Applied to today's world this would involve the efficient use of natural and social capital of society. To fulfil its endeavor, it is essential for economics to include the new insights of sciences and disciplines, in order to find a new vision regarding human development and nature. To do so, it is essential to set bridges with other disciplines that can contribute to its understanding of the current world. The latter can happen under the exercise of a transdisciplinary work, which goes far beyond the concept of interdisciplinarity.

The essence of transdisciplinary work is overpassing the boundaries of individual disciplines in the construction of new linguistics that set the background for the emergence of new concepts and ideas that can better answer the challenges that the current and future exercise of economics so urgently demands. In this book, many of these concepts, like ecosystem services, are breaking the ground of traditional economics with new interesting proposals. Certainly, this chapter has set a critic against the neoclassical monodisciplinary economic approach. It would not be fair to say that all what this discipline has reached is worthless. In fact, the reason that this discipline is still so hegemonic may partially be owed to the immaturity of an alternative school of economics.

This chapter is an exercise to understand the roots of the global crisis, but it is also a proposal on how to set new roads for an alternative school of economics to run through it. Such economics should have also new tools to deal with the qualitative domain of human existence, and a strong emphasis in relational, holistic and non-materialistic phenomena of life as a whole. In the search of objectivity, science tries to step out of any subjective, emotional or spiritual matters, and so did neoclassical economics. However, we need a school of economics that dares to work with the different dimensions of human life or, to put it in the words of Theodor Roszak to introduce the book "Small is Beautiful" (Schumacher 2010) we need "a nobler economics that is not afraid to discuss spirit and conscience, moral purpose and the meaning of life, an economics that aims to educate and elevate people (. . .)".

References

Aristotle (1995) Aristotle, 1995 Politik, 1. Buch. In: Aristoteles Phillosophische Schriften in sechs Bänden. Felix Meiner Verlag, Hamburg

Capra F (1997) The web of life; a new synthesis of mind and matter. Harper Collins Publishers, London

Fantuzzi G (2014) The sound of health. Front Immunol 5:351

Freudenthal G, McLaughlin P (2009) The social and economic roots of the scientific revolution: texts by Boris Hessen and Henryk Grossmann. Springer Netherlands

Fuders F (2017) Neues Geld für eine neue Ökonomie – Die Reform des Geldwesens als Voraussetzung für eine Marktwirtschaft, die den Menschen dient. In: Finanzwirtschaft in ethischer Verantwortung – Erfolgskonzepte für Social Banking und Social Finance. Springer-Gabler, Wiesbaden, pp 121–18

Fuders F, Nowak V (2019). https://link.springer.com/chapter/10.1007/978-3-030-25513-8_17

IPCC (2011) Special report renewable energy sources (SRREN) – summary for policymakers. Abu Dabi

Kuhn TS (1970) The structure of scientific revolutions, vol II, 2nd edn. The University of Chicago, USA

Maturana H (1988) Ontology of observing, the biological foundations of self-consciousness and the physical domain of existence. Conference workbook: texts in cybernetics, American Society for cybernetics conference, 18–23 October, Felton

Maturana H (1989) Language and reality: the constitution of what is human. Arch Biol Med Exp 22:77–81

Maturana H (1997) Metadesign. Instituto de Terapia Cognitiva INTECO. http://www.inteco.cl/articulos/006/texto_ing.htm. Maturana, Humberto

Maturana H, Mpodozis J (2000) The origin of species by means of natural drift. Rev Chil Hist Nat 73:261–310

Maturana HR, Varela FJ (1980) Autopoiesis and cognition. The realization of the living. Reidel, Dordrecht, p 13

Max-Neef M (2005) Foundations of transdisciplinarity. Ecol Econ 53:5–16

Max-Neef M (2010) The world on collision course and the need for a new economy. Ambio 39:200–210

Meadows DH, Meadows DL, Randers J et al (1972) The limits to growth; a report for the Club of Rome's project on the predicament of mankind (PDF). Universe Books, New York

Novak JD, Cañas AJ (2008) The theory underlying concept maps and how to construct and use them, technical report IHMC Cmap Tools 2006–01 Rev 01–2008, Florida Institute for Human and Machine Cognition, available at: http://cmap.ihmc.us/Publications/ResearchPapers/TheoryUnderlyingConceptMaps.pdf

Popper KR (1959) The logic of scientific discovery. Hutchinson & Co. London, UK

Röling N (2000) Gateway to the global garden: Beta/gamma science for dealing with ecological rationality.. Eight Annual Hopper Lecture. University of Guelph, Guelph

Schmidt U (2004) Official answer from January 5, 2004 to an inquiry of Rudolf Kraus (member of parliament), Bundestag, BMGS Leitungsabteilung Registry No 3538. Bundestag, Berlin

Schumacher EF (2010) Small is beautiful; economics as if people mattered. Harper Perennial, New York

Smith P, Max-Neef M (2011) Economics unmasked: from power and greed to compassion and the common good paperback. Green Books, UK

Varela F, Thompson E, Rosch E (1991) The embodied mind: cognitive science and human experience. MIT Press, Cambridge

von Bertalanfy L (1955) An essay on the relativity of categories. Philos Sci 22(4):243

Chapter 2
The 'Tragedy of the Commons' and the Role of the Money Interest Rate

Felix Fuders

> *(...) do not lend money at interest, and do not take a bribe against the innocent.*
>
> (Psalm 15:5)

2.1 Introduction

Determining the macroeconomic causes of natural resource degradation is key preventing further loss of global biodiversity. Many ecological economics, environmental economics or green economics texts indicate that the destruction of the environment and the overexploitation of natural resources can be partly explained by the "Tragedy of Commons" (Hardin 1968). In this metaphor, the overexploitation of nature results from natural resources and ecosystem services being freely accessible goods (*Open Access Regimes*). Classic economic theory conceptualizes this as a "market failure" that could be corrected by *assigning these goods private property rights* (Demsetz 1967; Cheung 1970; Hardin 1978; Ault and Rutman 1979; Barkley and Seckler 1972; Dales 1972; Smith 1981; Welch 1983; see also: Stevenson 1991; Common and Stagl 2008; Daly and Farley 2011). This argument is based on the assumption that only market goods (excludable and subject to competition) can be efficiently allocated by the market. To put it more simply, private actors will take care of their investment to maximize their economic benefit.

This chapter aims to demonstrate that the real reason for the overexploitation of natural resources cannot be ascribed to free access regimes. Rather, it originates from the *obligation to grow* the real economy. This obligation comes from the money interest rate as an opportunity cost of every productive investment and represents a pressure that applies to both private and common goods. Consequently, and contrary to conventional economic theory, the artificial privatization of natural resources that had always been freely accessible will not eradicate the underlying

---

F. Fuders (✉)
Economics Institute, Faculty of Economics and Administration, Universidad Austral de Chile, Valdivia, Chile
e-mail: felix.fuders@uach.cl

F. Fuders and P. J. Donoso (eds.), *Ecological Economic and Socio Ecological Strategies for Forest Conservation*, https://doi.org/10.1007/978-3-030-35379-7_2

problem or increase economic efficiency, but instead will even decrease it. The chapter will also explain that the allocation of property rights for natural resources that were originally freely accessible does not necessarily avoid an unsustainable management of those resources, and in fact does not even ensure that those resources will not be exploited until their total extinction.

The chapter starts with a classic Sustainable Yield curve analysis, used for ~100 years in natural resource management (Tsikliras and Froese 2019), as this is necessary to understand the argumentation line. Each section will end with a conclusion on which will be built each following section.

2.2 The *homo oeconomicus* Will Not Exhaust His Source of Income

2.2.1 Maximum Sustainable Yield

In order to understand why economic theory usually assumes that privatization helps to solve the problem that Hardin (1968) described as the "*Tragedy of the Commons*," let us imagine, by way of example, a fish population in a management area. Figure 2.1 shows a modified Sustainable Yield curve of that population. Although the curve refers to fisheries management, this model can be applied to *forests or any other renewable natural resource* (Maunder 2008).

The abscissa shows the fish stock and the ordinate shows the flow multiplied with the price of Fish P_f. The classic Sustainable Yield curve represents the stock's *net*

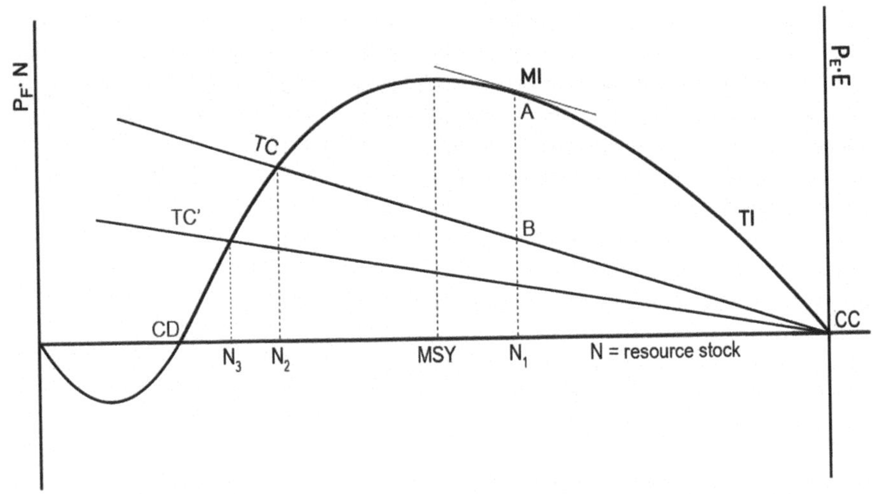

Fig. 2.1 Maximization of benefits (fishing example). (Source: own work, adapted from Daly and Farley 2011, p. 215)

growth rate corresponding to each stock level. It is called this way because if each year the same number of units is extracted as replenished (the distance between the abscissa and the curve), the stock remains the same (ceteris paribus).[1] Since Fig. 2.1 modifies the classic Sustainable Yield (SY) curve such that in the ordinate the quantity of harvested fish (N) is multiplied by the market price P_F, the Sustainable Yield (SY) curve becomes a sustainable Total Income (TI) curve.[2]

In the long term, any population of living beings (from fish to bacteria) will arrive at a point of stable equilibrium in which the reproduction rate and the death rate are equivalent: point CC on the graph (*carrying capacity*). The reason for this is that the habitat does not have resources for an ever-growing population; if the population grows, resources become scarcer. At point CC, the growth rate of the stock is 0.

Point CD (*critical depensation*) represents the critical stock level – the point at which the population cannot sustain itself and begins to decrease on its own. This is the *minimum viable population*. After this point negative feedback loops can result in population extinctions. For example, this could occur in fish if eggs were not fertilized because males are not able find them. In the case of a forest, if there are very few trees left, soil erosion and nutrient leeching might result in seeds unable to germinate. In addition, when there are few species left poor genetic diversity makes populations more susceptible to hereditary diseases, illnesses, parasites, etc. Although many authors leave the CD point out from Sustainable Yield curves (e.g. Turner et al. 1994; Maunder 2008; Tsikliras and Froese 2019), it is important to understand that very small populations are at risk of extinction even without further human harvesting (Daly and Farley 2011).

The specific shape of the curve tells us that it might be profitable to decrease the stock in order to increase the annual net regeneration. For example, a forest can be made more profitable by removing some trees, so that the others get more sunlight and generate more energy via photosynthesis, and thereby grow faster. Nevertheless, the owner should not have an interest in decreasing the stock by more than the *Maximum Sustainable Yield* (MSY). We will see that they would have even less interest in doing so by taking into account selling prices and harvesting costs.

[1]There is much uncertainty about what this Sustainable Yield curve looks like in the real world. It is not precisely known at what rate a given population reproduces, since this depends on many unpredictable factors, for example: time, predator-prey cycles, illnesses, etc. In addition, the exact size of a population is often unknown. Moreover, habitat destruction, air and water contamination, climate change, etc. all influence the shape of the curve (Daly and Farley 2011).

[2]The shape of the curve does not change as compared to the classic Sustainable Yield curve, since it is multiplied by a constant. However, the fish price is not necessarily a constant and depends, among other things, on the amount of fish captured. Here a constant fish price is assumed to simplify the analysis.

2.2.2 Maximization of Economic Benefits

The economic benefit $\prod$ is the difference between the Total Income (TI) (sometimes also called Total Revenue) and the Total Cost (TC):

$$\prod = \text{TI–TC} \quad (2.1)$$

The Total Cost refers to the *costs of harvesting a sustainable amount* (the amount that is reproduced). These costs include labor, equipment, etc. If we define these costs as our "effort," then our costs are the price of effort P_E multiplied by the effort E:

$$P_E E \quad (2.2)$$

If we harvest a fish population at point CC in Fig. 2.1, then our costs increase as the fish population is reduced. This is because catching fish will require more effort as their density decreases. In order to facilitate this analysis, let us suppose there is a linear relationship between the total costs and the quantity of fish caught. But this relationship is not necessarily a straight line. Rather, one could imagine that as the population decreases, the difficulty of catching an additional fish increases at a growing rate, and, as a result, the curve would have a steeper slope as we move to the left (Daly and Farley 2011).

For the fisherman who wants to maximize his economic gain, it makes sense to exploit the fish stock until it is at the point where the difference between the Total Income (TI) curve and the Total Cost (TC) curve is greatest (Fig. 2.1, point N_1). Here, the slopes of both curves are equal and are represented on the graph by parallel tangents.[3,4] This idea can be explained intuitively: if there is an abundance of fish, it is easy to catch them and the costs of catching each fish are low. It is not beneficial to decrease the stock beyond point N_1, even though the fisherman could obtain a large one-time income, since doing so would make fishing more expensive in the future. This last observation is *only valid if we leave the money interest rate outside of our analysis*. Later, we will see how taking the interest rate into account changes the situation.

[3]The distance between two curves is greatest at the point at which their slopes are equal. Given that the slope of one curve at a given point is equal to the slope of a tangent that touches the curve at that point, the two curves TI and TC have equal slopes where their tangents are parallel.

[4]The slope of the Total Cost (TC) curve is the Marginal Cost (MC), that is, the cost of catching one more fish. The slope of the Total Income (TI) curve is the Marginal Income (MI), that is, the return from harvesting the next fish. We see that the classical rule of maximization of benefits MC = MI (e.g. Frank 2015) also applies to the exploitation of natural resources.

2.2.3 Conclusion

The classic conclusion drawn from what has been outlined so far is that the *homo oeconomicus,* in seeking to maximize benefits, will not even reduce the stock beyond its MSY, much less reduce it to the point of putting the resource at risk of extinction.

2.3 The Case of Free Access Regimes

2.3.1 The "Tragedy of the Commons"

In the previous analysis, we assumed a private owner of the fish population (management area). In reality, oceans are essentially free access regimes with rival goods (fish). Here, what Hardin (1968) described as the "Tragedy of the Commons" applies. Instead of avoiding reducing the stock beyond N_1, new fisherman will enter the market as long as it is possible to achieve some gain; that is, as long as the Total Income (TI) is greater than the Total Costs (TC). In other words, the amount harvested will be greater than the amount that is reproduced annually (Stevenson 1991; Daly and Farley 2011). As a consequence, the fish stock will start to decrease.

New fisherman will enter the market until economic benefits can no longer be obtained; that is, until the Total Cost (TC) curve and the Total Income (TI) curve cross (see stock N_2 in Fig. 2.1). If harvesting costs are also reduced, e.g. because a more efficient way of catching fish is discovered, then at this point the stock will be even smaller (represented by the line TC' in Fig. 2.1). The stock that remains gets dangerously close to that at the CD point, threatening the viability of the population (Daly and Farley 2011).

2.3.2 Conclusion

According to what has been said so far, the conclusion is: the overexploitation of the ocean occurs because it is a free access regime. If, on the other hand, the ocean was privatized (by allocating private property via management areas or exclusive exploitation rights), it would not be overexploited because the private owners would attempt to preserve their stock of natural capital. This is the crux of the belief that privatization of nature would prevent or reduce unsustainable resource exploitation. The next section will demonstrate that in reality, the ocean – or other freely accessible natural resources– are not overexploited because they are free access regimes. On the contrary, it will be shown that it may be economically desirable that they be freely accessible.

2.4 Free Access to Natural Resources Is Economically Desirable

2.4.1 The Natural Dividend Is a Monopoly Rent

In the theoretical model of "Perfect Competition", competition between producers pushes prices down until they reach the minimum point on the Average Total Cost curve in the short and long term (SATC; LATC in Fig. 2.2). Gains that are greater than the opportunity cost of the invested productiion factors do not exist[5] (Frank 2015). If a producer was able to achieve a greater economic benefit, then this would indicate that the benefit is not justified by the investment of production factors; that is, the gain would not be based on the producer's effort and would be generated because of a *market failure* (monopoly, collusion, unfair competition).[6]

Note that this desired outcome of competition is not generated by the "Invisible Hand" of egoism, as some have misinterpreted Adam Smith's (1952) famous metaphor (e.g. Frank 2015), but *even though* we behave only thinking of ourselves. Even though we only think of our own benefit, competition forces us to produce good products at good prices to avoid losing competitiveness. In other words: competition limits our self-interest and induces us to behave as if we loved one

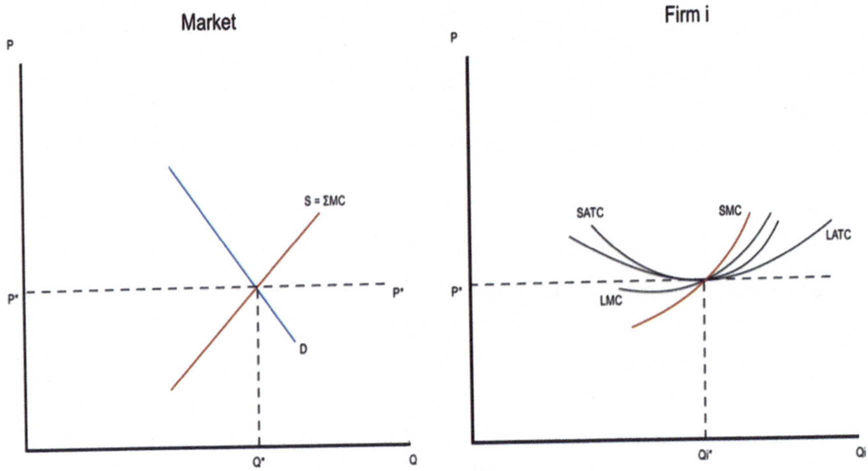

Fig. 2.2 Long-term equilibrium Perfect Competition. (Source: own work. Explanation: The market price P_* is at the lowest possible level (minimum Short-term Average Total Costs curve [SATC] and Long-term Average Total Cost curve [LATC]))

[5] Here, opportunity costs are included in the Total Costs.

[6] This is the reason why Competition Law usually prohibits collusion, unfair competition and the abuse of a dominant market position (Fuders 2009b).

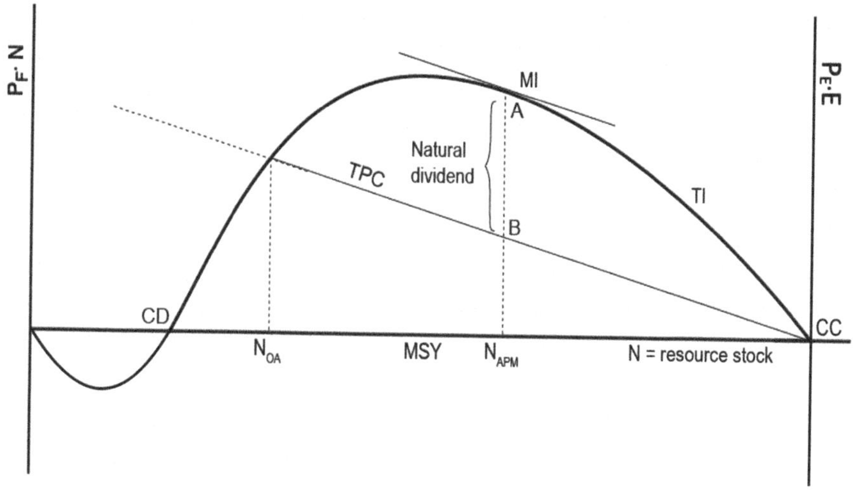

Fig. 2.3 Natural Dividend. (Source: adapted from Daly and Farley 2011, p. 226)

another (Fuders 2017; Fuders and Nowak 2019). We can see what undesired outcomes egoism causes if we take a look at any monopoly, which are contrary to perfect competition.

Monopolists can take advantage of their dominant market position to, for example, sell at a higher price than their production cost (including the opportunity costs of production factors invested). The resulting income is known as *monopoly rent.* Such an income is not explained by their efforts (the high quality of their product), but rather by their abuse of the power associated with the dominant market position. The term "rent" in economics emphasizes the fact that the payment is not justified by the producer's effort: it is a gain for nothing in return. The exploitation of natural resources can also generate gains that are greater than the mere compensation for effort; in Fig. 2.3, it is the difference between points A and B. This gain is often called a "natural dividend" (Daly and Farley, 2011: 226) or "natural resources income" (Acquatella, et al. 2013; Conrad and Clark 1995: 89, 94; Neher 1990: 51), since it originates from the reproductive force of nature, not from the exploiter.

This gain is only possible if exploitation is limited to a *single owner*, whether private or public. In contrast, as outlined above, in a free access regime exploitation by an increasing number of people would lead to stock N_2 in Fig. 2.3, where the Total Costs (TC) = Total Income (TI); that is, where there is no gain greater than the remuneration for the invested production factors equivalent to their opportunity cost.

Given all of this, we can argue that the "natural dividend" is really a *monopolistic rent*, and accordingly can be viewed as an *unjust* profit. All earnings that go beyond the recuperation of Total Costs –which include opportunity costs, that is, the profit that one could have earned by expending the same effort on the best alternative opportunity– are income that is not based on the producer's effort. In a perfect competition situation, this earning does not exist.

2.4.2 *Free Access Ensures Efficiency*

If we consider that a natural dividend can be better characterized as monopolistic income, then one can argue that free access regimes are actually desirable, since their result corresponds to the result produced by Perfect Competition. When there is no artificial restriction on access, ever more competitors (here fishermen) enter the industry (here the ocean). This increases the fish supply on markets and pushes prices to the lowest possible level, that is, to the point at which they only cover the Total Costs. *This is a desired situation*. In this situation, Price equals Marginal Income and Marginal Costs, and we can achieve what in economics is called "efficient allocation of resources" or "allocative efficiency" (Frank 2015: 346; Common and Stagl 2008: 310).

This is precisely the situation that existed for centuries: an undefined number of small fishermen (suppliers) fishing without any restriction; a large supply of fish in relation to the demand; and, consequently, low fish prices. The price of fish used to be so low that even today, many Christians eat fish on Fridays and during the Holy Week instead of fasting because fish was considered such a low value food that eating it was similar to fasting. We can see this as an example of Perfect Competition that functioned for thousands of years, without significant exploitation of the ocean. That natural resources should not be privatized has already been emphasized by John Stuart Mill (1885), Henry George (1935) and Silvio Gesell (1949), with the same argument as this text does: to ovoid monopoly profits for one individual who owns the exploitation rights.

2.4.3 *Conclusion*

Free access to natural resources is economically desirable, since it ensures allocative efficiency. In contrast, the artificial restriction of access to resources that have always been freely accessible and are produced by nature allows for monopolistic incomes, which Ecological Economics texts sometimes characterize as a natural dividend (or income). In addition to generating the described negative effects (monopoly rents), privatization does not necessarily solve the problem of overexploitation, given that it does not address the real reason why the ocean is overexploited, as will be explained in the next paragraph.

2.5 The Interest Rate as Opportunity Cost

2.5.1 *Gigantism: Just Greed?*

One could argue that the fact that natural resources are overexploited is not due to their nature as freely accessible goods, but rather that it is a problem of *scale*. To continue with the fishing example, if we were to continue fishing with small artisanal boats, such as those used 100 years ago, there would be no overexploitation *despite it being a free access regime*. With traditional technology, the TC curve in Fig. 2.3 would be so steep that the intersection with the TI curve would be far away from the CD point. There would be no danger of overfishing. The fact that Hardin's "Tragedy of the Commons" today is more relevant than ever is because the world has become relatively small in relation to exploitation technologies and not because population has grown, as Hardin suggested. Throughout human history, the ocean has been freely accessible, but in former times it was not "empty" or even close to being overexploited as it is today. One might say, God foresaw the world with such an abundance that normally, if we used traditional fishing technologies, it is unlikely that we could ever exploit the ocean to the point of emptying it, even if the worldwide population doubled. Rather, we can contend that the ocean is being emptied by "*gigantism*", to use the words of Max-Neef (1986): 58, 63, 136, 149, 151 f., 184. If we fished at a *Human Scale* (Max-Neef et al. 1991), then we would probably never empty the ocean.

One could ask what, then, is the reason for gigantism? Is it rooted in human greed? The answer is that it does not (only) have to do with greed. The owner of a modern fishing ship that currently exploits the ocean by periodically catching large hauls of fish does not necessarily earn more than a fisherman 100 years ago.[7] This is because the fishing ship owner's costs are enormous and therefore a respectively large amount of fish must be caught in order to recover those costs. Why, then, are bigger and bigger fishing ships built? The answer is that the pursuit of bigger ships derives from the *obligation to grow economically*. For its part, this obligation *stems from an interest rate* that exerts a tacit but constant pressure on the real economy to produce more and more, a fact that, unfortunately, is poorly understood in economic thinking. Our current financial system has existed for approximately 700 years, and during that time the economy has only done well if it was able to constantly increase production.[8] The following paragraphs are dedicated to explaining this phenomenon.

[7] It might be that the fishing company is indebted to such an extent that it does not even have enough income to decrease its debt.

[8] Approximately 700 years ago, the "*Bracteate,*" a coin that was difficult to hoard and was interest free, was replaced by the "*Denarius perpetuus,*" the money that is used today (see: Walker 1959; Azkarraga et al. 2011: 54; Fuders and Max-Neef 2014b).

2.5.2 The Exponential Growth of Deposits

Depending on the interest rate, financial assets held in bank accounts double in approximately 15 years (through interest and compound interest). That is, if one starts with 100 €, in 15 years the amount in that deposit would increase to 200 €, and in 15 more years it would be 400 €. As a consequence, the amount being gained, increases at a faster and faster pace, relative to the initial amount invested. Even the most sophisticated computers in the world will at some point no longer be able to calculate the amount of money that is accumulated in this way, due to a lack of zeros. This is the logic of an exponential function (Kennedy 1990, 2011; Creutz 1993; Constanza et al. 2012; Lietaer et al. 2013).[9] It is the absurd and immanent logic of our current unnatural (Soddy 1934: 176) financial system, in which money is created with money, a fact that was criticized already by *Aristotle* (Aristotle, 1995: 23).

It is worth to mention that *all* bank accounts that pay interest grow exponentially if the interest is not withdrawn. Even if the rate is very small, the moment will come when the deposit will have doubled. As long as a value doubles at regular intervals, the value grows exponentially. Many bank clerks even use this argument to convince clients to invest with them and illustrate how the client's savings will grow over time.[10] For the economy as a whole, this is even true if the money is withdrawn from a bank account since it will most likely appear on some other bank account (this effect is called *money multiplier*) and generate interest there. This means, even though this exponential function might not be recognized on one specific bank account, all bank accounts of an economy taken together will show exponential growth over time. The exponential function can be observed in Fig. 2.4, which depicts the money supply M3 in the U.S. (composed of coins and bills, sight and short-term deposits). The same exponential function can be seen in most economies in the world.[11]

2.5.3 The Exponential Growth of Total Debt

At the end of the day, borrowers are the ones who pay for the interest that banks pay out to their depositors. Accordingly, not only do banks endeavor to replace each

[9]If, for example, 2,000 years ago one cent was deposited in a bank with an interest rate of 5%, this investment today would be the equivalent of approximately 707 billion planets earth of pure Fein gold (assuming that the price of gold remained constant) (Kennedy 2006: 243).

[10]In an advertisement a big German finance group explains the compounding effect with an illustrative example. Money invested in this institution multiplied like chickens. When chickens lay eggs, chickens hatch again from these eggs, which then again lay eggs, see: Deka 2019.

[11]A current overview is provided by (Wikipedia 2019). The fact that the money supply follows an exponential function in developed countries and developing countries alike is highlighted by Lietaer et al. (2013).

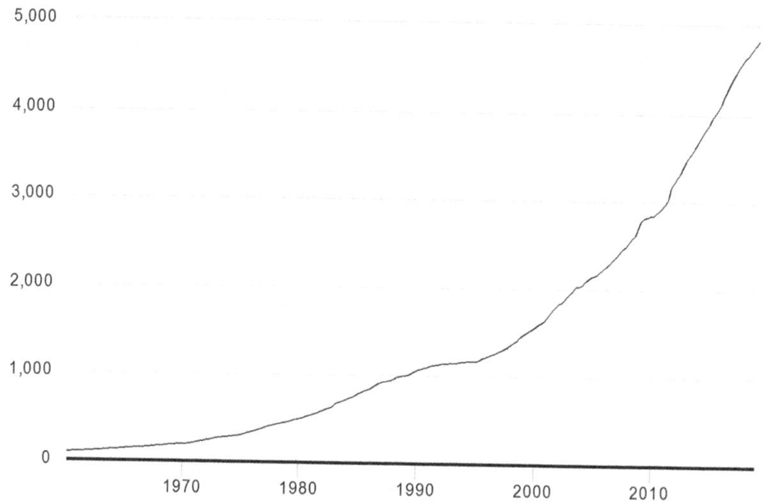

Fig. 2.4 Estimated money supply M3 U.S. (billions USD) (Since 2006, the Federal Reserve of the U.S. (Fed) has not published the money supply M3, which Central Banks throughout the world use as indicator to forecast inflation. Nevertheless, one can make an estimate by adding up the different components of the money supply M3 (Money base, cash, sight deposits and short-term deposits) that are still made public). (Source: own work. Based on data provided by OECD)

amortized loan with a new loan as soon as possible, to maintain interest payments functioning, but they must also constantly increase the volume of loans because the deposits also grow through interest and compound interest.[12] This is why the *total debt of a given country must grow in the same way* as the money supply does (Soddy 1934; Kennedy 1990, 2011; Farley et al. 2013; Constanza et al. 2012) (for the US economy see Fig. 2.5). As long as a deposit grows through the payment of interest, there must be more debt in the country, *since there is no interest without debt*. To use the words of Frederick Soddy (1934: 25) "For there is no credit without debt (. . .) money is a credit-debt relation from which none can effectually escape". This is a simple fact, and there is no way out of it. It seems to be necessary to emphasize this,

[12]The longer a financial system functions and the greater the volume of total debt in an economy is –that is, the more saturated the debts of economic agents (homes, companies and government) become– the more difficult it is for banks to continue increasing the debt volume. In their effort to find new borrowers, banks begin to loan money even to borrowers whose solvency is questionable and at low interest rates, as happened in the U.S. until 2008 in the mortgage loan market and as continues to happen today in Chile and also in Europe with loans to State governments; these are the loans most recently backed by the "rescue fund" ESM (European Stability Mechanism). Here, we can see why financial crises occur in every country at regular intervals, and why the so-called "sub-prime" crisis and "euro crisis" are actually the same crisis: the financial system is coming closer and closer to collapse. The system crumbles when banks cannot find enough solvent borrowers who can pay for the interest that the banks must pay on their deposits. The last time this happened on a large scale was in 1929, and it will probably happen again soon. See: Fuders 2009a, 2010a, 2016, 2017; Fuders, et al. 2013; Fuders and Max-Neef 2014a, b.

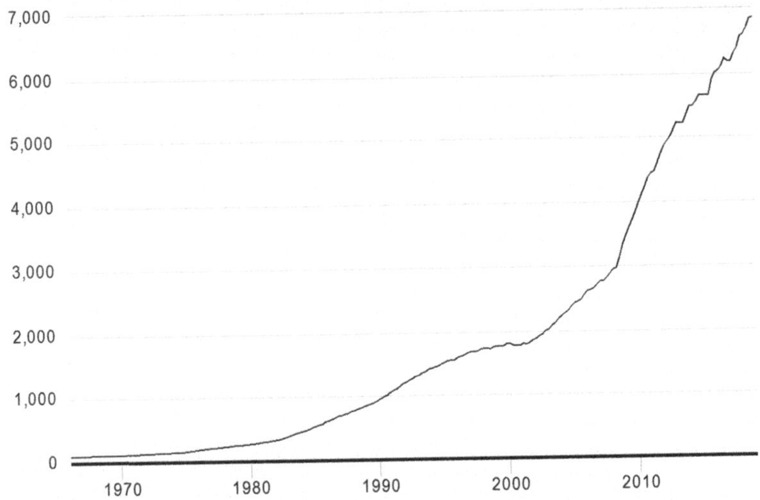

Fig. 2.5 Federal debt U.S. (billions USD). (Source: own work, based on data provided by the US department of treasury)

since some criticize this view arguing that simple explanations were to be mistrusted (Strunz et al. 2015). But is a complicated model always better than a simple one? In fact, the truth is usually simple. To use the words of Albert Einstein and Leopold Infeld: "Most of the fundamental ideas of science are essentially simple, and may, as a rule, be expressed in a language comprehensible to everyone" (Einstein and Infeld 1938: 29). Indeed, complicated "pseudo-mathematical" models increase the probability for errors (Keynes 1936: 297) and often make it difficult to focus on what is really essential. Here Goethe was probably right when he said: "What is hardest of all? That which seems most simple: to see with your eyes what is right before your eyes" (Goethe 1996: 230).

2.5.4 *The Obligation to Grow (Microeconomic View)*

We can now understand why our financial system is the reason for both neoclassical and Keynesian models always seeking economic growth. In political economics textbooks, the panacea for unemployment is economic growth (see, e.g.: Lachmann 2006; Fernández et al. 2006; Cuadrado et al. 2006). Indeed, empirical observations suggest that the economy (GNP) must grow three percentage points in order for unemployment to decrease by one point (Okun 1962). As a consequence, either the economy grows or the number of unemployed, and associated social costs, increase. This is why dominant political parties generally preach economic growth in their manifestos. Although we know that we have to grow in order to avoid unemployment, it is not well understood in economics why this is.

The obligation to grow finds its origin in the fact that deposits (and debt) grow by interest that, at the end of the day, someone has to pay. The problem is not that money is the counterpart of debt, as some might think, but that both money supply and debt grow exponentially because of interest (for a similar argumentation see: Kennedy 1990, 2011; Creutz 1993; Constanza et al. 2012; Löhr 2012; Lietaer et al. 2013). Governments of States, companies and households are ever more indebted. This can be confirmed by comparing companies' balance sheets today with those some 20 years ago: one is likely to observe more and more borrowed capital. But even those lucky producers that are not financed with borrowed capital are not free from pressure to expand. This is because their financial loads also increase since in most countries the national debt is growing, and thus the government is obliged to increase taxes on citizens and companies. In addition, workers want salary increases to compensate for inflation, which eventually is also the result of the growing money supply (Fuders et al. 2013; Fuders and Max-Neef 2014a, b).

There is a fourth and perhaps most important reason for which the productive economy must grow: *interest as an opportunity cost*. The interest rate that banks pay is a reference point against which the success of any productive investment is measured (e.g. Copeland et al. 2008), even in Ecological Economics texts (Common and Stagl 2008). If a company does not achieve at least an equal return on investment, its existence becomes senseless (see also Suhr 1988). In other words, even if a company did not have borrowed capital on its balance sheet and the government did not increase taxes and workers did not require a readjustment for inflation, the company would still have to grow because of the money rate of interest as an opportunity cost. Hence, the interest rate imposes the rhythm to which the productive economy must "dance" (Fuders 2017: 127 f.). As a result, economic growth under our current financial system can be seen *as an obligation*[13] *– not simply a "fetish"* (Hamilton 2003).

If companies do not manage to grow by expanding sales of their products, they will try to save costs so that their rentability would not fall below the opportunity costs. One frequently used cost reduction measure is replacing the human work, which is usually the heaviest cost burden, with machines, that is, increasing the productivity of labor. We can conclude that there is a constant and continuous pressure on companies to *either increase sales or replace human workers with machines* in order to reduce the steadily growing costs. This is why we see this well-known (but not well understood) dichotomy everywhere in which either the economy's output or unemployment grow. Since any politician fears unemployment and its associated social unrest, everything is done to foster economic growth. Here

[13]This is true if the sole purpose of the business is to make money. However, entrepreneurs could, of course, also start their business because they wished to make use of their talents in order to produce products that contribute to the common good, i.e. to the quality of life of people, and not only for financial gains. If all entrepreneurs understood this we could come close to what might be called an 'economy of neighborly love' (Fuders 2017; Fuders and Nowak 2019). In that case, entrepreneurs would care less about interest as an opportunity cost, and it might be sufficient for them to earn enough to live off it.

we understand that degrowth strategies –even though highly necessary to avoid further overexploitation of our natural environment– will cause unemployment and deep economic depression if the actual financial system is maintained.

2.5.5 The Obligation to Grow (Macroeconomic View)

What we have just seen from the microeconomic (supply side) point of view, we can also observe from the macroeconomic (demand side) point of view. We are obliged to grow on the one side while on the other side *interest also makes this growth possible*. We can envision this from three perspectives.

The higher the money interest rate, the faster deposits grow and therefore the higher will be the total money supply (composed of cash plus deposits). The role of the money interest rate in the exponential growth of the money supply can be seen in the U.S. (Fig. 2.4) and can be found similarly in any other countries. A higher money supply in turn means more monetary units circulating and, as more money is spent, aggregate demand grows. In other words, the money interest rate not only forces the producer constantly to achieve an equivalent (or higher) rentability, but also makes economic growth possible since it increases money supply and aggregate demand. That is to say, even if all loans in an economy were repaid, the money supply would not return to its original level since deposits would have acquired interest. We might call this effect "*interest money creation*" (Fuders et al. 2013). However, in classic monetary theory, the effect the money interest rate has on money supply and economic growth is seen exactly the other way around. Central Banks actually put interest rates down if they wish to increase economic activity.

That the money interest rate makes aggregate demand grow can also be explained using the so-called *money multiplier* model (sometimes called *checkbook money creation*). This is a model Central Banks and conventional monetary theory apply to explain the effect that the money interest rate exerts on money supply. According to this concept total money supply grows through the expansion of credits (e.g. Galbraith 1983; Stiglitz 1998; Larroulet and Mochón 2003; Mankiw and Taylor 2014). It is the reason why Central Banks try to increase interest rates if they fear inflation and wish to calm down economic growth as this has the effect of disincentivizing the granting of loans (e.g. Cuadrado et al. 2006; Fernández et al. 2006). However, the expansion of credits will most probably slow down *only initially* after raising the interest rates. In the medium and long terms, higher interest rates will result in more people depositing money into bank accounts, and also these deposits will increase even faster if interest rates are higher. Yet, money deposited on bank accounts *will not be out of circulation for a long time*. The growing deposits put an enormous pressure on the bank to increase the volume of money loaned out, since someone has to cost the interest that banks pay on deposits, otherwise the bank would go bankrupt soon. As outlined above, there is no interest without debt.

Hence, money deposited in bank accounts will circulate in the economy via loans.[14] The more money is deposited, the more (and not the less) loans the bank will have to grant. This means, also applying the *money multiplier* explanation higher interest rates will eventually lead to an increased credit expansion and with it money supply, aggregate demand and economic growth, and vice versa (Fuders et al. 2013; Fuders 2017; Stehlík et al. 2017), which is why it can be agreed upon that conventional monetary policy should be completely reconsidered (Lee and Werner 2018).

The third and simplest explanation why higher and not lower interest rates foster aggregate demand is that the money interest rate is a "reward for parting with liquidity" (Keynes 1936: 167).[15] The higher the interest rate, the higher the incentive to lend money (or to save it in a bank which lends it for you). Hence, the higher the interest rate, the easier money circulates in the economy and vice versa.

The aforementioned can be verified empirically. Notwithstanding the possibility of strong short-term fluctuations, the money interest rate and GDP growth rate are clearly positively correlated with each other when viewed over long time periods. Fig. 2.6 depicts this relationship for the US economy (for other economies see Lee and Werner 2018). If the classic interpretation of the effect that the money interest rate has on money supply and GDP was correct, then interest rate and GDP growth rate should instead be negatively correlated (Fuders et al. 2013; Stehlík et al. 2017).

2.5.6 Clear Distinction Between Money Interest Rates and Real Economic Profits

The fact that in both economics and finance frequently all types of earnings are labelled as "interest", must not lead to confuse returns driven out of productive investments with interest that is charged for loans.[16] Only the money interest rate

[14]Here we can also question if the so-called "Keynesian multiplier" really exists. In the view of post-Keynesian economics, the government can attenuate economic cycles via anticyclical spending and saving. That is, in times of economic boom and high tax incomes the government saves up money in funds that later can be spent if the economy drives into a recession to attenuate the economic downturn. However, money that is saved on bank accounts will keep on circulating in the economy via credits. Does it make a difference if the money is put into circulation by the government or the banking system?

[15]This is an excellent definition. However, many textbooks state that the money interest rate was a "prime to abstain from consumption" (e.g. Süchting 1995: 437; Engelkamp and Sell 2005: 166). However, money is only a voucher, or accounting unit to facilitate barter. Money is not a real good and does not have any intrinsic value, from which consumption one might abstain. Only if the voucher is interchanged for a real good, such as a car, and the owner decides to lend it to someone instead of using it himself, only then he abstains from consumption. And only then it makes "sense" (Steiner 1979: 50) and it seems to be justified to charge a rent (on the moral-ethic aspects of interest see Fuders 2010b, 2017).

[16]Interestingly, for Muslims the Quran prohibits this equalization of productive gains with interest for lending money, cf. Surah 2.275. The Holy Bible also prohibits the money interest rate (e.g.

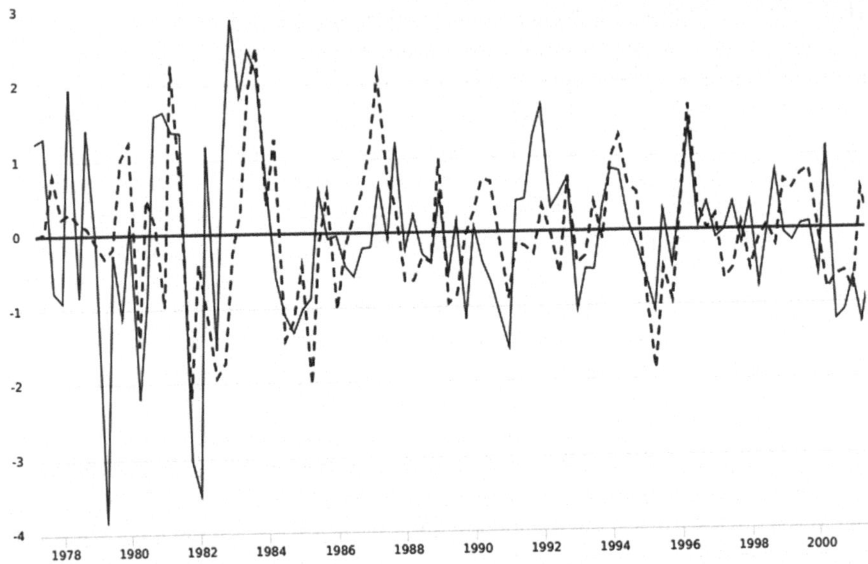

Fig. 2.6 Interest Rate and real GDP growth (% change from year ago). Solid line: real GDP growth; dashed line: 10-Year Treasury Constant Maturity Rate. (Source: own work, based on data provided by Federal Reserve Bank of St. Louis)

generates interest constantly and independently of economic production, and in this way creates the unnatural, exponential expansion of money supply and in this compels the productive economy to keep pace. Of course, also in a zero-interest environment there are possibilities to make a profit via productive investments. However, this profit would then be based on real economic output, i.e. on the real value that has been produced. This would not only make a mayor difference to the current system from a moral-ethical perspective (see Fuders, 2010b, 2017), but would also create a productive economy that is not compelled to constantly try to catch up with growing interest burdens.

Today we have quite some difficulties in distinguishing the money interest rate from real economic profits since we use the terms "money" and "capital" usually as synonyms, as if money was a production factor like capital. Yet money is no production factor, does not work (as banks sometimes try to make us believe) and does not have offspring (Aristotle 1995: 23). Money also is not a real good and, thus, does not have any intrinsic value (Galbraith 1983: 78; Keynes 1983). Money is only a medium to facilitate the exchange of real goods and services. Any increase of monetary units by interest (and debt) means that someone has to work for them and therefore increase production.

Deuteronomy 23,19; 23,20–21; 24,10; Exodus 22,25; Leviticus 25,35–37; Ezekiel 18,8; 18,13; 22,12; Psalm 15,5; 112, 5; Proverbs 28,8; Luke 6,30; 6,35; 1 Timothy 6, 9; see also Fuders 2017).

That real economic gains are also possible in an economic system with zero money interest rate, is sometimes used as an argument for interest being something natural and hence impossible to avoid. This argumentation is based on the *confusion of capital with money*. For example, a blacksmith lends one of his plows to a farmer for a fixed period of time and receives a sack of potatoes as a reward from the harvest. The sack of potatoes represented the interest for leasing the plow (Rösl 2006). However, the plow is real capital and not money. The sack of potatoes is a real economic gain. Something was really produced. In this example the blacksmith did not make money with money, which is why his gain is not interest but the lending fee for a real good (real capital). Note that this affirmation would even be true if the Blacksmith in this example was not paid with potatoes but with money that the farmer had gained from selling the sack of potatoes to someone else. Money flows from one person to another: total money supply does not increase. Because money is not earned with money, the financial system cannot decouple from real economy. If all loans worked like this, there were no money (and price) bubbles, and no obligation to keep up a certain growth rate.

In this context some interesting questions arise. In South America we still find groups of indigenous people that were able to maintain their traditional lifestyle and live in a more or less self-sufficient manner. If all people lived in such a way natural resources would surely not be over exploited. So, can we learn from them? How were they able to preserve this way of life? Is it because they are better connected to nature? Yes, this could be possible. But it also might be that the reason for being more connected to nature is because they do not have a financial system that obligated them to grow the last 700 years.

2.5.7 Conclusion

A positive money interest rate obligates to grow but also makes growth possible. That is, until the debt bubble bursts when too many debtors default and banks go bankrupt. At this point people lose confidence in the banking sector and start keeping money under the pillow. Then, money does not circulate anymore via loans, the money multiplier constricts, and deflation and deep economic depression prevail (as occurred during the Great depression – see Dornbusch et al. 2009). Since this is far from a desired outcome, for centuries world economies have grown or aspired to grow. The exponential growth of real productive capacity –inculcated by our financial system– also affected the size of fishing boats, to return to that example. That is, the growth imperative triggered not only the economic development but also the technological evolution over centuries.

Our financial system exerts constant pressure for economic growth, but just as there "ain't no such thing as a free lunch" (Dolan 1971: 14), there is no such thing as "sustainable growth". As the first law of thermodynamics tells us, we cannot create something out of nothing. The more that is produced, the more resources are needed, but many of them are finite (Maedows et al. 2004; Daly and Farley

2011; Azkarraga et al. 2011; Constanza et al. 2012; Farley et al. 2013). That endless physical growth is not possible is an issue that was already raised by Aristotle (1995: 17). Steady economic growth must lead to overexploitation of nature. This is even true for renewable natural resources, since steady economic growth means that at some point, we will start reducing the resource at a faster pace than it reproduces. This leads us to the next point.

2.6 Privatization Does Not Solve the "Tragedy of the Commons"

2.6.1 Privatization Does Not Necessarily Avoid the Overexploitation of Resources

Interest brings with it the need to grow, which gradually leads us to the *gigantism* that, when applied to free access regimes, leads to overexploitation. Yet, using the example of the ocean, even if this common resource was privatized (via management zones with a *single owner* per zone), overexploitation might still occur, even to the point of threatening population viability. That even a limited number of appropriators (owners) might increase the quantity of resource units they harvest so that they either dissipate all potential rents or totally destroy the resource has been pointed out also by Ostrom (1990). The reason is, again, the interest rate: the greater the interest rate, the smaller the Net Present Value (NPV) of future harvests.[17] If the NPV of future harvests is less than the amount I can obtain by selling all of my fish stock (or timber in case of a forest) today, it is economically profitable to sell the entire stock today instead of managing harvests sustainably and each year only harvesting the amount that is reproduced (Ackermann 1994). Or, in other words: if the interest rate paid by the bank is high enough, it is worth it to harvest the entire stock and let the money "work" in the bank. Envisaging such a situation Herman Daly once remarked that anything in nature that does not reproduce at the pace of the interest rate is potentially in danger of extinction (Daly et al. 1989).

2.6.2 Privatization Does Not Necessarily Avoid Unsustainable Resource Management

We should mention that the interest rate places a value on time ("time is money"). This is probably the main reason why renewable resources are often managed in

[17]The NPV of a future earning is the value that would have to be deposited in a bank today, at the current interest rate, in order for the deposit to reach the expected future value of the earning.

an unsustainable way. An example is the forest landscape in south-central Chile, which is dominated by exotic, fast growing species while native forests are generally secondary and degraded (see also Chaps. 4, 5 and 8 in this book). Why do forest companies prefer to plant monoculture plantations of eucalyptus or pine instead of native species? A pine tree grows quickly, but it also sells cheaply. By contrast, a native tree grows more slowly but sells at a higher price. If, for example, a native tree needs two times as much time as a pine tree (*Pinus radiata*) before it can be harvested, but would also sell for about three times the price, as is the case with Chilean raulí – *Nothofagus alpina* (Cubbage et al. 2007), one might think that the plantation owner would prefer to plant raulí, since the return in relation to the time needed until it can be harvested is higher than that of a pine tree.

Yet this is not the case. Instead, forest companies prefer to plant exotic species, thereby also increasing biodiversity loss, soil degradation and the risk of forest fires. The reason for this is that their profit calculation is distorted by the fact that *time is not a neutral factor*, but rather interest places a value on time. Accordingly, a species that grows faster often appears to be more profitable than a species that grows slowly, even if it sells at a much higher price in relation to the time that is needed until harvesting. In other words, depending on the interest rate, the NPV of the price of a tree that is harvested in 30 years might be less than the NPV of a species that is harvested in 15 years, even though it sells for a price twice as high. This 'time-is-money-pressure' applies to both privately as well as to commonly owned resources. Here, Elinor Ostrom is right when she concludes regarding Common-Pool Resources (CPR) and tying on insights from Game Theory that "the higher the discount rate, the closer the situation is to that of a one-shot dilemma in which the dominant strategy of all participants is to overuse the CPR" (Ostrom 1990: 91).

There are two obvious avenues to avoid monocultures of exotic species. The first is that the government imposes rules that restrict such actions, as is done in many parts of the world. The second would be to have a financial system in which the interest rate did not distort how economic actors calculate profits. This would allow for long term investments in natural resources and, hence, result in the planting of more native forestry.

2.6.3 Conclusion

The current system forces a choice between unsustainable growth and unemployment and so misery as Constanza, et al. (2012: 42 f.) correctly put it. As a result, environmental degradation is occurring across the globe at record levels and we cannot continue in this way (Maedows et al. 2004; WWF 2018). It should be mentioned that the interest rate paid for lending money is a very *unnatural construct* invented by human beings (Aristotle 1995: 23; Kennedy 2011: 19 f.) and contradicts the laws of physics (Soddy 1934). It is possible to change the financial system and move toward one that permits a sustainable development (Kennedy 2011; Azkarraga et al. 2011; Fuders 2009a, 2010a, 2016, 2017; Fuders and Max-Neef 2014a, b). In the next paragraph it will me mentioned how this could be achieved.

2.7 Summary, Outlook and Final Considerations

Allocating private property rights to originally freely accessible natural resources might produce short-term benefits as *homo oeconomicus* will not seek to deplete their source of income. However, privatization as a strategy for environmental protection does not attack the root of the problem and also permits monopoly rents (*natural dividend*) for some people, and therefore generates unjust incomes, inequality and allocative inefficiency. It was further illustrated that since privatization does not get at the root of the problem, it does not ensure sustainable resource management and, in fact, does not even guarantee that those resources will not be exploited until their total extinction.

Chile is an excellent example. Even though its coastal zones (management areas) and forests are widely in private hands, the biodiversity in these ecosystems is increasingly degraded. This chapter defends the hypothesis that environmental degradation is not solely a product of classic market failure but results from the fact that the economy is obliged to grow. *This is the primary and transversal failure* of our economic model, though it is not recognized as such in conventional economic texts. The growth imperative does not (only) come from greed and economic agents' interest in increasing their profits, but rather originates in the financial system in which money is the counterpart of debt and both the money supply and the total debt of a given country grow constantly through a mathematical logic based on the interest rate. This exerts a constant tacit pressure on the economy to grow at least at the same rate. In other words, regardless of whether or not natural resources are freely accessible goods, they will be overexploited in the long term because of this logic of growth.

Economic schools should critically analyze conventional economic theories in order to formulate a new model for a market economy that is not perverted by the need to grow. We could probably learn a lot from the German-Argentine economist Silvio Gesell (1949) and his work, "The natural economic order". Gesell proposed a currency that cannot be hoarded eternally, and thus circulates without interest as "reward for parting with liquidity" (Keynes 1936: 167) being necessary. This interest free currency, Gesell called it "*Freigeld*" (free money), would serve solely as a means to facilitate the interchange of goods and services and not to store wealth. Consequently, it would truly comply with the concept that conventional economic theory usually calls "monetary neutrality" (Stiglitz 1998: 187), but which cannot be sustained under today's financial system (Fuders 2017). Today, money is not neutral. Rather, it generates a constant increase in a country's total debt (and, in fact, also in inequality since deposits and debt grow in a likewise manner); and generates the obligation to grow, and with it contributes significantly to the destruction of the planet.

We can envision, that without the obligation to grow and without the possibility of accumulating large amounts of *virtual wealth* (Soddy 1933, 1934: 36) in bank accounts that reproduce "on their own" (which even ancient Aristotle (1995: 23) found to be unnatural), the abuse of nature might decrease significantly. In any event, what can be affirmed is that with the current financial system, a sustainable future is not possible (Azkarraga et al. 2011). Hence, environmental protection politics that do not take this system into consideration can be considered a farce (Fuders and Max-Neef 2014a).

It is worth mentioning that development does not necessarily require economic growth. Growth is quantitative and refers to the increase of the total output of an economy while development is qualitative and depends on the subjective perception of the satisfaction of human needs. We can develop infinitely without having to steadily increase the production of goods and services. For example, we could improve our computers, making them more efficient and effective without necessarily increasing the number of units sold. If we want an economy that satisfies fundamental human needs, that is Development at a Human Scale (Max-Neef et al. 1991),[18] if we want to turn our world away from the current "collision course" (Max-Neef 2010) then we have to reform our money first.

Finally, we might also need to rethink private property laws. From an economic perspective, private property makes sense when applied to man-made products but not necessarily when applied to natural resources. Surely nobody would produce something that he later cannot label "his own". But this argument does not apply to natural resources, as e.g. primeval forests, since they are already supplied by nature; and that a private owner better protects a natural resource than the government is not necessarily true as has been argued in this chapter. We have seen that also under private control the money rate of interest leads to unsustainable management of the forests and fishing industry, allowing for monopoly incomes. More detailed thoughts on private property law and the management of natural resources are outlined in Chap. 3.

References

Ackermann F (1994) The natural interest rate of the forest: macroeconomic requirements for sustainable development. Ecol Econ 10:21–26

Acquatella J, Altomonte H, Arroyo A, Lardé J (2013) Rentas de Recursos naturales no renovables en América Latina y el Caribe. Cepal, Santiago de Chile

[18]Human needs in the Human-Scale-Development-Approach are seen as ontological, i.e. stemming from the condition of being human and can be characterized as few, finite, classifiable and do not vary through all human cultures and across historical time periods, in contrast to the notion of what economics define as "wants", which are infinite and insatiable. What do change over time and between cultures are the strategies by which these needs are satisfied. An index that measures the subjective perception of the satisfaction of fundamental human needs is the "Human Scale Development Index" (Fuders et al. 2016).

Aristotle (1995) Politik, 1. Buch. In: Meiner F (ed) Aristoteles Phillosophische Schriften in sechs Bänden.. (Bd. 4). Felix Meiner Verlag, Hamburg
Ault D, Rutman G (1979) The development of individual rights to property in tribal Africa. J Law Econ 22(1):163–182
Azkarraga J, Max-Neef M, Fuders F, Altuna L (2011) La evolución sostenible II. Lanki (Mondragon Unibertsitatea), Eskoriatza
Barkley PW, Seckler DW (1972) Economic growth and environmental decay – the solution becomes the problem. Hartcourt Brace, New York
Cheung S (1970) The structure of a contract and theory of a non-exclusive resource. J Law Econ 13(1):49–70
Common M, Stagl S (2008) Introducción a la Economía Ecológica. Reverté, Barcelona
Conrad J, Clark C (1995) Natural resource economics – notes and problems. 5th ed. Cambridge University Press, Cambridge
Constanza, R., Alperovitz, G., Daly, H., Farley, J., Franco, C., Jackson, T., ... Victor, P. 2012. Building a sustainable and desirable economy-in-society-in-nature. New York: United Nations Devision for Sustainable Development
Copeland T, Weston J, Shastri K (2008) Finanzirungstheorie und Unternehmenspolitik. Konzepte der Kapitalmarkorientierten Unternehmensfinanzierung, 4a edn. Pearson, Munich
Creutz H (1993) Das Geldsyndrom. Wege zu einer krisenfreien Marktwirtschaft. Wirtschaftsverlag Langen Müller/Herbig, Munich
Cuadrado JR, Mancha T, Villena JE, Casares J, González M, Marín JM, Peinado ML (2006) Política Económica - Elaboración, objetivos e instrumentos, 3rd edn. McGraw-Hill, Madrid
Cubbage F, Mac Donagh P, Sawinski Junior J, Rubilar R, Donoso P, Ferreira A et al (2007) Timber investment returns for selected plantations and native forests in South America and the southern United States. New For 33(3):237–255
Dales JH (1972) Pollutions property and prices, 4a edn. University of Toronto Press, Toronto
Daly H, Farley J (2011) Ecological economics – Principles and applications, 2da edn. Island Press, Washington, USA
Daly H, Cobb J, Cobb C (1989) For the common good: redirecting the economy toward community, the environment, and a sustainable future. Beacon Press, Boston
Deka (2019) Youtube. [Online] Available at: https://www.youtube.com/watch?v=CSmBku1qZsg. Accessed 30 July 2019
Demsetz H (1967) Toward a theory of property rights. Am Econ Rev 57(2):347–359
Dolan EG (1971) TANSTAAFL – the economic strategy for environmental crisis. Holt, Rinehart and Winston, New York
Dornbusch R, Fischer S, Startz R (2009) Macroeconomía, 10th edn. McGraw-Hill, México
Einstein A, Infeld L (1938) The evolution of physics. Cambridge Uiversity Press, London
Engelkamp P, Sell F (2005) Einführung in die Volkswirtschaftslehre, 3rd edn. Springer, Berlin
Farley J, Burke M, Flomenhoft G, Kelly B, Murray F, Posner S et al (2013) Monetary and fiscal policies for a finite planet. Sustainability 5:2802–2826
Fernández A, Parejo JA, Rodríguez L (2006) Política Económica, 4th edn. McGraw-Hill, Madrid
Frank R (2015) Micoreconomics and behaviour, 9th edn. McGraw-Hill, New York, USA
Fuders F (2009a) Die natürliche Wirtschaftsordnung als Option nach dem Zusammenbruch. Aufklärung & Kritik 16(2):128–145
Fuders F (2009b) EG-Wettbewerbsrecht. SVH, Saarbrücken
Fuders F (2010a) Alternative concepts for a global financial system – an answer to the present world financial crisis. Estudios Internacionales 166:45–56
Fuders F (2010b) Warum der Zins auch moralisch nicht zu rechtfertigen ist. Humane Wirtschaft, Issue 2:26–29
Fuders F (2016) Smarter money for smarter cities: how @regional currencies can help to promote a decentralised and sustainable regional development. In: Dick E, Gaesing K, Inkoom D, Kausel T (eds) Decentralisation and regional development – experiences and lessons from four continents over three decades. Springer, Cham, pp 155–185

Fuders F (2017) Neues Geld für eine neue Ökonomie – Die Reform des Geldwesens als Voraussetzung für eine Marktwirtschaft, die den Menschen dient. In: Krämer G (ed) Finanzwirtschaft in ethischer Verantwortung - Erfolgskonzepte für Social Banking und Social Finance. Springer-Gabler, Wiesbaden, pp 121–183

Fuders F, Max-Neef M (2014a) Local money as solution to a capitalistic global financial crisis. In: Pirson M, Steinvorth U, Largacha-Martinez C, Dierksmeier C (eds) From capitalistic to humanistic business. Palgrave Macmillan, Hampshire

Fuders F, Max-Neef M (2014b) Dinero, deuda y crisis financieras. Propuestas teórico-prácticas en pos de la sostenibilidad del sistema financiero internacional. In: Fernández J (ed) Economía Internacional. Claves teórica-prácticas sobre la inserción de Latinoamérica en el mundo. Latin, Guayaquil

Fuders F, Nowak V (2019) The economics of love: how a meaningful and mindful life can promote allocative efficiency and happiness. In: Steinebach C, Langer Á (eds) Enhancing resilience in youth – mindfulness-based interventions in positive environments. Springer, Cham, pp 259–277

Fuders F, Mondaca C, Azungah Haruna M (2013) The central banks' dilemma, the inflation-deflation paradox and a new interpretation of the Kondratieff waves. Economía 36:33–66

Fuders F, Mengel N, Maria del Valle B (2016) ´indice de desarrollo a escala humana: propuesta para un indicador de desarrollo endógeno basado en la satisfacción de las necesidades humanas fundamentales. In: Pérez Garcés R, Espinosa Ayala E, Terán Varela O (eds) Seguridad Alimentaria, actores Terretoriales y Desarrollo Endógeno. Laberinto, Iztapalapa, pp 63–106

Galbraith JK (1983) El Dinero - De dónde vino /Adónde fue, 2nd edn. Hyspamerica, Buenos Aires - Madrid

George H (1935) Progress and poverty – an inquiry into the cause of industrial depressions and of increase of want with increase of wealth – the remedy. 50th anniversary edition Ausg. New York: Robert Schalkenbach Foundation

Gesell S (1949) Die natürliche Wirtschaftsordnung durch Freiland und Freigeld. Rudolf Zitzmann, Lauf

Goethe J (1996) Goethe – Werke, Kommentare und Register, Hamburger Ausgabe in 14 Bänden – Band 1: gedichte und Epen I, 16th edn. C.H. Beck, Munich

Hamilton C (2003) Growth fetish. Allen & Unwin, Crows Nest

Hardin G (1968) The tragedy of the commons. Science 162:1243–1248

Hardin G (1978) Political requirements for preserving our common heritage. In: Bokaw HP (ed) Wildlife and America. Council on Environmental Quality, Washington, pp 310–317

Kennedy M (1990) Geld ohne Zinsen und Inflation - Ein Tauschmittel, das jedem dient. Permakultur-Verlag, Steyerberg

Kennedy M (2006) Geld ohne Zinsen und Inflation - Ein Tauschmittel, das jedem dient, 9th edn. Goldmann, München

Kennedy M (2011) Occupy money. Kamphausen Verlag, Bielefeld

Keynes J (1936) General theory of employment, interest and money. Hartcourt, New York

Keynes J (1983) Economic articles and academic correspondence. In: Moggrigde D (ed) The collected writings of John Maynard Keynes. Macmillan, Cambridge, p 402

Lachmann W (2006) Volkswirtschaftslehre 1. Springer, Berlín-Heidelberg

Larroulet C, Mochón F (2003) Economía, 2nd edn. Mac Graw-Hill, Santiago

Lee KS, Werner R (2018) Reconsidering monetary policy: an empirical examination of the relationship between interest rates and nominal GDP growth in the U.S., U.K., Germany and Japan. Ecol Econ 146:26–34

Lietaer B, Arnsperger C, Goerner S, Brunnhuber S (2013) Geld und Nachhaltigkeit – Von einem überholten Finanzsystem zu einem monetären Ökosystem. Ein Bericht des Club of Rome/EU-Chapter. Europa Verlag, Vienna

Löhr D (2012) The euthanasia of the rentier – a way toward a steady-state economy? Ecol Econ 84:232–239

Maedows D, Randers J, Maedows D (2004) The limits to growth – the 30-year update. Chelsea Green Publishing, Vermont

Mankiw GN, Taylor MP (2014) Economics, 3rd edn. Cengage Learning EMEA, Hampshire
Maunder M (2008) Maximum sustainable yield. In: Encyclopedia of ecology. Elsevier, Amsterdam, pp 2292–2296
Max-Neef M (1986) La Economía Descalza – Señales desde el mundo invisible. Nordan, Estocolmo
Max-Neef M (2010) The world on a collision course and the need for a new economy. Ambio 39:200–210
Max-Neef M, Elizalde A, Hopenhayn M (1991) Human scale development - conception, application and further reflections. Apex, New York
Mill J (1885) Principles of political economy. Appleton, New York
Neher P (1990) Natural resource economics - conservation and exploitation. Cambridge University Press, Cambridge, New York
Okun A (1962) Potential GNP: its measurement and significance. In: I. A. Association (ed) Proceedings of the business and economic statistics section. American Statistical Association, Alexandria
Ostrom E (1990) Governing the commons - the evolution of institutions for collective action. Cambridge University Press, Cambridge
Rösl G (2006) Regionalwährungen in Deutschland – Lokale Konkurrenz für den Euro?. Diskussionspapier Reihe 1: Volkswirtschaftliche Studien Nr. 43. Frankfurt, Bundesbank
Smith A (1952) An inquiry into the nature and causes of the wealth of nations. Great Books of the Western World, No. 39. W. Benton, Chicago
Smith R (1981) Resolving the tragedy of the commons by creating private Propeny rights in wildlife. CATO 1:439–468
Soddy F (1933) (Reprint 1961). Wealth, virtual wealth and debt – the solution of the economic paradox. 2nd. London: Briton
Soddy F (1934) The role of money – what it should be contrasted with what it has become. Routledge, London
Stehlík M, Helperstorfer C, Hermann P et al (2017) Financial and risk modelling with semicontinuous covariances. Inf Sci 394–395:246–272
Steiner R (1979) Die soziale Grundforderung unserer Zeit – In geänderter Zeitlage, Zwölf Vorträge, gehalten in Dornach und Bern v. 29. 11. bis 21.12.1918. 2 ed. Dornach: s.n
Stevenson G (1991) Common property economics – a general theory and land use applications. Cambridge University Press, Cambridge
Stiglitz J (1998) Macroeconomía. Ariel, Barcelona
Strunz S, Bartkowski B, Schindler H (2015) Is there a monetary growth imperative? Helmholtz-Zentrum für Umweltforschung, Leipzig
Süchting J (1995) Finanzmanagement, Theorie und Politik der Unternehmensfinanzierung, 6th edn. Gabler, Wiesbaden
Suhr D (1988) Alterndes Geld – Das Konzept Rudolf Steiners aus geldtheoretischer Sicht. Novalis, Schaffhausen
Tsikliras A, Froese R (2019) Maximum sustainable yield. In: Encyclopedia of ecology, 2nd edn. Elsevier, Amsterdam, pp 108–115
Turner R, Pearce D, Bateman I (1994) Environmental economics – an elementary introduction. Harvester Wheatsheaf, New York
Walker K (1959) Das Geld in der Geschichte. Rudolf Zitzmann, Lauf
Welch W (1983) The political feasibility of full ownership property right: the cases of pollution and fisheries. Policy Sci 16:165–180
Wikipedia (2019) [Online]. Available at: https://en.wikipedia.org/wiki/Money_supply. Accessed 10 May 2019
WWF (2018) Living planet report – 2018: aiming higher. Gland (Switzerland): s.n

Chapter 3
Allocative Efficiency and Property Rights in Ecological Economics: Why We Need to Distinguish Between Man-Made Capital and Natural Resources

Felix Fuders and Roberto Pastén

So neither the one who plants nor the one who waters is anything, but only God who gives the growth.
(1 Cor. 3: 6–7)

3.1 Introduction

3.1.1 *Privatization of Natural Resources Does Not Protect Against Over Exploitation of Nature*

Chapter 2 outlined why, despite the *homo oeconomicus* having a vested interest in preserving future sources of income, assigning property rights to natural resources does not necessarily prevent nature from being excessively exploited. This hypothesis runs counter to conventional economic thinking in which assigning property rights is presented as a solution to overexploitation. Instead, it was argued that the lack of sustainability in the management of natural resources has its origins in our monetary system where the money interest rate (as the opportunity cost of any productive investment) exerts a constant pressure on businesses to achieve a rentability at least as high, which is why the economy has to grow at least as the same pace. In other words, *regardless* of whether or not natural resources are freely accessible goods, they will be over-exploited in the long term due to this logic of constant economic growth. In addition, the privatization of the originally freely accessible natural resources allows for so-called monopoly rents (often described as "*natural dividend*") and, thus, an unjust income, i.e. an income not based only

F. Fuders (✉) · R. Pastén
Economics Institute, Faculty of Economics and Administration, Universidad Austral de Chile, Valdivia, Chile
e-mail: felix.fuders@uach.cl

F. Fuders and P. J. Donoso (eds.), *Ecological Economic and Socio Ecological Strategies for Forest Conservation*, https://doi.org/10.1007/978-3-030-35379-7_3

on the owner's effort.[1] Chile provides a case in point. Although the coastal areas are divided in exclusive management zones (*áreas de manejo*) and the forests are predominantly privately owned, neither aquaculture nor forestry are sustainably operated (regarding aquaculture see: Sommer 2009; Allsopp et al. 2008; regarding forestry: see Chaps. 4 and 5). Here, unsustainable resource management cannot be explained with the free access to resources.

3.1.2 Regulation of Privately-Owned Resources?

Although unsustainable practices are common in south American forestry (for instance, planting monocultures of non-native species and the practice of clear cutting), governments could regulate such practices through legislation. For example, in Germany the federal forest law seeks to balance the interests of forests owners and the common good (§ 1, No 3 Bundeswaldgesetz). Forest owners are obliged to manage their forests in a proper and sustainable manner (§ 11 Bundeswaldgesetz). The respective federal state laws in Germany build on this premise by obligating regional and local governments to establish forest use plans that try to maintain the protective and recreational functions of forests as well as their biodiversity (e.g. Art. 6 Bayerisches Waldgesetz). Similar regulations enacted across Europe have resulted in marked improvements in forest management. For example, the clear cutting is now very uncommon in many countries (Paletto et al. 2008; Lundmark et al. 2017).

Even better than regulation, however, would be to establish a financial system that does not force the productive economy to constantly grow and in which the money interest rate does not bias the profit calculations of economic agents as explained in Chap. 2. Assigning private property rights and artificially restricting the access to natural resources and in doing so allowing for monopoly rents would then be unnecessary. As was already outlined (Chap. 2), such a monetary system was suggested 100 years ago by Silvio Gesell (1949).[2] Gesell proposed a currency (*Freigeld*) that circulates in the economy without interest as "reward for parting with liquidity" (Keynes 1936: 167) being necessary. This currency would serve solely as a means to facilitate barter and not to store "virtual wealth" (Soddy 1933). It would probably truly comply with the concept that in economic theory usually is known as the "neutrality of money" (e.g. Stiglitz 1998: 187), but which cannot be hold in the current financial systems (Fuders 2017).

It is quite conceivable that without the obligation to grow, an equilibrium could be achieved, that comes close to the "*steady-state-economy*" described by Herman Daly (1991), but without the need for the government to artificially regulate input

[1] Economic "rent" is a term that defines an amount of money earned that exceeds that which goes beyond the opportunity costs of production factors invested. Economic rents often arise from market inefficiencies.

[2] See: Fuders 2009, 2010, 2016, 2017; Fuders and Max-Neef 2014a.

and throughput of production processes (Fuders 2016).[3] Unfortunately, it is unlikely that such a reform of our financial system will occur in the near future. It should also be noted that technological innovations (possibly themselves triggered by the growth imperative) cannot be withdrawn. Consequentially, even if such a reform occurred, we would not return to using axes to cut trees or small artisanal fishing boats to catch fish. Indeed, when viewed through the lens of Hardin's (1968) "*Tragedy of the Commons*" metaphor, technological advancements have made huge swathes of natural resources as accessible for exploitation as the shared pasture ground in a village.

This begs the question, how can we protect our natural resources meanwhile the current system keeps functioning the way it does, i.e., meanwhile all productive processes have to grow at least at the rhythm of the money interest rate as opportunity costs of any productive investment? Given our current world financial system would it not still be recommendable to restrict access to natural resources *combined with strict regulations* (e.g. prohibition of clear-cuttings and monocultures of exotic species) in order to achieve a sustainable management, even though this allowed for monopolistic incomes? These questions will be explored in the next paragraphs.

3.2 Natural Resources Belong to All People

3.2.1 Distinction Between Man-Made Products and the Produce of Nature

Since the excessive exploitation of nature does not result from free access to resources, but from the growth imperative that the money rate of interest exerts since about 700 years over our economies worldwide, it was concluded in Chap. 2 that natural resources, especially land and soil, mineral resources, water and air, should remain freely accessible. From the perspective of allocation theory, free access (in the absence of the growth imperative) is most efficient and results in prices equaling marginal costs.

We need private ownership of man-made goods, otherwise no one would manufacture them. For example, let us imagine a coffee loving country, whose government declared drinking coffee as a human right and therefore prohibited private ownership of coffee machines. Would coffee machines be produced or sold in this country? Probably not. Anyone who produced a coffee machine would not be able to sell it, since in the moment it is produced, the producer would lose the ownership. It also seems fair that someone who produces goods is rewarded for the

[3] At the same time this monetary system would abolish a powerful force behind the steadily growing income inequality, avoid inflation and deflation as well as financial crisis. On this see: Fuders 2009, 2010, 2016, 2017; Fuders and Max-Neef 2014a, b.

creation of the added value, i.e. for his labor or work performance. Private property has always existed, even though it was not always explicitly anchored in laws. The biblical commandment "Thou shalt not steal", for example, would be superfluous if private property rights were not thought to be necessary to assure a living together in justice, peace and harmony. However, one might wonder whether it is right and just to consider 'pure natural resources', that is goods that are produced and provided by nature only, as private property. These are goods that humans did not create or add any value such as land, water (lakes, oceans, groundwater basins) and resource stocks that provide ancient forests, original fishing grounds or mineral resources. Thus, private property is not necessary to provide an incentive so that these goods are produced, since nature does this for us. It makes sense, therefore, to distinguish between goods made by man and those produced by nature, and assign *private property only to man-made goods*. This would result in pure natural resources becoming freely accessible to all humans and animals. This idea is not completely new, but almost completely forgotten. It has been envisioned 150 years ago by Henry George (1935) in his in-depth inquiry into the fairness aspects of such a distinction. Fifty years later Silvio Gesell (1949) incorporated similar thoughts into his model of a "Natural Economic Order".

3.2.2 Free Access for Everyone to 'Pure Natural Resources'

Free access to 'pure natural resources' is not only required to achieve allocative efficiency (Chap. 2) but also recommendable from the point of view of justice. After all, they were created for everyone alike (George 1935). This would truly be a natural property order since tribes of aborigines all over the world do not recognize property rights on nature but they do respect belongings of each other, which are the produce of man having applied labor to transform natural resources, even though this 'property law' might not be written down anywhere but exists only as a sort of customary law.

Incidentally, in biblical times there was no private property on land, mineral resources or water below the king, i.e. the government of State (Leviticus 25: 23). The biblical commandment "Thou shalt not steal" can therefore *only* refer to goods produced or harvested by man. This aligns with St. Paul's first epistle to the Corinthians: "I planted, Apollos watered, but God gave the growth. So, neither the one who plants nor the one who waters is anything, but only God who gives the growth" (1 Cor. 3: 6–7). Today, however, we often harvest that which we have not even planted or watered yet, without work, should we claim private ownership of such natural products? Only if man himself performs work to achieve the end product is it just to claim private property (George 1935). It is therefore worth asking whether the effort of harvesting (without applying any finishing) justifies claiming property of those resources?

It might be argued that harvesting is, indeed, sufficient effort to justify property to the harvested resources. Timber or fish would then belong to the person that felled the tree or set the net, respectively. But even then, the *access* to primeval forests and fishing grounds or to other resources made by nature and exploited by humans have to be *possible for everyone*. This is a powerful argument that pure natural resources should not be privatized. Had man also tried to privatize land, water and, more recently, air (emissions trading can be considered as a form of privatization of the air; Lohmann 2006) already in Moses times, maybe God would have specified in the Ten Commandments that the commandment not to steal is not applicable to land, water and air, which in any case can be regarded as gifts to all mankind and, thus, equally belong to everyone (Psalm 115; George 1935).

3.2.3 Potential Access for Everyone to Plant and Harvest

What has been said so far does not apply to agricultural and forestry products that have been planted by man himself. These should not remain freely accessible, since effort must be rewarded. Here, the value added to the product from the effort of planting and watering justifies private property. Interestingly, this aligns the, already cited, epistle to the Corinthians: "The one who plants and the one who waters have a common purpose, and each will receive wages according to the labor of each" (1 Cor 3: 8). Today, however, this is not always the case as the owner of land can earn more than a corresponding pay (more than the opportunity costs of the production factors invested) for his efforts. This additional remuneration has been described as a "*natural dividend*" but, as explained in Chap. 2, constitutes a monopoly rent and results precisely from the allocation of soil, mineral resources or water to a restricted number of private owners. So, on the one side, access to land needs to be restricted so that the one who planted can also collect a harvest. On the other side, the assigning of property rights that enable a small number of persons high monopolistic incomes should be avoided. And from a moral point of view: must not all people have to have an equal right to access the land used for agriculture and forestry?

How can we grant the access to land or other resources for everyone but at the same time (1) assure a sustainable use and (2) grant those who plant, water and harvest a just remuneration according to their efforts, i.e. production factors invested? Here is justified the demand for equal rights to *potentially* access the land and water so that everyone has the same opportunity to plant and water and later harvest. In other words: While private property on land and water as pure nature-made products should not exist, the produce that is gained with land and water as well as any edifications should still be private. The next lines will propose three policies to get out of this dilemma, depending on the resource.

3.3 Policies

3.3.1 *"Pure" Nature-Made Products: Concession or Quotas*

Not only do we not solve the environmental degradation problem through private property of biotic or abiotic resources purely produced by nature. Rather, our private property system becomes *highly unfair*. How can it be, e.g., that in many South American countries a significant proportion of total land tenures belong to only a few families. With what right do private companies, in their name (and purse), exploit copper in Chile, even though the Chilean constitution correctly states that copper belongs to the Chilean people[4]?

3.3.1.1 Concessions

If the state does not want to exploit the natural resource, it may give concessions to private firms to extract and market the resource on *its behalf*. Copper mining in Chile operates on the basis of concessions. However, the companies market the copper not on behalf of the state but for themselves, requiring a relatively small royalty fee of ~4–9% levied. This tax could certainly be much higher. Actually, the return of the sale should be transferred *fully* to the state, who could then use part of the income to pay the mining company a sum equivalent to the opportunity costs of production factors invested. With the remaining income, the State can finance its budget. Thus, indirectly *all citizens* would benefit from natural resource exploitation, as intended by the Chilean constitution. The same could apply to the exploitation of original fish stocks or the cutting of old-growth forests (if it was decided to exploit them). Concessions allow the State to impose rules and supervise exploitation ensuring sustainable extraction. However, the current model for concessions in Chile allowing private companies to exploit *and sell* the natural resources in its own name could be interpreted as a form of camouflaged privatization of goods that belong to all.

The suggestion that the exploitation of pure natural resources, where humans have not added any value (in form of planting and watering) should be carried out by private enterprises on behalf of the State, *must not be mistaken with Marxism*. This text argues that *only* commodities purely produced by nature should be common property, not all means of production, as suggested by Karl Marx (Marx 1872). This corresponds to the *natural*, millennia-old practice, it is, as stated above (Chap. 2), to be advocated from the point of view of allocation theory and appears fair and, by the way, also Christian. It has already been mentioned that in biblical times private

[4]"Disposición Tercera Transitoria", Constitution from 1980 in conjunction with "Disposición 17ª Transitoria", Constitution from 1925. On might even argue that natural resources belong to the whole world, and not only to the people who by chance live on, over or nearby the territory where the resource is exploited (Gesell 1949).

ownership of land, water and air did not exist. All land (and other natural resources) was tenured by the respective King. This system was mirrored in Europe were, until the Early Middle Ages, only the King and his nobles owned land (a power structure with parallels to current federal governments). Everyone else was a tenant, paying rents, or 'tributes', from the goods they produced. The more land a family would want to use for agricultural production the more tribute they had to pay. This system has been described as the origin of the obligation to pay taxes (Pfeifer 1993).[5]

In order to assure that a concession is truly a concession, and not a hidden form of privatization, concessions should comply with three policies: Firstly, they should be re-called after a set time period. Secondly, they should be assigned to as many individuals or companies as economically reasonable. These two policies increase the *potential opportunity* for everyone to access. Thirdly, the concessionaires are required to submit the exploited resource directly to the State or, alternatively, sell the resource in the name of the State and transfer the *total return* to the State. They would then be paid a customary return corresponding to their cost structure and production factors invested.

3.3.1.2 Quotas

For the exploitation of primeval old-growth forests as well as fishing ground also *quotas (contingents)* can be assigned to individuals or firms. A main difference between quotas and concession is that quotas can be given to a relatively high (but limited) number of individuals. Furthermore, in contrast to concessions, quotas do not require the negotiations of complicated contracts between concessionaire and the respective government. Concessions therefore are particularly suitable assigning exploitation rights that require high initial investments such as those that are reliant on economies of scale (e.g. mining activities). This is not the case for the exploitation of primary forests or fishing grounds. Quotas should only be implemented when effective monitoring can be ensured. Strict monitoring is one of the eight design principles for efficient governance of commonly owned resources that Elinor Ostrom proposes (Ostrom 1990). If the government manages to effectively control quotas (e.g. by fining those who exceed them), this will lead to the dominant game theoretical strategy where all cooperate (Ostrom 1990), i.e. a situation where all stakeholders win.

[5]Only later this governmental structure was watered down when Kings in Europe gave land titles to ever more nobles, who themselves also gave land titles to nobles of inferior categories (landlords). Although this resulted in more private people (nobles and landlords) owning land tenures, the vast majority of the citizens still did not own any land. In consequence many felt discriminated, the divine precedence of the governmental structure was questioned, and social class struggles emerged. These led into land reforms in many European countries and the registration of land ownership in cadasters where every person could own land.

Regarding quotas some challenging questions arise:

- Should the quotas be sold or assigned freely?
- If they were to be sold, how would they be priced?
- If they were to be free, to whom should they be given and who takes the decision?
- Should they be transferable (marketable) or not?

Any price that a government sets up is unlikely to be the market equilibrium price, that is to say, the *Pareto-optimal* price that balances demand and supply. But, a government (or any local decision taker in self-governed commonly owned resources) could assign quotas via an auction where they are assigned to those individuals who offer the most. This way the price can function as rational to assign the resource to those who most value them, e.g. those who perceive the highest utility from harvesting the timber or fish. In addition, selling has the advantage that those, who did not receive a contingent, would indirectly benefit as the revenue from sales could mean that the State needs charge less taxes elsewhere.

Here, there is a problem, not commonly discussed in the mainstream economic literature, that is worth mentioning: the marginal utility of money (MU_M), that is the utility we perceive from earning the next unit of money, depends on our "virtual wealth" (W_v) (Soddy 1933), i.e., the units of money we already possess: $MU_M = f(W_v)$. The more monetary units someone possess the less they will value earning the next unit (Fig. 3.1). This is not a specific problem with quotas but a general problem regarding market equilibrium that brings along extreme wealth inequality (overview in Piketty 2014)[6] and that makes any market equilibrium imperfect. The classic definition of efficiency, i.e. the cost of the last unit produced equals the value the consumer attributes to this good (marginal costs equal price), is perverted by the fact that the monetary units used to express costs and prices do not have the same perceived value (utility) for each individual consumer. A certain price can

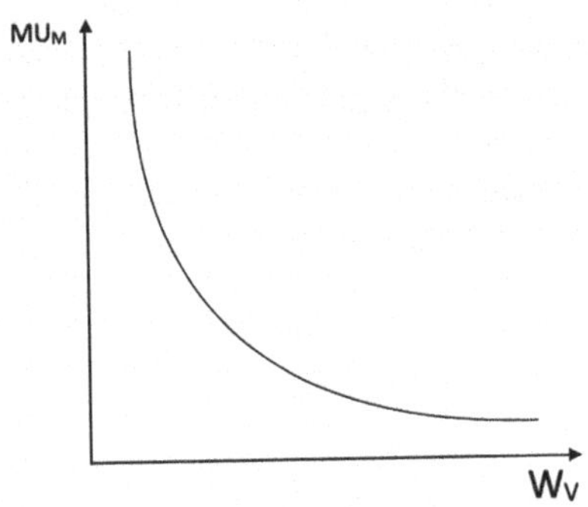

Fig. 3.1 $MU_M = f(W_v)$. (Source: own work)

[6]Piketty has shown through tax records that the capital-income ratio β has risen from approximately 2.5 to over 4 worldwide since the end of World War II. He forecasts that β will have risen to more than 6.5 by the end of the century.

be unaffordable for one person and extremely cheap for another. Hence, marginal production costs might equal the price, but not necessarily the marginal utility of the consumer.

Alternatively, quotas could be free. People who are less interested in exerting the exploitation right could opt to sell it if quotas were to be transferable. Then, someone who has a particularly strong interest in using the natural resource might buy additional quotas from others with less interest in exploiting the resource (such as those who attribute a higher value to the money they can receive selling their contingent instead of exploiting the resource themselves). Here, again the price functions as rational to assign the resource to those individuals who most value them, something to be desired from perspective of allocative efficiency. Nevertheless, a strong argument exists for quotas to be non-transferable, or at least that the number of quotas that one individual can buy be limited. Otherwise a rich person could buy all quotas and monopolize the originally commonly owned recourse. This is not to be desired from the point of view of justice (the resource was created for everyone) or allocative efficiency, since the individual who obtains a mayor proportion of quotas would be granted a monopolistic position, that per definition is not efficient (price unequal marginal revenue) permitting monopoly rents. In other words: allowing one or few individuals to obtain majority portions of exploitation rights can be seen as camouflaged privatization, comparable to a concession where the concessionaire is allowed to sell the extracted resource in its own name (and purse).

Regarding the question of who should be in charge of assigning exploitation rights, there is a strong argument that this should be done as much on a local level as possible (although this might not always be possible, e.g. in case of oceanic fishing grounds). All four examples of a successful sustainable use of commonly used resources that Elinor Ostrom describes in her masterpiece "*Governing the Commons*" had in common that the use of the common-pool resource was locally regulated. Better than by a central government on the local level it is possible to (1) clearly define exploitation rights (boundaries), (2) allow local people to participate (collective choice agreements), to (3) monitor the compliance of established rules and to (4) apply sanctions when needed. These are four of the eight design principles Ostrom (1990) proposes for a sustainable common usage of resources. Here, a federal governmental structure, such as found in Switzerland or Germany, has an advantage compared to a more centralized governmental system. Nevertheless, it is also possible that strongly centralized governments, such as Chile, could try to decentralize the allocation of quotas by the regional dependencies of the central government (*regiones*) who could then form locally based governing entities to govern common-pool resources making participate local stakeholders.

3.3.2 Leasing of Land and Costal Management Areas

As an alternative to granting concessions or quotas, the state could also lease land and management areas in coastal zones of the ocean. This model is probably best suited to natural resources that are *planted and harvested by man* such as agricultural produce. This way we can comply with the above defined rule that pure nature-made resources (here land and water) are to be hold by the public hand while the products that humans make using the natural resources (here the forestry or agricultural products) should be privately owned. The leasing rent level could be based either on the hectares leased or on the profits of the forestry (or aquaculture) company. Since the land belongs to all people, the lease should be designed in a way to make it worthwhile to lease, but without monopoly rents being possible. The ground on which self-used houses are built could be exempted from the leasing rent. As in the case of granting concessions, leasing of land allows the State to impose rules. For example, it would only be leased to those individuals or companies that demonstrate sustainable forestry, agriculture or aquaculture. Such a solution would require the State to buy back all land properties from private owners.

This approach has already been proposed by Silvio Gesell (1949), who demanded next to an interest free money (*Freigeld*) a land reform, where all land is in public hand but can be leased, while buildings are privately owned. Gesell called this land "*Freiland*" (free land) and he argued, like this text does, that land is a gift by nature to all people. In a civilized society regarding the access to land there is another aspect to be considered: prices are particular high in those regions where the infrastructure is good. However, infrastructure is usually not created by the owner of the land, but rather constitutes a joint service of the society (George 1935; Daly and Farley 2011; Löhr 2013, 2018). This is a further argument that the land itself should not be privately owned (unlike the planted products or the buildings on it). Furthermore, soil represents a universal basis of life and, as such, has a special position within the natural resources (similar to air and water): every person needs land to live on. However, land on a finite planet is limited (inelastic supply). Land tenure conflicts between indigenous people and immigrants (see Chap. 7) would be significantly reduced.

3.3.3 *Private Property and Levying Monopoly Rents*

Another possibility is to continue to *allow private ownership* of natural resources but to try to levy monopoly rents through taxes. Regarding real estates a tax based on the value of the land surface only, i.e. excluding the value of buildings or crops as proposed by Henry George (1935) could bring the desired effects. This is similar to a current proposal for a real estate tax reform presented by the initiative "*Grundsteuer: Zeitgemäß*!" (For a contemporary land tax!) in Germany (Löhr 2013, 2018; Kriese 2019). Again, this initiative proposes to exclude the value of buildings from an assessment of the property tax and to tax only the land value, i.e. from the value of the real estate as a whole, the value of any edification would be subtracted. Such a plan is contrary to the current taxing systems applied in Germany, Chile and probably most other countries in the world. In the conventional system, the basis for the assessment of real estate taxes is the value of the property as a whole, composed of the land value and the value of any buildings.

The proposed form of taxation respects the distinction between the land as a pure nature-made natural resource that belongs to all, and the work performed on this land by man who should be given the right to usufruct his efforts. The value of the land alone (excluding the value of buildings and crops) can be regarded as a monopolistic gain for the owner, since it derives from the assignation of property rights (land titles) to him and not from his effort. It seems therefore *fair and just* to levy this increase of value by the State who then can redistribute it to everyone (George 1935). In practice, and recalling the proposal of John Stuart Mill (Mill 1885), a fair estimate could be made of the market value of the land, and future additions to that value, not due to the improvements of the proprietor, could be absorbed by the state. Note that this solution does not differ drastically from the discussed solution of leasing land to citizens. While in one case citizens pay a leasing rate to the State to usufruct *public* property, in the other, they pay taxes on the value of *their* property.

We consider that the proposed form of land taxation is more just compared to the current property tax also for another reason. As discussed above, real estate prices are especially high in regions with good infrastructure, but this value is most often created by the community, not the landowner (George 1935). In this way the current taxing system *privatizes common efforts*. A reformed real estate tax would also prevent the situation, common in many cities worldwide, where the best locations remain fallow for long periods of time. As the current taxing system taxes the whole value of the property (including land and buildings) land owners are disincentivized to invest in expensive, quality and sustainable constructions, since this would require paying higher taxes. On the contrary, fallow land currently incurs in no or relatively low taxes. This generates a powerful incentive in cities to leave plots idle, at least as long as real estate prices rise steadily for speculative reasons.[7] Increasingly, investors will buy prime real estate sites for speculative reasons and not to actually develop or to plant on it causing inflated prices. Local residents who need land for homes are then unable to afford to buy in such areas.

A tax based only on land value would levy the increase in land prices and destroy speculative rents and this way reduce speculative investments (George 1935). At the same time, it would generate an elegant incentive to build in prime locations with high-quality and ecological standards as the tax burden of the property owner would not increase compared to less sophisticated constructions. That is to say, the owner would not be punished for constructing long lasting, sustainable and architecturally beautiful buildings, as is currently the case. Here we also see an advantage of the pure land value tax compared to the above-mentioned solution of leasing of land: Unlike a tenant, a landowner has a greater incentive to build with higher quality standards and, in doing so, will generate positive external effects that benefit all. A

[7]This can be seen as a symptom of our monetary system that generates an exponentially growing money supply that seeks to be invested (Fuders 2016, 2017; Fuders Max-Neef 2014a, b; Fuders et al. 2013).

tax of pure land value would also mean that land monopolies, as seen in Chile and Brazil, would be less profitable, since fallow land would generate tax, but no income for owners.

Note that the total tax burden of a person will not necessarily increase since the government of the State could use the income to lower other taxes ("tax shift"). Henry George was convinced that the real estate tax *alone* could be set in a way that all expenses of the government would be financed (George 1935). In fact, as mentioned above, the origins of taxes have roots in the tributes paid by citizens to the King as a sort of leasing rate for the use of his land. If the land value tax is the only existing tax it would then not be necessary to tax production, commerce or labor as taxing systems in most countries do today, putting a pressure on the productive economy, employment, our creative potential and finally, wealth. Hence, a pure land value tax would not compromise the incentive to improve or lessen the production of wealth (George 1935). Indeed, the *international competitiveness* of the country's economy could be enhanced by such a system.

3.4 Summary

This text builds on the insight from Chap. 2, that access to natural resources should remain free. This would not only be recommendable from the point of view of allocation theory, but also because it promotes a fairer system since natural resources were created for all humans (and animals) alike. However, this solution is not sustainable while maintaining the existing monetary system as this is a powerful driver of the economy's growth imperative. Given this current financial system, this chapter invites to reflect about a reform of resource property law that allows to protect natural resources and, additionally, to promote justice. Three concrete proposals for such a reform depending on the type of resource have been described:

1. Concessions and quotas
2. Leasing of land and costal management areas
3. Private property but levying monopoly rents

Concerning the land tenure, the reform proposal number three is particularly promising as it could tie into existing legislation and therefore has the greatest chance of being implemented. In addition, unlike a tenant, the private landlord has a stronger incentive to build according to high-quality and sustainability standards and, thus, to generate positive externalities which are beneficial to the common good. From this perspective the proposal of a pure land value tax would be preferable to the lease-solution. However, for the exploitation of pure nature-made natural abiotic or biotic resources such as copper, coal, rare earths, original fish stocks and primeval forests reform proposal number one (concessions and quotas) seems to be most appropriate.

In the classic liberal view all production factors, including natural resources, are privately owned. On the contrary, following a Marxist perspective all production factors are commonly owned. The purpose if this chapter, however, was to invite to reflect on a scheme of defining property rights that depends on the origin of the product itself as it had been proposed first by Henry George. Only for the produce of humans private property should exist while for pure nature-made products (land, soil, water, air, mineral resources, virgin forests, oceanic fishing grounds), where no man has added any value, private property should not exist. Such a system would not only be fairer, but it would improve allocative efficiency. The text then discussed, in reference to specific examples, regulations and policies that would help to enact and facilitate this system. However, such a sea change in property rights would not be without difficulties and would require a more detailed inquiry than can be provided here.

References

Allsopp M, Johnston P, Santillo D (2008) Challenging the aquaculture industry on sustainability. Greenpeace, Amsterdam

Daly H (1991) Steady-state economics, 2nd edn. Insland Press, Washington

Daly H, Farley J (2011) Ecological economics-principles and applications, 2da edn. Island Press, Washington

Fuders F (2009) Die natürliche Wirtschaftsordnung als Option nach dem Zusammenbruch. Aufklärung & Kritik 16(2):128–145

Fuders F (2010) Alternative concepts for a global financial system – an answer to the present world financial crisis. Estudios Internacionales 166:45–56

Fuders F (2016) Smarter money for smarter cities: how regional currencies can help to promote a decentralised and sustainable regional development. In: Decentralisation and regional development – experiences and lessons from four continents over three decades. Springer, Cham, pp. 155–185

Fuders F (2017) Neues Geld für eine neue Ökonomie – Die Reform des Geldwesens als Voraussetzung für eine Marktwirtschaft, die den Menschen dient. In: Finanzwirtschaft in ethischer Verantwortung – Erfolgskonzepte für Social Banking und Social Finance. Springer-Gabler, Wiesbaden, pp. 121–183

Fuders F, Max-Neef M (2014a) Local money as solution to a capitalistic global financial crisis. In: From capitalistic to humanistc business. Palgrave Macmillan, Hampshire, New York

Fuders F, Max-Neef M (2014b) Dinero, deuda y crisis financieras. Propuestas teórico-prácticas en pos de la sostenibilidad del sistema financiero internacional. In: Economía Internacional. Claves teórica-prácticas sobre la inserción de Latinoamérica en el mundo. LATIn, Guayaquil

Fuders F, Mondaca C, Azungah Haruna M (2013) The central' banks dilemma, the inflation-deflation paradox and a new interpretation of the Kondratieff waves. Economía (U de los Andes) 36:33–66

George H (1935) Progress and poverty – an inquiry into the cause of industrial depressions and of increase of want with increase of wealth – the remedy, 50th anniversary edn. Robert Schalkenbach Foundation, New York

Gesell S (1949) Die natürliche Wirtschaftsordnung durch Freiland und Freigeld. Rudolf Zitzmann, Lauf

Hardin G (1968) The tragedy of the commons. Science 162:1243–1248

Keynes JM (1936) The general theory of employment, interest and money. Hartcourt, New York

Kriese U, Löhr D, Wilke H (2019) Grundsteuer: Zeitgemäß! – Der Reader zum Aufruf. Verlag Thomas Kubo, Münster
Lohmann L (2006) Carbon trading – a critical conversation on climate change, privatisation and power. Dev Dialogue 48:1–159
Löhr D (2013) Prinzip Rentenökonomie – Wenn Eigentum zu Diebstahl wird. Metropolis, Marburg
Löhr D (2018) Boden – die verkannte Umverteilungsmaschine. Zeitschrift für Sozialökonomie (ZfSÖ) 198/199:3–19
Lord T (1993) Holy bible. Thomas Nelson, Nashville
Lundmark H, Josefsson T, Östlund L (2017) The introduction of modern forest management and clear-cutting in Sweden: Ridö State Forest 1832–2014. Eur J For Res 136(2):269–285
Marx K (1872) Das Kapital – Kritik der politischen Oekonomie. Buch 1 Der Produktionsprozess des Kapitals. Otto Meissner, Hamburg
Mill JS (1885) Principles of political economy. Appleton, New York
Ostrom E (1990) Governing the commons – the evolution of institutions for collective action. Cambridge University Press, Cambridge
Paletto A, Sereno C, Furuido H (2008) Historical evolution of forest management in Europe and in Japan. Bull Tokyo Univ For 119:25–44
Pfeifer W (1993) Etymologisches Wörterbuch des Deutschen, digitalisierte und von Wolfgang Pfeifer überarbeitete Version im Digitalen Wörterbuch der deutschen Sprache. [Online] Available at: https://www.dwds.de/wb/wb-etymwb. Accessed 04 07 2019
Piketty T (2014) Capital in the 21st century. Harvard University Press, Cambridge, USA
Soddy F (1933) (Reprint 1961). Wealth, virtual wealth and debt – the solution of the economic paradox. London: Briton
Sommer M (2009) Acuicultura Insostenible en Chile 10(3):1–23

Part II
Chile

Chapter 4
Subsidizing Green Deserts in Southern Chile: Between Fast Growth and Sustainability of Forest Management

Roberto Pastén, Nicolás Nazal, and Felix Fuders

4.1 Introduction

Local conservationism is often criticized by limiting the expansion of agriculture or resource extraction especially in poor regions (Yergeau et al. 2017; Adams et al. 2004) Conservationists on the other hand argue that conservation can alleviate poverty by supplying ecosystem services, chief among them tourism services while at the same time protecting the environment (Grimm et al. 2008). Thus, 'win-win' scenarios may be possible in which ecosystems and their services are protected and poverty alleviated. Nevertheless, for local entrepreneurs and public officials alike, ecosystem service provision, particularly tourism is not enough to reduce poverty and some level of extraction activities (e.g. forest exploitation) should be allowed. These critics highlight the role that protected areas can play in limiting agricultural development and the exploitation of natural resources. It has been shown the protagonist position that protected areas can play in supplying ecosystem services, promoting tourism and improving infrastructure. For example, empirical studies have found that protected areas, on average, are effective in reducing deforestation, although not as much as advocates may have expected (e.g. Cropper et al. 2001; Andam et al. 2008; Pfaff et al. 2009; Sims 2009).

For some authors, the goal of conservation and poverty alleviation are conflicting, for others, the establishment of protected areas resulted in an increased welfare for the population (see Yergeau et al. 2017 for references). In lieu of the vision that

R. Pastén (✉) · F. Fuders
Economics Institute, Faculty of Economics and Administration, Universidad Austral de Chile, Valdivia, Chile
e-mail: roberto.pasten@uach.cl

N. Nazal
Centro de Desarrollo de Negocios Valdivia, Universidad Austral de Chile, Valdivia, Los Rios, Chile

F. Fuders and P. J. Donoso (eds.), *Ecological Economic and Socio Ecological Strategies for Forest Conservation*, https://doi.org/10.1007/978-3-030-35379-7_4

a strong conservation policy would not be an option to alleviate poverty and that some extraction activities are necessary to maintain a minimum welfare level, Chile stablished at the beginning of the 1980s subsidies to help to develop a strong forest sector. These subsidies were granted mostly independently of the species planted.

This article examines how this subsidy policy (negative tax) on marketed forest products may affect cash income, the delivery of ecosystem services and under what conditions this negative tax is welfare reducing rather than welfare enhancing using the case of the forest sector in Chile as an example. This article proceeds highlighting the permanent conflict between the extraction of a natural resource (e.g. forest) and its conservation and critically analyzes the effect of a negative tax for forest plantation. From the case study we draw some lessons for the response to forest subsidy policies of the regional economy and its effects on distribution of land between resource extraction and conservation. In the following section, this theoretical model is developed and solved, followed by a discussion of the model and its predictions, of the research questions and policy implications that are raised. A final section of conclusions is included.

4.2 Case Study

4.2.1 Context/Background

In Chile, there are currently 101 units of Wildlife Areas protected by the State, 36 of them correspond to national parks, 49 to national reserves and 16 to natural monuments. Altogether, these state-protected areas preserve a total of 14.6 million hectares, which corresponds to 20% of the national territory (CONAF 2019).

In a world where protected areas reached 14.81% of the total land territory (Worldbank 2014), growing significantly from 8.2% in 1990, the effort of some nations to increase the territory under protection has been substantial. However, of all the plant, amphibian, reptile, bird and mammal species that have gone extinct since the sixteenth century, 75% were harmed by overexploitation or agricultural activity or both in recent years (WWF 2018). Such statistics may suggest that the conservation efforts are not yet sufficient to accomplish the goal of preservation and sustainability of terrestrial ecosystems.

In Chile, for instance, less than 100,000 ha of Alerce (*Fitzroya cupressoides*) are left. This species, also known as Lawal or Lahuen by the Mapuche indigenous people, is unique in its kind, the oldest in the southern hemisphere and second oldest on the planet after the Sequoia (*Sequoia sempervirens*). This resource was reduced due to the overexploitation of the past century, by the use of its valuable wood or its replacement with exotic plantations (Torrejon et al. 2011). Despite of the conservation efforts of the Chilean State and despite the fact that this tree was declared a natural monument in 1977, the Chilean protected areas system was not able to protect this iconic and unique species of the southern geography. This takes on greater relevance if we consider that a large proportion of this remaining 100,000 ha have been guarded by private initiatives of NGOs such as The Nature Conservancy (TNC) and the Pumalín Foundation (Rivera and Vallejos-Romero 2015).

Chile serves as a case of study of the application of structural reforms implemented by the government since the 1970s (Klein 2007). Among these reforms, we find the implementation of governmental policies for boosting industrial forestry in the country (see also Chaps. 5 and 6 in this book). These policies promoted the conversion of native forests from south-central Chile into industrial monoculture plantations of fast-growing exotic species, in particular pines (*Pinus radiata*) and eucalyptus (*Eucalyptus globulus*, *E. nitens* and others).

4.2.2 Economic Life Cycle of Forestry

In order to analyze the life cycle and economic consequences of forest plantations it is necessary to refer to each of the stages of the production process of wood products as well as their effect on the environment. When the land is used for the first time for plantation of monoculture plantations, the process begins with a stage of 'cleaning' and preparation of the soil. This requires few and unskilled labor to carry out the extraction of other tree species or simply a burning method, no longer allowed but vastly used until the decade of 1990. If the soil was being used for agricultural purposes this stage is not necessary and it goes straight to the plantation. After this stage, the plantation takes place followed by forest management (thinning and pruning for *radiata pine*). Continuing with the production cycle, harvesting and transformation of the timber product is carried out with greater necessity of human work force and capital than in the earlier stages. In this stage the work force is mostly required especially during the harvesting process, which takes place every 15–20 years. The harvested wood is mainly transformed into paper pulp and other products which generated exports of 5377 million dollars in 2017 (DIRECON 2017).

4.2.3 Employment

The forest industry generates around 300,000 direct jobs which corresponds to approximately 3% of the total occupation of the Chilean work force (Lignum 2019). According to the national socioeconomic survey CASEN (2016), in areas eminently suited for forestry, the level of unemployment is consistently higher than the national average. In areas where the main activity is forestry, the unemployment ranges between 1.5% and 3% higher than the national average and is constantly above the macro region where these communes are situated. A reason for this situation could be the fact that most of these communes are covered in forest plantations with a crop rotation of more than 15 years and the harvest and re-forestation activities do not generate sufficient long-lasting direct employment and added values. While in former times the families living in these areas used to make up their livelihood with the traditional agricultural activity during the whole year, they now are mostly working in tasks related to the long forest cycles. However, these conclusions should be taken with caution since we do not know whether or not without the industrial forestry activities employment rates would be lower.

4.2.4 *Income and Poverty*

The reduction of poverty has been a macroeconomic and social success in Chile during the past three decades, certainly due to the work of all industries in the 'entrepreneurial ecosystem'. Nevertheless, when overviewing the contribution of the forest industry, we observe that in spite of the reduction of poverty in communes where it displays its operations, poverty is still persistently higher in communes with forest industry compared with the rest of communes in Chile (CASEN 2016). Here, the same caveat as before applies since we cannot compare this with the counterfactual situation, which is why we cannot disregard evidence that without forest poverty could have been even higher.

Regarding household incomes during the same period and according to the same survey, they are significantly lower in communes where forestry plantations are extensive. In consequence, the socioeconomic impact of the forest industry measured by indicators of employment, monetary incomes and poverty rates in the range of operation of the forest companies seems to be clear.[1]

4.2.5 *Environmental Liabilities of the Forest Industry*

The expansion of agroforestry like every economic activity generates an environmental cost, especially in matters of biodiversity. In Chile, as it happened much earlier in great part of Asia and Europe, agriculture was extended in the second half of the twentieth century from soils more suited and easier to grow to areas of more fragile soil, which until then had not been much affected by human activity. The expansion of agriculture and then forestry towards pristine ecosystems caused

[1] However, this may also respond to the 'cycle of copper.' Copper is the other important product that Chile exports next the forestry products. When copper is in an economic boom phase, the regions with mining activity increase their incomes and employment and reducing poverty indicators, whereas regions with agricultural and forest industries experience the contrary. This may be seen as a sort of curse for economies based in the extraction of natural resources. When the mining sector is depressed, the effect goes the other way around. The study takes into account the mining boom period but not the period when copper price was depressed.

important environmental liabilities, especially related to the loss of native forest and a dramatic decrease of biodiversity (Stephens and Wagner 2007).

The Intensive forest activity in Chile, whose consequences we are analyzing, started in the middle of the past century, in the first place for energetic usage (wood and coal) as fuel for homes and industries, later for the exploitation of native forest chips that were exported entirely to Asian markets (especially Japan) and finally with the replacement of hundreds of thousands of hectares of natural forests by crops of exotic tree species. Native forests were replaced with monoculture of introduced (nonnative) species, mainly pine and eucalyptus for the production of cellulose and timber products, activities subsidized by governmental incentives (Law DL 701, see next section).

Despite of the scarce documentation of the situation of forestry environmental liabilities and their difficult measurement and assessment, among the most important liabilities of the forest industry we may find:

- Loss of biodiversity
- Reduction of streams in areas with plantations
- Landscape fragmentation and cultural uprooting
- Conflict between communities and companies transformed into a national issue between the State of Chile and groups of indigenous peoples (Mapuche).

The debate on social and environmental consequences of the forest industry started in Chile in the middle of the 1980s. The requirement of social and environmental responsibility certifications can be seen as a result of this debate.

4.2.6 DL 701, an Incentive to the Replacement of Native Forests

In 1974 the Chilean government signed the decree DL 701 which granted a bonus of 75% of the cost of Pine and Eucalyptus plantations. After 23 years in force, in 1998 the DL 701 was extended for 15 more years with a retroactive section since 1996. In this occasion, the law was reformulated to redirect the benefit to small and medium producers.

In the years 2012 the Chilean government renewed the DL 701 again for 2 more years. In 2015, the law was submitted to the Congress for another extension this time despite of the resistance of social and environmental groups. The argument for a new extension of this subsidy was that boosting forestry in the country would help reduce the greenhouse gases emission. Basically, the idea was that through the plantation of more Pines and Eucalyptus Chile would improve its balance of carbon. The new Law of Forestry Development No. 19.561 redefined small and medium forest producers as main beneficiaries of the bonus.

According to figures compiled by (Gysling et al. 2017) between 1974 and 2015 plantations for a total of 1,250,347 ha were subsidized. The afforestation implemented during this period was mainly with monoculture of Pine and Eucalyptus. Between the years 1976 and 2016, the Chilean State allocated nearly 15 million USD per year in benefits derived of the DL 701. According to (Gysling et al. 2017) in total over 566 million USD have been granted by the Chilean State for forestry development due to this law, therefore subsidizing monoculture plantations of nonnative species, mostly in favor of large forest companies.

The native forests of south-central Chile located between the eighth and tenth region were mostly ecologically deteriorated in the period between 1980 and 1996 because of the replacement of old-growth forest by plantations of nonnative rapidly growing species. Because of the DL 701 subsidy, plantations increased by 77.583 ha/year, reaching a maximum of one million hectares in 1984 (Gysling et al. 2017). The cold evergreen forest did not look so strongly affected since the arrival of settlers in the area in the early nineteenth century, only that this time it was not a survival reason that led to the burn and clear-cutting of thousands of hectares of virgin forest. Despite the profitable business of forestry exports, companies were largely subsidized with the alleged intention of restoring degraded soils or low productivity areas with forest.

4.2.7 Environmental Impacts

The so called rapid-growth plantations with species such as Pine and Eucalyptus can reach crop rotations (period between seeding and harvesting) from 12 to 25 years depending on the characteristics of the soil, water and temperature conditions (Frêne and Núñez 2010). In contrast, in plantations of native forest species, the rotation is much slower and may range from 35 to 50 years for some *Nothofagus* species.

Some of the most significant impacts of industrial forestry, understood as intensive monoculture plantations of mainly exotic (non-native) species, are those which directly affect the soil where plantations are scattered. The Association of Forest Engineers for the Native Forest has identified some of the key negative consequences such as compaction, removal, erosion and depletion of nutrients (Astorga and Burschel 2017). Concerning the effect in the availability of water in plantation areas, scientific studies have determined as consequence of Eucalyptus plantations that moisture reserves of the soil rapidly degrade: after only 4–6 years the water consumption during the year is similar to the one observed in a mature native forest. Therefore, the water consumption of this type of plantation is such that in order to reach a fast development, it generates a hydric imbalance and affects the physicochemical quality of waters. Studies carried out by researchers from Universidad Austral de Chile established that the amount of water necessary to produce a cubic meter of forest volume in *Pinus radiata* plantations at south-central Chile ranged from 241 to 717 m^3. In the case of Eucalyptus, the transpiration rates depending on the species fluctuate around 20 and 40 l per tree and day (Huber and Trecaman 2000).

Due to this high consumption of water, the forest plantations leave behind a lower availability of this resource for the activities of the population living in the surrounding areas which affects negatively their quality of life (Donoso and Otero 2005). While native forests are an integrated ecological system dominated by local tree species and natural accompanying vegetation as well as animals, fungi and soil microorganisms, the forest plantations are artificial ecosystems of a single fast-growing species, planted in homogeneous blocks of the same age, with a very low presence of other flora and fauna species. Better that plantations, original forests generate a series of ecosystem services that are fundamental for the life and welfare of human beings such as water purification, climate regulation, soil formation, CO_2 absorption and oxygen release, in addition to other cultural, recreational and tourism services and being home of people and ancestral communities that have developed complex balanced systems to inhabit these places in continuous co-adaptation (Chazdon et al. 2008).

4.3 The Model

We develop and solve an analytical theoretical model to describe people's response to an introduction of a subsidy (negative tax) on marketed forest products. The model is based on Anthon et al. (2008) where the authors use a two-sector model to study the effect of a tax on forest products marketed by the poorest people. These authors do not consider environmental effects in the model, but they do include leisure in the utility function that can easily be adapted to the environmental case. However, the model presented here has several main differences with Anthon et al. (2008). While in a two-sector model labor is required in both sectors, in our model environment is not necessarily affected by two sectors. In the model presented here land is distributed either for use in extractive uses (e.g. mining, agriculture, forest, etc.) or for conservation. But, while conservation produces a flow of ecosystem services, extraction activities preclude those ecosystem services to be provided, and this is the second and main difference with Anthon et al. (2008) that will prove to be critical in the assessment of results presented here.

We use the framework of a regional representative agent model and assume that total land T is either destined to conservation z_C or for extractive activities z_F. For the purpose of this model we identify the extractive activity with timber extraction, but it could also be any other activity (e.g. mining, agriculture, commercial, fishing etc.) where a product is marketable and ecosystem services provision total or partially impaired. The local economy maximizes aggregate utility u of ecosystem services provision q and flow of consumption c. Cash income for consumption is obtained from forest production and any payment able to be captured from ecosystem service provision provided by conservation, e.g. tourism. A minimum of consumption is required not possible to be fulfilled exclusively through revenues from conservation (i.e. tourism or payment from ecosystem services in general), mostly owed to the impossibility to capture financially all ecosystem services provided by conservation. Income from forest activity is pz_F and income from payment for ecosystem services, such as tourism or other activities associated with conservation, that is wz_C, where p and w are the price of marketed forest products and payment for ecosystem services respectively. Conservation activities produce ecosystem services given by $q = z_C^{\theta}$, if there is no forest extraction, total ecosystem services are given by $\overline{q} = T^{\theta}$. Forest activities prevent the provision of the ecosystem services in the amount $T^{\theta} - (T - z_F)^{\theta}$. The price w represents the price of every parcel of land conservation that is possible to capture as payment for ecosystem services but not the whole social value of ecosystem services. The households´ problem to be solved is[2]:

$$\max_{c,q} u(c, q) = c^{\beta} q^{(1-\beta)} \tag{4.1}$$

$$\text{s.t.} \quad c = pz_F + wz_C \tag{4.2}$$

$$T = z_c + z_F \tag{4.3}$$

$$q = z_C^{\theta} \tag{4.4}$$

$$c \geq wT \tag{4.5}$$

$$c, q, T, z_C, z_F > 0 \tag{4.6}$$

$$0 < \beta < 1; 0 < \theta < 1 \tag{4.7}$$

The households maximize utility under the budget constraint (4.2), a land constraint (4.3) and an environmental constraint (4.4) and a minimum consumption constraint (4.5). For simplicity, we build a static model, i.e. cash income is equal to consumption. The price of consumption is normalized to 1.

4.4 Results

The result section is structures as follows. First, we look at the optimal land distribution, consumption and ecosystem services provision. We then introduce a negative tax (a subsidy) in the forest sector to see the effects on the optimal values

[2]The meaning of s.t.: important constraint that the functions must follow.

of the control variables. Second, we observe the effects of changes in the price paid for ecosystem services and its effects particularly in cash income for consumption. Finally, we discuss some welfare effects associated to the negative tax.

4.4.1 Basic Results

We first take a look on the optimal land allocation, ecosystem services provision and consumption. The concentrated problem can be defined as:

$$\max_{z_C} u = (p(T - z_C) + wz_C)^{\beta} {z_C}^{\theta(1-\beta)} \tag{4.8}$$

Which gives the following agent's land allocation:

$$z_C^* = \frac{(\alpha - \beta)\, pT}{\alpha\,(p - w)} \tag{4.9}$$

$$z_F^* = \frac{T\beta p - \alpha wT}{\alpha\,(p - w)} \tag{4.10}$$

$$\alpha = \beta + \theta\,(1 - \beta) \tag{4.11}$$

Under Eq. (4.5), $z_C^*, z_F^* \in [0, T]$ as the appendix shows.

Both, land destined to conservation and land destined to forest extraction are complex nonlinear functions of the price of forests p and the price of Ecosystem Services w, so it is not straightforward to say what the effect will be. This is the goal of the following section. The results from Eqs. (4.9) and (4.10) can be used to calculate the provision of ecosystem services q^* and total cash income for consumption c^*

$$q^* = \left(\frac{(\alpha - \beta)\, pT}{\alpha\,(p - w)}\right)^{\theta} \tag{4.12}$$

$$c^* = \frac{T\beta p}{\alpha} \tag{4.13}$$

The optimal level of ecosystem service provision in our model depends exclusively on land destined to conservation z_C and how productive in terms of the provision of ecosystem services this land is θ. Using Eq. (4.3), q^* can also be determined in terms of land for extraction z_F rather than for conservation, i.e. $q^* = (T - z_F)^{\theta}$, showing that ecosystem services decrease as land for extraction activities, e.g. timber extraction, increases. The optimal households' cash income for consumption is directly proportional to the price of the extractive resource but surprisingly does not depend on the price of ecosystem services.

4.4.2 *Effects of a Negative Tax (Subsidy)*

Now we turn to the effect of a negative tax, i.e. the effect of a subsidy. This subsidy is modeled as an increment in the price p per unit of land used for forest exploitation:

$$\frac{\partial z_C^*}{\partial p} = \frac{-\alpha(\alpha-\beta)wT}{(\alpha(p-w))^2} < 0 \tag{4.14}$$

$$\frac{\partial z_F^*}{\partial p} = \frac{\alpha wT(\alpha-\beta)}{(\alpha(p-w))^2} > 0 \tag{4.15}$$

$$\frac{\partial q^*}{\partial p} = -\theta z_C^{\theta-1}\frac{\alpha(\alpha-\beta)wT}{(\alpha(p-w))^2} < 0 \tag{4.16}$$

$$\frac{\partial c^*}{\partial p} = \frac{T\beta}{\alpha} > 0 \tag{4.17}$$

As expected, the results show that a subsidy per forested area, increase land destined to resource extraction, decreases land for conservation and, thus, decreases the provision of ecosystem services. However, the subsidy increases regional cash income.

4.4.3 *Effects of Payment for Ecosystem Services*

Now we turn to the effect of an increase in the price of the ecosystem service w. One of these ecosystem services is tourism, if this is the setting, w is the price per unit of land destined to conservation and where the unique activities allowed are those of tourism. As stated in Eq. (4.5), it is not possible to obtain minimum levels of consumption only based on income from these ecosystem services. This is more than an assumption; it is a requirement for the model to be consistent. It is consistent in addition with the fact that most of the ecosystem services provided by the ecosystem are public goods without possibility to assign private property rights and in this they cannot be remunerated, except for a few cases where payment for ecosystem services have been implemented but, as shown above, these instruments have been only partially successful at least as a broad mechanism of funding conservation.

$$\frac{\partial z_C^*}{\partial w} = \frac{\alpha(\alpha - \beta) pT}{(\alpha(p - w))^2} > 0 \tag{4.18}$$

$$\frac{\partial z_F^*}{\partial w} = -\frac{\alpha(\alpha - \beta) pT}{(\alpha(p - w))^2} < 0 \tag{4.19}$$

$$\frac{\partial q^*}{\partial w} = \theta z_C^{\theta-1} \frac{\alpha(\alpha - \beta) pT}{(\alpha(p - w))^2} > 0 \tag{4.20}$$

$$\frac{\partial c^*}{\partial w} = 0 \tag{4.21}$$

As expected from the results above, as response to an increase of w land allocated to z_C^* as well as ecosystem services provision q^* increase while land allocated to z_F^* decreases. Surprisingly, there is no effect of an increase of w on regional income c^*. The intuition, as elaborated in the appendix, is that even though revenue from conservation increases, this is totally offset by the increase in conservation of land and the reduction in land assigned for resource extraction while the latter is more productive (in terms of generating cash income) than the former.

4.4.4 Welfare Effects

The indirect utility function is given by

$$u = \left(\frac{T\beta p}{\alpha}\right)^{\beta} \left(\frac{(\alpha - \beta) pT}{\alpha(p - w)}\right)^{\theta(1-\beta)} \tag{4.22}$$

It is possible to show that in our setting an increase of the return of timber extraction is welfare enhancing if the condition $p > \frac{\alpha w}{\beta}$, or equivalently, if the income from payment for ecosystem services is not enough to satisfy the minimum standard of consumption, i.e. if $c > wT$. However, in a more general setting higher returns in the extraction sector can be welfare reducing if

$$\frac{p}{w} < \left(1 + \theta \frac{1 - \beta}{\beta}\right) \tag{4.23}$$

The subsidy is welfare reducing if the expression in Eq. (4.23) holds, i.e. if the relative return of conservation to extractive activities w/p is high; if the environment is highly valued relatively to consumption (i.e. a high $(1 - \beta)/\beta$); or if the land destined to conservation is highly rich in terms of the quantity and quality of ecosystem services it provides (i.e. θ is high).

4.5 Discussion and Policy Implications

The fact that the subsidy reduces welfare when the land destined to conservation is highly rich in terms of the quantity and quality of ecosystem services it provides (i.e. θ is high) is particularly interesting. Natural sciences are still (and most likely will always be) far away from completely understanding all ecosystem services and, above all, the interactions and connections of the different ecosystem services with each other and with the fauna/flora and human beings (Daly and Farley 2011). We are therefore unable to assess neither the "quantity" nor the "quality" of ecosystem services that a forest provides (i.e. whether θ is high or not). We can conclude that such a generalized subsidy that promotes any kind of reforestation (including monocultures of nonnative species), as it has been implemented in Chile so far, should be applied very carefully.

The model does not distinguish between subsidies that are granted for extraction and reforestation of whatever species and those subsidies that could theoretically be granted for reforestation of only native species in heterogeneous and diversified plantations (no monocultures). In the latter case the loss of ecosystem services and the associated welfare loss would be minimized while at the same time welfare gains from extraction would not necessarily be less. Native species usually sell significantly more expensive than Pine tree or Eucalypts and some species do not even need so much longer to be harvested (Donoso 2014). A powerful incentive why investors prefer Pine or Eucalyptus plantations is the fact that the money interest rate gives time a (relatively high) value which makes the fast-growing species economically more attractive (on this see Chap. 2 in this book). So, if a subsidy would only be granted for planting native species and not, as was the case in Chile until now, for any type of reforestation, this could compensate for the welfare loss (which in fact can be interpreted as a loss of interest) generated by the increased time period to harvest. Since with reforestation of native species in heterogeneous cultures there would be no (or only a minimal) negative impacts on ecosystem services, such subsidy would enhance welfare. It is therefore recommendable to introduce subsidies that are only granted for the reforestation of native species and guaranteeing species diversity.

Our model shows (see appendix Eq. (A.9)) that the return provided by the ecosystem does not have any negative impact on the households' income but a positive effect on the conservation of the environment. One might think that this should lead to a total conservation of land. However, this is not the case, since our restrictions show that if total land is devoted only to conservation, it is not possible to achieve minimum levels of consumption only through the captured revenues from ecosystem services.

In our model a high level of consumption is welfare reducing through the loss of ecosystem services. On the contrary, in conventional economic thinking consumption is an indicator for welfare; the more we produce and consume the higher is supposed to be our total welfare. The "more is better" is seen as one of the characteristics of consumer preferences in Microeconomics (Frank 2008:

64). This is why GDP is still – although criticized (see e.g. Constanza et al. 2014) – frequently used as welfare indicator. Our model clearly shows that a steady increase in consumption in not welfare enhancing; we even see a trade-off between consumption and welfare. It is worth mentioning that we also recognize this trade-off in alternative development and Quality-of-life indexes, as many of those indexes use arbitrarily chosen artefacts or circumstances to measure what is believed by their authors to be an indicator for the quality of life. For example, the quantity of hospital beds or physicians per capita or the access to potable water are, among others, such indicators we usually find in these indexes. However, a certain national income (GDP) and with it the destruction of ecosystem services is required to be able to provide this infrastructure. Furthermore, it is questionable if a high number of physicians per capita really means that we live healthier. If we live, as still many people in Chile and Brazil do, self-sufficiently in the mountain countryside with fresh air, little stress (compared to living in big cities) and consuming small-scale-farming organic food we will probably get less ill and need less medical care, as living in a modern metropolis. We also do not necessarily need potable water if we have a spring or a clean cascade coming from the glaciers next to the house providing us with the purest water imaginable.

An alternative to measure welfare that does not promote the mentioned trade-off between GDP growth and destruction of ecosystem services is the "Human Scale Development Index" (Fuders et al. 2016) that measures the subjective perception of the satisfaction of fundamental human needs. This index is builds on the *Human-Scale-Development*-approach from Max-Neef et al. (1991). According to this concept human needs are not infinite but few, finite and classifiable and the same across cultures and historical time periods. What might change in different cultures or times is the form to satisfy these needs (Max-Neef et al. 1991). Many fundamental human needs, such as "affection", "protection", "creation", "leisure", "understanding" or "liberty", can most probably be easier satisfied living in the Chilean mountain cordillera as living in the metropolitan area, even though average GDP might be relatively low. The conservation of ecosystems increases the outcome of this index, as ecosystem service provision is key for most fundamental human needs to be satisfied. Using this or a similar index to measure development of a country or region could significantly enhance the understanding of the importance of ecosystem conservation in politics.

Appendix

Basic Results

The concentered model is

$$\max_{z_C} u = (p\,(T - z_C) + w z_C)^{\beta} {z_C}^{\theta(1-\beta)} \tag{A.1}$$

And the first order condition is:

$$\frac{\partial L}{\partial z_c} = -\beta\left(p\left(T - z_c\right) + wz_c\right)^{\beta-1}(p - w)\, z_C{}^{\theta(1-\beta)} + \left(p\left(T - z_C\right) + wz_C\right)^{\beta}\theta\left(1 - \beta\right) z_C^{\theta(1-\beta)-1} = 0 \tag{A.2}$$

Rearranging Eq. (A.2) we get that

$$-\frac{\beta\left(p - w\right)}{p\left(T - z_C\right) + wz_C} + \frac{\theta\left(1 - \beta\right)}{z_c} = 0 \tag{A.3}$$

and from Eq. (A.3)

$$z_C^* = \frac{\theta\left(1 - \beta\right)pT}{\left(\beta + \theta\left(1 - \beta\right)\right)\left(p - w\right)} \tag{A.4}$$

if we collect preference's parameters in

$$\alpha = \beta + \theta\left(1 - \beta\right) \tag{A.5}$$

Combining (A.4) and (A.5), and noting by (A.5) that $\alpha > \beta$, gives the optimal allocation to conservation in Eq. (4.9) in the text

$$z_C^* = \frac{\left(\alpha - \beta\right)pT}{\alpha\left(p - w\right)}$$

Note in the first order condition in (A.3) that the second term in the left-hand side gives the marginal benefits of increasing conservation, which does not depend on returns to each sector p, w, but only depends on the marginal damage of ecosystem services lost $(1 - \beta)$ and how productive the ecosystem is in terms of ecosystem services provision θ. The first term on the left-hand side shows the opportunity cost of conservation in terms of reduced income for forest production, which is positive if the return to the forest sector p is higher than the returns captured trough the ecosystem services provision w. The optimal allocation to conservation occurs where both effects cancel each other out.

The optimal allocation of land to the forest sector can be determined by inserting z_C^* in the total land restriction in Eq. (4.3)

$$z_F = T - z_c$$

$$z_F = T - \frac{\theta\left(1 - \beta\right)pT}{\left(\beta + \theta\left(1 - \beta\right)\right)\left(p - w\right)}$$

$$z_F^* = \frac{T\beta p - \left(\beta + \theta\left(1 - \beta\right)\right)wT}{\left(\beta + \theta\left(1 - \beta\right)\right)\left(p - w\right)},$$

This expression is equal to the optimal allocation of land to the extractive sector (i.e. forest) in Eq. (4.10) in the main text if Eq. (A.5) holds:

$$z_F^* = \frac{T\beta p - \alpha w T}{\beta (p - w)}$$

Note that $z_F^* \geq 0$ if

$$\frac{p}{w} \geq \frac{\alpha}{\beta} \tag{A.6}$$

Since $\alpha > \beta$, this condition requires that the return to the extractive sector (either forest, agriculture, mining etc.) must be higher than the collected value for ecosystem services payment, (tourism, use of fresh water, etc.). As we show below, this expression is equivalent to the restriction in Eq. (4.5) $c \geq wT$.

The total environmental quality q is calculated from Eqs. (4.4) and Eq. (4.9)

$$q^* = z_C^\theta$$

$$q^* = \left(\frac{(\alpha - \beta)\, pT}{\alpha (p - w)} \right)^\theta .$$

Cash income for consumption is calculated inserting the optimal values of land conservation Eq. (4.9) and forest land in Eq. (4.10) in the budget constraint (4.2):

$$c^* = p z_F^* + w z_C^*$$

$$c^* = p \frac{T\beta p - \alpha w T}{\alpha (p - w)} + w \frac{(\alpha - \beta)\, pT}{\alpha (p - w)}$$

$$c^* = \frac{T\beta p^2 - \alpha w p T + (\alpha - \beta)\, w p T}{\alpha (p - w)}$$

$$c^* = \frac{T\beta p^2 - w\beta p T}{\alpha (p - w)}$$

$$c^* = \frac{T\beta p\, (p - w)}{\alpha (p - w)}$$

$$c^* = \frac{T\beta p}{\alpha} \tag{A.7}$$

Note that if (A.6) holds, then

$$c^* \geq wT \tag{A.8}$$

This means that even if the whole land is destined to conservation activities that would not be enough to satisfy minimum levels of consumption and some extractive activities are necessary.

Effects of a Negative Tax (Subsidy)

$$\frac{\partial z_C^*}{\partial p} = \frac{(\alpha - \beta)\, T\alpha\, (p - w) - \alpha\, (\alpha - \beta)\, pT}{(\alpha\, (p - w))^2}$$

$$\frac{\partial z_C^*}{\partial p} = \frac{-\alpha\, (\alpha - \beta)\, wT}{(\alpha\, (p - w))^2} < 0$$

$$\frac{\partial z_F^*}{\partial p} = \frac{T\beta\alpha\, (p - w) - \alpha\, (T\beta p - \alpha w T)}{(\alpha\, (p - w))^2}$$

$$\frac{\partial z_F^*}{\partial p} = \frac{\alpha w T\, (\alpha - \beta)}{(\alpha\, (p - w))^2} > 0$$

$$\frac{\partial q^*}{\partial p} = \theta z_C^{\theta - 1} \frac{\partial z_C^*}{\partial p}$$

$$\frac{\partial q^*}{\partial p} = -\theta z_C^{\theta - 1} \frac{\alpha\, (\alpha - \beta)\, wT}{(\alpha\, (p - w))^2} < 0$$

$$\frac{\partial c^*}{\partial p} = \frac{T\beta}{\alpha} > 0$$

Effects of Payment for Ecosystem Services

$$\frac{\partial z_C^*}{\partial w} = \frac{\alpha\, (\alpha - \beta)\, pT}{(\alpha\, (p - w))^2} > 0$$

$$\frac{\partial z_F^*}{\partial w} = \frac{-\alpha T \alpha (p - w) + \alpha (T \beta p - \alpha w T)}{(\alpha (p - w))^2}$$

$$\frac{\partial z_F^*}{\partial w} = -\frac{\alpha (\alpha - \beta) p T}{(\alpha (p - w))^2} < 0$$

$$\frac{\partial q^*}{\partial w} = \theta z_C^{\theta - 1} \frac{\partial z_C^*}{\partial w}$$

$$\frac{\partial q^*}{\partial w} = \theta z_C^{\theta - 1} \frac{\alpha (\alpha - \beta) p T}{(\alpha (p - w))^2} > 0$$

$$c^* = \frac{T \beta p}{\alpha}$$

$$\frac{\partial c^*}{\partial w} = 0 \tag{A.9}$$

Result in Eq. (A.9) is intriguing. It shows that the return provided by the ecosystem does not have any negative impact on the households' income but does have a positive effect on the environment. At first sight it seems to indicate that it is possible to conserve the environment without any opportunity cost in terms of consumption only by increasing w, leading in the end to a total conservation of land. However, this is not the case. Our restrictions show that if total land is devoted only to conservation, it is not possible to achieve minimum levels of consumption only through the captured revenues from ecosystem services. Some level of extraction activity is necessary to sustain life.

Another way to see the null impact of w on aggregate income is trough total differentiation of the budget constraint

$$c^* = p z_F^* + w z_C^*$$

$$\frac{dc^*}{dw} = p \frac{dz_F^*}{dw} + z_C^* + w \frac{dz_C^*}{dw} \tag{A.10}$$

According to the land constraint

$$-\frac{dz_F^*}{dw} = \frac{dz_C^*}{dw},$$

rewritten Eq. (A.10)

$$\frac{dc^*}{dw} = z_C^* - (p - w)\frac{dz_C^*}{dw}.$$

It is possible to show that $z_C^* = (p - w)\left(dz_C^*\right)/dw$, and thus $(dc^*)/dw = 0$

$$z_C^* = \frac{(\alpha - \beta)\, pT}{\alpha\,(p - w)}$$

$$(p - w)\frac{dz_C^*}{dw} = (p - w)\frac{\alpha\,(\alpha - \beta)\, pT}{(\alpha\,(p - w))^2} = \frac{(\alpha - \beta)\, pT}{\alpha\,(p - w)}$$

Hence, the additional income obtained by the higher price of conservation z_C^* is totally compensated by increasing z_C^* and reduced z_F^* while the former is less productive by the latter.

Welfare Effects

According to Eqs. (4.1), (4.9) and (4.10), the indirect utility function is given by

$$u = \left(\frac{T\beta p}{\alpha}\right)^{\beta}\left(\frac{(\alpha - \beta)\, pT}{\alpha\,(p - w)}\right)^{\theta(1-\beta)}.$$

This expression in logarithms becomes:

$$\ln u = \lambda + \beta \ln p + \theta\,(1 - \beta) \ln p - \theta\,(1 - \beta) \ln\,(p - w)$$

with λ, a constant, given by

$$\lambda = \beta \ln\left(\frac{T\beta}{\alpha}\right) + \theta\,(1 - \beta) \ln\left(\frac{(\alpha - \beta)\, T}{\alpha}\right)$$

$$\frac{\partial \ln u}{\partial p} = \frac{\beta}{p} + \frac{\theta\,(1 - \beta)}{p} - \frac{\theta\,(1 - \beta)}{(p - w)} \tag{A.11}$$

According to Eq. (A.11), it is possible to show that an increase in the return of forest p is welfare enhancing if the condition $p > \frac{\alpha w}{\beta}$, holds, or equivalently, if the income from ecosystem services is not enough to satisfy minimum standard of consumption, i.e. it is required that $c > wT$. However, in a more general setting an increase in price and in this higher returns in the extraction sector can be welfare reducing if

$$\frac{p}{w} < \left(1 + \theta \frac{1-\beta}{\beta}\right) \tag{A.12}$$

If Eq. (A.12) holds, higher returns in the extractive sector (e.g. a subsidy) is welfare reducing if the left side of Eq. (A.12) is low or the right side is high. Thus, if the relation between the return to conservation to the return of the extractive sector w/p is high; if the relative valuation of the environment to the valuation of consumption $(1 - \beta)/\beta$ is high or if the richness of the ecosystems in terms of provision of ecosystem services θ is high.

References

Adams WM et al (2004) Biodiversity conservationand the eradication of poverty. Science 306:1146–1149
Andam KS et al (2008) Measuring the effectiveness of protected areas networks in reducing deforestation. Proc Natl Acad Sci U S A 105:16089–16094
Anthon S, Lund JF, Helles F (2008) Targeting the poor: taxation of marketed forest products in developing countries. J For Econ 14:197–224
Astorga L, Burschel H (2017) Avanzando en la propuesta hacia un Nuevo Modelo Forestal. Bosque Nativo 59:9–15
CASEN (2016) Resultados encuesta CASEN 2016. In: M. d. Social (ed) . Ministerio de Desarrollo Social, Santiago de Chile
Chazdon RL et al (2008) Beyond deforestation: restoring forests and ecosystem services on degraded lands. Science 320:1458–1460
CONAF (2019) Parques nacionales. Retrieved 1 29, 2019, from http://www.conaf.cl/parques-nacionales/parques-de-chile
Constanza R et al (2014) Time to leave GDP behind. Nature 505:283–285
Cropper M, Puri J, Griffiths C (2001) Predicting the location of deforestation: the role of roads and protected areas in North Thailand. Land Econ 77(2):172–186
Daly H, Farley J (2011) Ecological economics – principles and applications, 2nd edn. Island Press, Washington
DIRECON (Dirección General de Relaciones Económicas Internacionals), M. d. R. E (2017) Reporte anual comercio exterior de Chile. Ministerio de relaciones exteriores, Departamento de estudios, Santiago
Donoso PJ (2014) Ecología Forestal Bases para el Manejo Sustentable y Conservación de los Bosques Nativos de Chile. In: Donoso C, Gonzales M, Lara A (eds) Ecología forestal. Ediciones UACh, Valdivia, pp 505–526
Donoso PJ, Otero LA (2005) Hacia una definición de país forestal: ¿Dónde se sitúa Chile? Bosque 26(3):5–18
Frank R (2008) Microeconomics and behaviour. McGraw-Hill/Irwin, New York
Frêne Conget C, Núñez Ávila M (2010) Hacia un nuevo Modelo Forestal en Chile. Bosque Nativo 47:25–35
Fuders F, Mengel N, Maria del Valle B (2016) ´indice de desarrollo a escala humana: propuesta para un indicador de desarrollo endógeno basado en la satisfacción de las necesidades humanas fundamentales. In: Pérez Garcés R, Espinosa Ayala E, Terán Varela O (eds) Seguridad Alimentaria, actores Terretoriales y Desarrollo Endógeno. Laberinto, Iztapalapa, pp 63–106

Grimm NB, Faeth SH, Golubiewski NE, Redman CL, Wu JG, Bai XM, Briggs JM (2008) Global change and the ecology of cities. Science 319:756–760
Gysling Caselli A, Álvarez González V, Soto Aguirre D, Pardo Velásquez E, Poblete Hernández P, Bañados Munita J (2017) Anuario forestal. INFOR, Santiago
Huber A, Trecaman R (2000) Effect of a Pinus Radiata plantation on the spatial distribution on soil water content. Bosque 21(1):37–44
Klein N (2007) The shock doctrine: the rise of disaster capitalism. Alfred A. Knopf, Toronto
Lignum (2019) El sector forestal genera más de 300.000 empleos en Chile. Retrieved 1 29, 2019, from http://www.lignum.cl/2015/10/22/el-sector-forestal-y-la-generacion-de-empleos
Max-Neef M, Elizalde A, Hopenhayn M (1991) Human scale development – conception, application and further reflections. Apex, New York
Pfaff A, Robalino J, Sanchez-Azofiefa GA, Andam KS, Ferraro PJ (2009) Park location affects forest protection: land characteristics cause differences in park impacts across Costa Rica. BE J Econ Anal Policy 9(2):1–24
Rivera C, Vallejos-Romero A (2015) The privatization of conservation in Chile: rethinking environmental governance. Bosque 36(1):15–25
Sims KE (2009) Conservation and development: evidence from Thai protected areas. Amherst College – Dep. of Economics, Amherst
Stephens SS, Wagner MR (2007) Forest plantations and biodiversity: a fresh perspective. J For 205(6):307–313
Torrejon FG, Cisternas M, Alvial I, Torres L (2011) Colonial timber felling consequences of the alerce forests in Chiloé, southern Chile (18th and 19th centuries). Magallania 39(2):75–95
Worldbank (2014) Terrestrial protected areas. Retrieved 1 29, 2019, from http://data.worldbank.org/indicator/ER.LND.PTLD.ZS?end=2014&start=2014&view=bar&year=2014
WWF (2018) In: Grooten M, Almond RE (eds) Living planet report: aiming higher. WWF, Gland
Yergeau M-E, Boccanfuso D, Goyette J (2017) Reprint of: linking conservation and welfare: a theoretical model with application to Nepal. J Environ Econ Manag 86:229–243

Chapter 5
Land Use as a Socio-Ecological System: Developing a Transdisciplinary Approach to Studies of Land Use Change in South-Central Chile

Daniela Manuschevich

5.1 Introduction

The study of socio-ecological systems seeks to understand how nature intertwines with social, political, economic, and environmental dimensions of particular societies (Kates et al. 2001; Clement 2013). Applied to land use change, a socio-ecological approach examines interactions among political, economic, and social domains that contribute to land use change over time. In turn, land use change produces ecological and environmental consequences (Turner et al. 2007) which, consequently affect the long-term sustainability and the adaptive capacity of socio-ecological systems (Gallopin 2006; Turner and Robbins 2008).

The socio-ecological approach has become increasingly popular among ecologists who have come to understand that a sole focus on the "environment" is limited. In fact, the management of natural resources is the "management of people" (Berkes 2012). Environmental problems are inherently complex, where several dynamics are at play. Such problems can only be analyzed and ultimately addressed through the use of multiple disciplines and epistemologies (Holling 2001; Berkes et al. 2003). Yet, how should we integrate the insights of social and natural sciences? Ideally, it has been argued that transdisciplinarity, the integration of several disciplines, must be done from the very beginning, i.e. the conception of the research project (Norgaard and Baer 2005). The conception of the problem or core research question must explicitly consider multiple aspects of environmental problems. Although, the transdisciplinary approach promises a fertile ground both for problem solving and knowledge generation, it presents significant intellectual and methodological challenges – from research design to data collection and analysis.

D. Manuschevich (✉)
Geografia, Universidad Academia Humanismo Crisitiano, Santiago, Region Metropolitana, Chile
e-mail: dimanusc@syr.edu

F. Fuders and P. J. Donoso (eds.), *Ecological Economic and Socio Ecological Strategies for Forest Conservation*, https://doi.org/10.1007/978-3-030-35379-7_5

This chapter first presents the theoretical conception of transdisciplinary research in environmental sciences. Secondly, presents an empirical case: land use change and forestry legislation in Chile. Thirdly, is demonstrated how spatial modeling can be used to connect policy views and landscape outcomes. Finally, is presented the initial findings of 3 years of ethnography conducted in rural areas. As we shall see, the ethnographic approach yielded fresh insights that relevant for thinking in policy alternatives for sustainable development.

According to Norgaard and Baer (2005), that generates transdisciplinary knowledge requires that the conception of the core research question integrates multiple disciplines. That is to say that, transdisciplinary research is not merely the addition of several knowledge. Instead, transdisciplinary research means *trespassing the discipline itself* (Max-Neef 2005). The perspective proposed here is that transdisciplinarity is a movement, an attempt to transcend arbitrary borders and boundaries too often enforced within the academy. It seeks to examine human-nature relationships in detail, to address urgent problems, and apply knowledge in the pursuit of sustainable solutions. As a movement, transdisciplinary research does not seek to resolve the disciplinary barriers-once and for all, but rather, acts as a call for mobilization beyond (and unsettling) disciplinary knowledge barriers. Therefore, transdisciplinarity is an *open endeavor,* never final, and always open to new knowledge.

The differentiation between studies of social and natural phenomena is, in part, a historical byproduct of Western ontology (e.g., the Cartesian divide). For example, such a perspective views society as separate from nature. Therefore, humans are viewed as "disruptive" influences on environments, and environments are understood as "dangerous" for human populations (Binder et al. 2013). In this way, pressures – whether human or environmental – become naturalized and therefore seemingly impossible to change. In both cases, a lack of integration between ecological and social forms reduces possibilities for understanding the phenomena, and crucially, limits our ability to intervene and propose alternatives. If the human species are *only* a source of pressure and stress to natural systems, or conversely, if nature is *naturally* a danger, we are left with few options (Binder et al. 2013) – we must protect ourselves from nature and limit human population. At the same time, humans are viewed homogenously – as ravenous consumers and destroyers of the planet. Economic, social, ideological, and cultural differences exist and entail diverse relationships, understandings, and valuations of nature (Berkes 2012). Thus, under a transdisciplinary gaze, social and environmental dynamics are ontologically linked, meaning it is a well-suited framework for adequately addressing sustainability issues (Daly and Farley 2011).

Considering all the above-mentioned aspects, in Fig. 5.1 is presented a socio-ecological framework for addressing land use change as a socio-ecological system. As can be seen in Fig. 5.1, land use change is the result of economic and political process, how people use the land, discourses regarding land use policy and biophysical conditions. In turn, land use change modifies the environment and influences how people live both in materially and symbolically. The term socionatures, aims to push us beyond the nature-culture divide and create a more

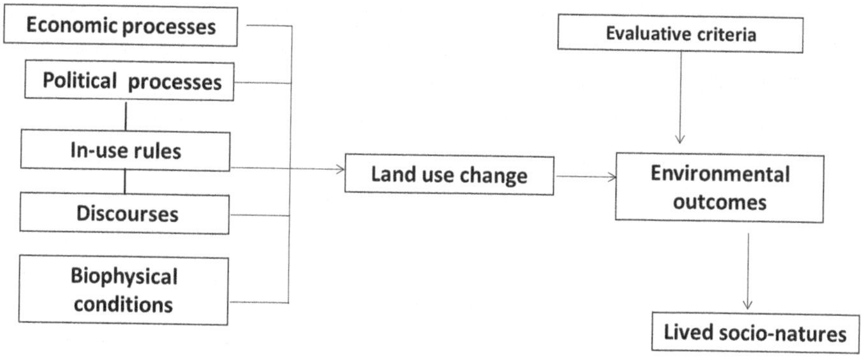

Fig. 5.1 A socio-ecological framework of land use change. Land use change is the result of the interaction among economic, political, and environmental variables. (Framework adapted from Clement and Amezaga (2013), Turner et al. (2007) and Swyngedouw (1999))

dynamic, holistic, and social vision of the natural world (Swyngedouw 1999; Castree and Braun 2001). Swyngedouw (1999, p. 443) provides a sharp account of the term socio-natures:

> It is not at all a denial of the material reality of those things we routinely call natural – be they trees, rivers, animals, or anything else. Rather, it's an insistence that the physical opportunities and constraints nature presents societies with can only be defined relative to specific sets of economic, cultural, and technical relations and capacities. In other words, the same 'chunk' of nature – say the Amazon rainforest - will have different physical attributes and implications for societies, depending on how those societies use it. In this sense, the physical characteristics of nature are contingent upon social practices: they are not fixed.

Over the last 30 years, critical geographers, human ecologists, political ecologists, anthropologists, and ecological economists have developed a very large body of knowledge demonstrating the social underpinnings of environmental problems. For example, Blaikie and Brookfield's seminal study (1986) documented the social roots of land degradation. They argue that geological processes only become problematic in relation to human needs. Watts and Bohle (1993), amidst the international discussion on famines and hunger, countered notions that famines were caused by a lack of productivity and analyzed the structural conditions which produce crop failure and hunger. For them, famines were not the simple result of a lack of food production, but rather related to entitlement (economic capability), empowerment (social and political power within the household, workplace, and public) and highly influenced class-based appropriation of surplus and production.

In a study of gendered agricultural practices, Rocheleau and Ross (1995) found that women favored the production of food, while men sought to plant non-native "miracle trees" which they credited with the ability to rectify land degradation. The women in this study were concerned about increased dependence on the cash economy and concomitant threats to food security. They felt increasingly marginalized within their households (as food crops were less valuable "companion species" than male tree plantations). The afore-mentioned studies highlight the eminently social nature of environmental issues, including land use change. The limits of simplistic common-sense solutions, and the need to embrace the complexity of socio-ecological interactions.

5.2 Timber Farm Expansion in Southern Chile

Over the last 40 years, Chile has witnessed a spectacular landscape transformation. In 1973, only 330,000 ha were planted with timber farms (Camus 2006; Armesto et al. 2010). By 2016, that figure increased 713%, up to 2,414,000 ha (INFOR 2016). These tree farms are generally of exotic species, monocultures of Radiata pine (*Pinus radiata*) or Eucalyptus spp. The plantations are cultivated in accordance with the logic of industrial management (e.g., agrochemical inputs, high plantation density, extensive clear cuts), mainly to produce paper pulp. Clear-cuts are done on average every 15–20 years, depending on the species.

The landscape of south-central Chile has much to tell us about striking environmental transformations. Since the arrival of the Spaniards, the southern cone of Latin America has suffered a process of deforestation and territorial change (Lara et al. 2012). Deforestation was initially driven by mining and export-oriented wheat farming (Camus 2006). By 1940, poor soil management and deforestation contributed to erosion problems (Elizalde 1970). As a result, in 1931, the first Forest Law[1] was enacted and aimed to curb deforestation. However, without proper implementation deforestation continued in frontier regions (e.g., mountainous areas in the Andes, coastal areas, and Patagonia (Camus 2006).

By the beginning of the twentieth century, radiata pine was recognized for its rapid growth rate, reaching maturity in a mere 20 years. It was therefore used in schemes to control soil erosion. As a result, the National Timber Association promoted and legitimized afforestation with non-native species. The industrial afforestation model complemented and reinforced import substitution models, which predominated across Latin America in the aftermath of the 1929 economic crisis and World War II. In Chile, the timber sector was privileged as a source of raw materials to be exported elsewhere (Camus 2006). The expansion of tree farms intensified during the 1960s, enabled by the proliferation of public-private reforestation agreements and the creation of the afforestation corporation (COREF, later called CONAF). Later, under Allende's administration (1970–1973), state policies led to the development of plantations on approximately 40,000 ha (Camus 2006, p. 242). However, the most profound transformations took place only after 1974 (on this see also Chap. 6 in this book).

[1] Decree law 4363, published on 31st of July, 1931.

5.2.1 Economic Drivers

In September 1973, the military-led government seized control of the Chilean government by force, remaining in power until 1990. The military regime sought to re-engineer the national economic model, imposing free market reforms and promoting commercial timber plantations (Clapp 1995). Relevant policies in this new economic model, which directly contributed to land use change in south-central Chile over the next 40 years, include:

1. *Decree of Law (DL) 701 of 1974.* DL 701 paid between 75% to 90% of reforestation costs within a year after a plantation was developed, effectively subsidizing growers of planted species. Expenses were readjusted in accordance with the Consumer Price Index in response to inflationary pressures. Timber investments were quickly recovered. DL 701 also granted tax exemptions and presumed income and declared forested land plots as unsuitable for expropriation (Camus 2006; Niklitschek 2007).
2. *Loans of liaison.* One of the barriers to afforestation is its high up-front cost. As a result, credit packages were created that provided landowners with access to capital with very low interest rates (Camus 2006).
3. *Removal of restrictions on timber exports.* As part of the import substitution model, before 1973, it was illegal to export raw materials, such as logs. The military-led government eliminated these restrictions and allowed timber-derived exports in any stage production (Niklitschek 2007).
4. *Elimination of wheat price bands.* With the elimination of wheat price bands, and as the profitability of wheat farming declined, timber farming became increasingly attractive (Niklitschek 2007).
5. *Removal of import tariffs for industrial goods.* Without import tariffs, it was cheaper to obtain machinery needed to process timber (Niklitschek 2007).
6. *DL 600.* This law fixed income tax rates for foreign investors and allowed foreign companies to exchange foreign debt for domestic capital. In 1982, many small forestry companies went bankrupt. They were later bailed out by the state and then handed over to foreign investors or their subsidiaries in Chile. This process also facilitated land concentration, as large companies survived while others went bankrupt (Moguillasky and Silva 2001; Camus 2006).
7. *Openness to international markets.* Finally, the free trade agreements pursued by the *Concertación*[2] governments (1990–2010) opened access to new international markets, further increasing the demand for forest products, and with it, the profitability of timber plantations (Niklitschek 2007).

---

[2]*Concertacion de Partidos por la Democracia* (Caucuss of parties for democracy): a center-left political coalition that lead the national referendum aiming to terminate the military-led government. This coalition was in power from 1990 to 2010 (5 4-year periods).

Collectively, these policies ensured Chilean timber plantations to become among some of the most profitable in the world (Sedjo 1983; Cubbage et al. 2007). This, of course is not due to some magical property of these trees, but rather, due to the productive and capital accumulation strategy that was assembled around these species. The policies that led to the ascendance of the Chilean commercial forestry sector were implemented without consulting local populations, and under the condition of a repressive authoritarian military rule.

5.2.2 *Politics: Reactions from the Social and Political World*

Although civil dissidence was officially prohibited during the military dictatorship and tightly controlled during the gradual transition to democracy (Ulianova and Estenssoro 2010), the substitution of native forest has been a source of civil society concern since the 1980s (Donoso 2012). The first government of *La Concertación* developed a new law to protect the native forest[3] (Biblioteca del Congreso Nacional de Chile 2011). Discussions began in 1992, but only resulted in legislation 16 years later, in 2008 (Biblioteca del Congreso Nacional de Chile 2011). This law aims to ensure the "protection, recovery and improvement of native forests, in order to ensure forest sustainability" (2008, sec. 22). Clearly, the fate of Chile's forests is still a contentious issue. The bill's first draft came out of the Chamber of Deputies in 1994 and stipulated seven articles to regulate the use of the native forest, and in some cases prohibited the substitution of native forest for non-native species. It was rejected by the forestry and agricultural associations, represented by the Chilean Timber Corporation (CORMA) and by the Agriculture Society (SNA), as they claimed it violated their property rights, enshrined in the 1981 Constitution (Manuschevich and Beier 2016). As a result, the final version of the bill was very weak (Biblioteca del Congreso Nacional de Chile 2008; Manuschevich 2014). Under the administration of Ricardo Lagos (2000–2006), the Agricultural Ministry decided to proceed with the forestry policy, on the condition that exotic plantations were excluded ("Worktable Agreements" (2006). As a result, substantial discussions centered on payment for native forestry management and conservation. Some sought of a fourfold increase in subsidies compared to what was finally approved (Pizarro and Zolezzi 2003), but this idea was ultimately rejected (Library of the National Congress of Chile 2008). Civil society actors anticipated that financial incentives for conservation and management would be inadequate, but as government representatives put it: "You need a minister and a half to fight for a larger budget" (Manuschevich 2014).

This reflects the peculiarity of the Chilean Constitution, which accords the President exclusive power over budget allocation for any policy. Currently, the NFA has been ineffective and regressive. Farmers are not compensated for management

[3] Law number 20.283.

expenses, and even gaining access to the native forest subsidy involves navigating a veritable bureaucratic obstacle course (Cruz et al. 2012). Furthermore, in Chile, planting native trees is challenging – for technical, ecological, and financial reasons (e.g., summer drought, shortage of affordable seedlings, non-native wild herbivores, lack of knowledge and technical assistance, and the fact that most forestry research focuses on management techniques for non-native species (Donoso and Otero 2005).

5.2.3 *Environmental Outcomes*

The direct substitution of native forests with plantations is largely documented (Echeverria et al. 2006; Echeverría et al. 2007; Miranda et al. 2015, 2016; Zamorano-Elgueta et al. 2015). Moreover, recently Heilmayr et al. (2016) demonstrated that tree farms expansion is resulting in a forest transition dominated by tree farms, rather than forest, as it has happened in industrialized countries. Native forest and tree farms are different land cover, both in terms of ecosystem structure and function, thus these two ecosystems are different in the provision of goods (Chazdon et al. 2016). As mentioned, tree farms are monocultures managed to maintain the maximum of timber production per unit of area. In contrast, native forest encompasses 12 different forestry types with at least 80 tree species. Although recognized for its biodiversity (Myers et al. 2000), native forest are still diminishing in the region, despite the passage of native forestry legislation (Reyes and Nelson 2014; Miranda et al. 2016). Currently, the vast majority of Chilean timber farms are located in regions that harbor the greatest known diversity of vascular plants (Bannister et al. 2011; Pliscoff 2015).

Soil properties, such as carbon content, sediment nutrient retention are negatively affected by tree farms (Cisternas et al. 2001; Oyarzun et al. 2007, 2011; Soto et al. 2019). In addition, several studies have shown that forest plantations reduce summer waterflow (Little et al. 2009; Huber et al. 2010; Stehr et al. 2010; Jullian et al. 2018). Summer waterflow is a critical element for Mediterranean ecosystems. During summer season, marked by high temperatures and little to no rainfall. Tree farms also increase wild fire risk (Carmona et al. 2012; McWethy et al. 2018). In 2017 nearly 11% of the national total of area planted with tree farms burnt in wildfire that lasted for 3 weeks (CONAF 2017). After the wildfire the soil is exposed, which further increases soil degradation and water scarcity.

In summary, political and economic factors have undermined the feasibility of native forest conservation and has fostered environmental degradation. Considering the discussion above, how might we best understand and propose alternatives for the conservation of native forests? In the following section, two examples of different research techniques are provided. Each technique seeks to provide a different angle on the alternatives for forest conservation.

5.3 Socio-Ecological Integration: Framework and Research Tools

Land use change in Chile has been fueled by the high profitability of timber plantations, in a functional dance with the public policies. Attempts to regulate land use change were made, with little success due to the inner political dynamics presented in Sect. 2.2. This chapter could finish here. However, the aim of this chapter is not only to describe land use change though transdisciplinary lenses, but also to explore research venues that can enlighten change towards a more sustainable path. Following this logic, two research tools are presented in the following section. One is the use of scenarios and land use modeling in connection to the policy analysis. The second is the use of ethnographic methods to understand from a material and symbolic perspective what it as meant to live in areas that have changed dramatically in a lifetime. In other words, how the socio nature has been transformed due to land use change.

5.3.1 *Scenario Modeling*

Scenario modeling can be particularly useful as a tool to explore different policy alternatives from a broader perspective. Policy alternatives can help us to visualize the consequences of different policies before they happen or, teach us lesson of the consequences of the decisions taken in the past.

Manuschevich and Beier (2016), using DYNA-CLUE (Verburg and Overmars 2009) software, developed three scenarios of land use change based on the political analysis. These scenarios were converted into land use maps for the Malleco-Vergara watershed, in the Araucanía region. DYNA-CLUE uses physical, social and environmental information to estimate where each land use is most feasible, connecting social and environmental variables. In this way, these maps are can become images of alternative pasts and possible futures. These scenarios were based on the policy views, namely

1. Negotiated scenario: NFA enters into force in 1994 instead of in 2008.
2. Industrial scenario: the law as proposed by the Chilean Wood Corporation and the National Agriculture Society.
3. Conservation scenario: the law as proposed by academics and NGOs interested in conservation, that is, prohibiting substitution.

Manuschevich and Beier (2016) did not find major differences in the change in land use between scenarios for the Malleco-Vergara basin, Araucanía. However, it was found that in high elevation areas of this basin, increased subsidy packages and legislative changes might help to promote the recovery of native forest. The impact of the subsidy would depend on the landowner's sensitivity to a larger subsidy, which in turns depends on the landowner disposition, expectations and the easiness of the subsidy process.

Interestingly, land plots owned by indigenous communities were significantly less likely to be covered by tree farms. Moreover, areas in conflict among the state of Chile and Mapuche people, consistently deviated from model prediction. This evidence suggest that land tenure is a relevant variable on land use change in Araucania. The work done by Manuschevich and Beier (2016) helped to understand whether the subsidy proposed would have made a large difference in land use change. Their work also shed light on systemic variables such as land tenure and the importance of a good policy design, both in terms of subsidy amount and application process.

Spatial modelling can be used to explore a wider range of scenarios, which then can be assessed in terms of ecosystem services provision. Manuschevich et al. (2019), created four scenarios of land use change for the entire region. Those prospective scenarios (made for 2030), combined regulation and high and low areas (demand) for native forests and timber farms. Then each scenario was assessed in terms of ecosystem services provision, using InVEST. They found that land regulations are important for ecosystem service provision in any scenario. For example, in Fig. 5.2, is shown the potential differences in sediment production among a scenario of A. Business as usual: land use continues has it has happened

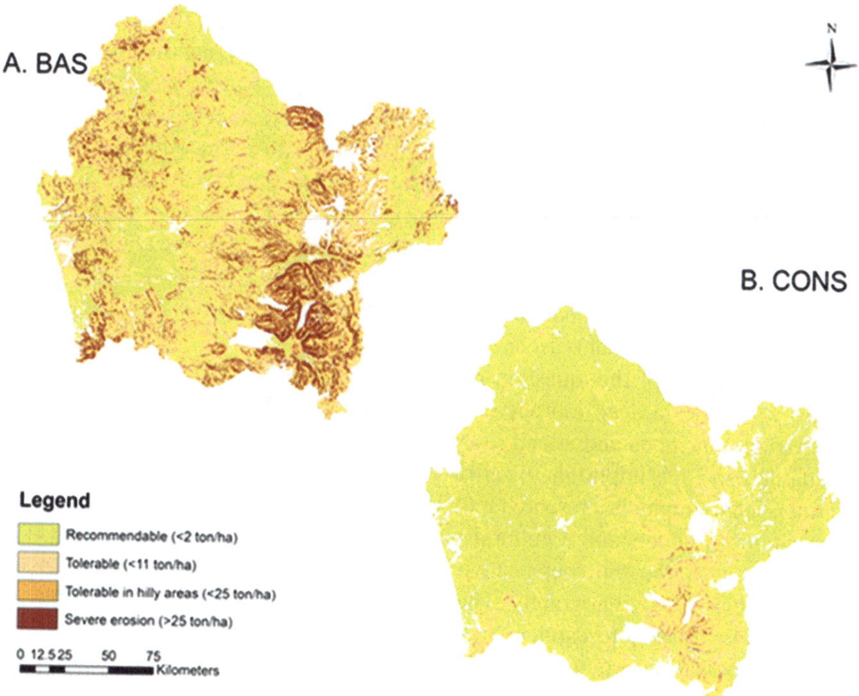

Fig. 5.2 Absolute differences in erosion rates among scenarios. (**a**) BAS scenario (**b**). Optimistic conservation scenario Absolute values are a per pixel value in ton ha-1 year-1. Classes of erosion tolerance are based on Ruhoff et al. (2006)

to B. Forest conservation, where enough subsidy is paid to cover the average opportunity cost of forest conservation. A conservation scenario would reduce erosion to tolerable levels in almost the entire region, compared to maintain the current trend in land use change. In this way, land use modeling in combination to ecosystem service modeling can inform and help to visualize outcomes of alternative paths to land use policy.

Overall, land use and ecosystems services modeling can be used in multiple ways to explore different policy alternatives and inform about the possible consequences. However, land use modeling and maps is just one of the approaches that can be used to understand land use change.

5.3.2 An Ethnography of the Campesino de Montaña

Undoubtedly, global trade plays a vital role in driving land use change (Meyfroidt et al. 2010; Verburg et al. 2014). However, not all landscapes are equally permeable to the global demands for commodities, not all landscapes are easily domesticated to supply the global needs for commodities. Previous sections of this chapter have exposed the political economy of land use change, commercial timber expansion, and native forest conservation. It was sought to explain, from a national perspective, how particular policies contributed to land use change. Nevertheless, these phenomena operate in a different way at different scales. At a local scale, what does it mean to live and witness the replacement of the native forest by standardized timber monocultures? How does one perceive, understand, and make sense of such profound transformations? Environmental modeling may help us to predict accelerated erosion risks associated with industrial farming, but how does it affect the ways people inhabit the land and make sense of life in the shadows of the plantations? Under which conditions a farmer would be interested in recovering the native forest. For what reasons?

An exploration of this question requires a set of research techniques. In other words, it requires a deep journey into social sciences. To explore the importance of agrarian transitions and attend to experiences of environmental degradation and standardization, it is important to gain a sense of the quotidian lives of those who call it home. Ethnography is a research tool, based on a detailed observation of everyday life, aiming to understand a cultural phenomenon point of view of the subject of the study, this is, the *emic* perspective. The information presented in the following section is based on ethnography. These are the preliminary results of 3 years of qualitative field-based research conducted with rural people in south-central Chile. Ethnography was used to understand what the native forest means for peasants and see under which conditions a forest policy could foster native recovery.

Identifying details such as names, dates, and specific locations have been changed, to ensure anonymity. The work presented here is based on participant observation conducted for 3 years, 40 semi-structured interviews and 3 participatory-mapping exercises, complemented with literature review and census

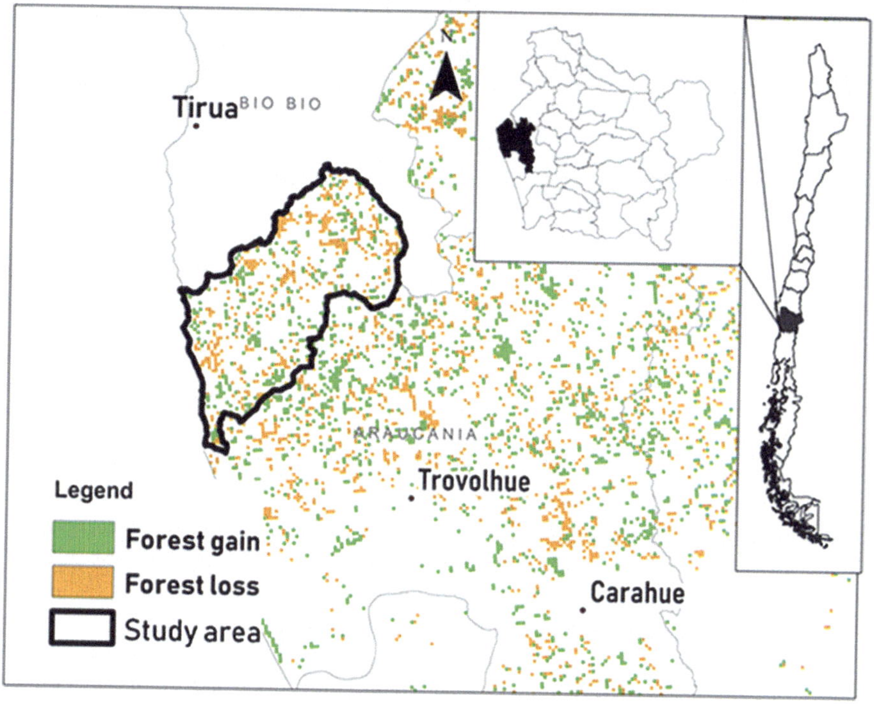

Fig. 5.3 Area where the ethnography was conducted. Green indicates forest gain, while orange indicates forest loss. (Data from 2007 to 2014 of the forest service (CONAF 2016))

data. Overall the ethnography is more than each separate technique of data collection, is a method of inquiry focused on everyday details of the peasant's life and allows a holistic comprehension of the material and symbolic aspect of the peasants' life. Ethnographies are often communicated as a narrative based on the information collected from interviews, observations and a documentation. The entire narrative constitutes the results of the work.

The ethics committee at Universidad Academia Humanismo Cristiano and Universidad Bernardo O'Higgins oversaw the research. Qualitative data was analyzed using the computer aid coding software ATLAS.ti and spatial information was analyzed with Arc GIS 10.3. The fieldwork was conducted in a mountainous coastal area (from 500 up to 750 m.a.s.l), located around 38°S, in south central Chile (Fig. 5.3). The climate is Mediterranean, with oceanic influence, due to the elevation, with some snow falling in the winter (Sarricolea et al. 2017). The soil in this area is rather poor and eroded, thus, it is suitable for forestry activities (CIREN 2012). That being said, local people engage in pastoral activities, herding sheep and cows.

The area encompasses 9019 ha, with remnants of evergreen forest (CONAF 2016). Between 1978 and 2014, 38.5% of the native forest area was lost due to deforestation and wildfires (CONAF 2016; Zhao et al. 2016). Of the forest lost, since

1978, 54% was planted with tree farms in 2014 (Zhao et al. 2016). A major part of this study took place in a community where a major wildfire destroyed some 1000 ha in 2011, caused by an unattended charcoal oven. The residents have been coping with serious difficulties in the aftermath of environmental devastation – namely the lack of water, especially during summer.

The rural areas of the municipality have lost a significant part of the population in the last 35 years. According to the 2017 census, the entire census district population decreased by 26% (INE 2017a). However, according to more detailed data form the local hospital, people registered there decreased by 47% between 1976 and 2018. The hill dwellers had lower educational attainment levels (when compared with the national average). Men slightly outnumber women. The population is aging, meaning for every 100 people of working age, 62.5 are either elders or children (INE 2017b). People employ pluri-active livelihood strategies by producing charcoal, firewood, potatoes, cattle, and sheep for market (along with chickens and vegetable gardens for familial consumption). This diversification is typical of campesino livelihoods. Both men and women occasionally supplement their agricultural activities by obtaining temporary positions (contracted on a monthly basis) with local forestry contractors, as cooks, security guards, or crossing guards. Forestry operations are transient, rarely lasting more than a few months.

Most of the hill dwellers arrived in the 1930s from nearby towns, and gold mining settlements (Jara et al. 2006). During this time, the state was awarding land grants to Chileans, specifically 10 ha per male child. Poor and landless families who previously worked as sharecroppers in the *Fundos (large farms)* were eager to finally own a piece of land. At that time, land ownership in Chile was extremely concentrated and opportunities for ownership were rare. In 1935 in central Chile around 4.1% of the population owned 83.5% of the land; while 79.6% of the population owned only 3.9% of the land area (Bauer 1994). Life was very simple; people lived in *ranchas*, where the top roof was made of chunks of bromeliads. House construction was precarious, with dirt floor, no running water and little clothes, no electricity, hospital or school. Fire was used to open space to cultivate wheat, vegetables, and have some herding, such as oxen. A yoke of oxen is still a very important means of doing farm work such as plowing, and transporting wood. Roads were only a small path transited by hand-made carts that would often fall apart before getting to the destiny. Getting to Carahue would take 4 days in caravans where people slept below the cart in sheep leathers. In Carahue, peasants sold their products and brought mate, sugar and salt, and went back for another 4-day trip.

Today, firewood from native forest represents at least 50% of the house hold income (Jara et al. 2006). From field observation, an active household can produce 70–100 m^3 of firewood per year, reaching \$1500–\$2250. Firewood in produced during the summer to buy food and fodder for the winter. Firewood production is very dependent on the working capacity, since only males conducted this type of work, and many are elders, they often cannot produce even is prescribed in management plans provided by CONAF. However, for those who do not have management plan- often because their do not legally have all the paperwork- life is very precarious and often live from charcoal or work in temporary jobs at the forestry operations or minimum-wage jobs funded by the state. In some cases, the firewood income is complemented with recollection of a native nut (*Gevuina avellana*). It takes between 1 and 2 days to collect a full sack o nuts that is then sold at $30–$40 dollars. Productive households can reach up to 70 sacks (working everyday Mach to May, when the nuts are mature). This can represent an income of up to $2800 per year.

Aerial images and elders accounts indicate that this area was fully covered by evergreen forests. These forests were initially used for dye extraction from tree barks. These tree barks were then, transported to the nearest town called. Once the dye industry closed, people started to saw valuable tree species such as *Weinmannia trichosperma,* to be sold in Trovolhue, Carahue and Tirúa. As the most valuable woods diminished, people started to produce charcoal, and continued to sell remaining forest as firewood. By the end of the 1980s timber firms started to buy land, while in the 1990s and the 2000s CONAF contractors came to peasant's houses to offer tree farms planting using the DL701.

Peasant are very conscious of the environmental degradation, several mentioning that before water was never scarce and several species were lost. For example, before there were *moscardones naranjos* (*Bombus dalbomii*), el león (*Puma concolor*), camarones de agua (*Cryphiops caementarius*), and a great diversity of birds, such has carpintero (*Campephilus magellanicus*). Several remember the honey they extracted from the forest and the soil being much more fertile than now, potatoes grew easily. After the wildfire the landscape changed profoundly and, as of today, the area burnt looks like the picture in Fig. 5.4.

The history of land ownership, settlement, environmental degradation, land use change and emigration are relevant to understand how meaning is constructed around land and forest. Based on the ethnographic work, two different views were identified in regard to the land and forest. One sees the land as means of living, it implies a sense of freedom and belonging, while the other sees the land as means to make money. In the first view land ownership, meant to have a way to sustain their families and not having to work for a wage, therefore for many peasants their land means freedom. With land, they were able to define themselves and gain a measure of personal autonomy, as these following quotes illustrate:

> If a farmer does not have a piece of land, he is nothing. What would he do outside of the farm? He would suffer the misery of being hungry, which would not happen in the farm.

Fig. 5.4 Picture of the landscape 6 years after the 1000-ha wildfire

While other commented that:

I have family in Santiago [the capital of Chile] and they say: when I get free of all of this, I'm going to live the countryside.

This sense of freedom also connects to the sense of belonging and connection to their parent and grand parents

We are like salmons ... we come back to where we are from.

For those who stayed, land means the sacrifice their parents made to have a piece of land, raise their children and live. As mentioned before, life conditions were very hard. Thus, for many elders, their land does not have a monetary value, as this 71-year old male farmer expresses.

It must have been so hard for my father to get all of this! I think it would be very ugly to take this and give it away... They would come and pay for it, but he won't pay what is really worth.

Many elders didn't want to leave their homes in the mountains, but some must do it for health reasons. For a sick 80-year-old man, this move was traumatic.

It was terrible to leave my father's land. I wanted to live there, so that my father's sacrifice won't be lost.

However, not everyone assigns such positive, non-financial values to land. As tree farms expands and environmental degradation progress, and becomes naturalized into the landscape, tree farm becomes the future that some farmers foresee, in particular among farmers at the age of 40 years, about the time when the DL701 was decreed. In this future, their grandchildren won't live there. When asked about the future, whether they want to see more native forest and people living in the area, some responses were rather bleak.

What for? Our children only need the land, to sell it to the forestry companies, and make money. Nobody will remember that we lived here.

In line with this ideal, land, animals and forest, are only valued in terms of the money that can produce, therefore environmental degradation is not bad as long as in opens new possibilities for commodity production. These commodities can be animals, wood, or potatoes, as this excerpt illustrates:

I think the wildfire was for better. Because we can now have many animals, which are like a bank for the farmer. Seated, you watch the herd grow and you make money.

You must take something out of the land, to have some tranquility.

Peasant, when asked about whether they would be any policy that would conserve the native forest, the DL 701 stands up as a benchmark for decision making. A law modification done un 1998[4] reoriented the law towards medium and small landholders, and implied that forestry contractors visited peasants' houses offering them to conduct the timber plantation in exchange for some or the entire subsidy

[4]Law number 19,561, which modified the DL701, published on May 16th, 1998.

money. If the peasant agreed the contractor conducted the entire application and planting then, either the farmer or the contractor received the money. This system finished in 2015 when the DL701 was left with no funding in the House of Representatives, due to a corruption case. Peasants participated in the DL701 system and would only be willing to participate in forest conservation or planting if it is as easy and profitable as the DL701. For example, when asked whether he has planted native species, a peasant responded:

> We have not done it here, because the forests had to be improved . . . .and CONAF offered to pay us ... just as they had paid before when one planted.

While one of the leaders of the *comités* commented:

> with the same [money] amounts . . . . we would all plant native people because [with what] it was paid in the 701 . . . they threw it down almost the majority of the native forest and plant pine and eucalyptus on it . . . . because . . . . That was money.

The DL701 shaped the profits expectations. Today peasants expect that the subsidy for native forest planting operates in a similar way to the DL70, otherwise they are not interested in planting. This evidence speaks to one of the key variables mentioned in Sect. 3.1. If the NFA would increase the subsidy amount, would people plant more? From the evidence collected in the interviews it becomes clear that the application procedure of the DL701 created a benchmark for peasants. Any effective law would need to consider that.

5.4 Conclusion

In this chapter a combination of historical and political economic analysis, modeling techniques, and qualitative research results was presented. The combination of multiple research paradigms and research techniques enables a deeper and richer understanding of the land use change in Chile. A transdisciplinary endeavor necessarily combines multiples disciplines generating new knowledge, and a new conception of the problem.

In the last 45 years, political and economic changes promoted the expansion of tree-farms, which were favored by the state. Despite the attempts to have an effective NFA, due to the inner political dynamics, it was not possible. Nevertheless, land use modeling can be used to explore the effect of different policy proposals. Land use modeling complements very well with qualitative methods. Ethnography provides a rich and deep understanding of what it meant and means today to live in the fringes of timber expansion and environmental degradation. However, material conditions are symbolically interpreted. Land and environmental degradation are interpreted in different ways. However, the state has proven itself to be unwilling to make effective commitments at forest conservation. Drawing on the insights of ecology and the social sciences, we can help to shed light and model alternative futures that by combining different disciplines provided a more complete view of land use as a socio-ecological system.

In conclusion, a socio-ecological approach enables a more holistic –and humane– vision of the land use changes witnessed in Chile in last 70 years. Only the combination of both social and natural science, in equal footing, will allow a deeper comprehension of the causes and impacts f tree farms expansion in Chile, as a socio-ecological system.

Acknowledgements The author would like to acknowledge the financial support of CONICYT+FONDECYT+ 11150281, as well as the contributions and suggestions of Dr. Melida Gurr.

References

Armesto JJ, Manuschevich D, Mora A et al (2010) From the Holocene to the Anthropocene: a historical framework for land cover change in southwestern South America in the past 15,000 years. Land Use Policy 27:148–160

Bannister JR, Vidal OJ, Teneb E, Sandoval V (2011) Latitudinal patterns and regionalization of plant diversity along a 4270-km gradient in continental Chile. Austral Ecol 37. https://doi.org/10.1111/j.1442-9993.2011.02312.x

Bauer AJ (1994) La sociedad rural chilena: desde la conquista española a nuestros dias. Andres Bello

Berkes F (2012) Sacred ecology, 3rd edn. Routledge, New York

Berkes F, Colding J, Folke C (2003) Navigating social–ecological systems: building resilience for complexity and change. Cambridge, UK: Cambridge University Press

Biblioteca del Congreso Nacional de Chile (2008) (ed) Historia de la Ley No 20.283. Ley sobre recuperación del bosque nativo y fomento forestal

Biblioteca del Congreso Nacional de Chile (2011) (ed) Historia de la Ley No 20.488. Prorroga vigencia del decreto Ley N° 701, de 1974, y aumenta incentivos a la Forestación

Binder CR, Hinkel J, Bots PWG, Pahl-Wostl C (2013) Comparison of frameworks for analyzing social-ecological systems. Ecol Soc 18:26. https://doi.org/10.5751/ES-05551-180426

Blaikie P, Brookfield H (1986) Land degradation and society. Land degradation and society

Camus P (2006) Ambiente, Bosques y Gestion Forestal en Chile: 1541–2005. LOM, Santiago

Carmona A, González ME, Nahuelhual L, Silva J (2012) Efectos espacio-temporales de los factores humanos en el peligro de incendio en Chile mediterráneo. Bosque (Valdivia) 33:321–328

Castree N, Braun B (2001) Social nature: theory, practice and politics. Massachusetts: Wiley

Chazdon RL, Brancalion PHS, Laestadius L et al (2016) When is a forest a forest? Forest concepts and definitions in the era of forest and landscape restoration. Ambio 45:538–550. https://doi.org/10.1007/s13280-016-0772-y

CIREN (2012) Estudio Agrologico: Descripción de suelos materiales y simbolos IX región. Centro de Información de Recursos Naturales. Santiago, Chile

Cisternas M, Araneda A, Martínez P, Pérez S (2001) Effects of historical land use on sediment yield from a lacustrine watershed in Central Chile. Earth Surf Process Landf 26:63–76

Clapp RA (1995) Creating competitive advantage: forest policy as industrial policy in Chile. Econ Geogr 71:273–296

Clement F (2013) For critical social-ecological system studies: integrating power and discourses to move beyond the right institutional fit. Environ Conserv 40:1–4. https://doi.org/10.1017/S0376892912000276

Clement F, Amezaga JM (2013) Conceptualising context in institutional reforms of land and natural resource management: the case of Vietnam. Int J Commons 7:140–163

CONAF (2016) Sistema de Informacion Territorial – CONAF 2016. In: Sistema de Informacion Territorial. https://sit.conaf.cl/. Accessed 19 Apr 2018

CONAF (2017) Análisis de la Afectación y Severidad de los Incendios Forestales ocurridos en enero y febrero de 2017 sobre los usos de suelo y los ecosistemas naturales presentes entre las regiones de Coquimbo y Los Ríos de Chile. Sanitago de Chile

Cruz P, Cid F, Rivas E, et al (2012) Evaluación de la Ley N°20.283 sobre recuperación del bosque nativo y fomento forestal. Subsecretaría de Agricultura

Cubbage F, Mac Donagh P, Sawinski Junior J et al (2007) Timber investment returns for selected plantations and native forests in South America and the southern United States. New For 33:237–255. https://doi.org/10.1007/s11056-006-9025-4

Daly HE, Farley J (2011) Ecological economics. 2nd edition, Principles and applications. Washington, D.C., USA: Island Press

Donoso C (2012) Una mirada a nuestros Bosques Nativos y su defensa. Andros, Santiago

Donoso P, Otero L (2005) Hacia una definición de país forestal: ¿Dónde se sitúa Chile? Bosque 26:5–18

Echeverria C, Coomes D, Salas J et al (2006) Rapid deforestation and fragmentation of Chilean temperate forests. Biol Conserv 130:481–494

Echeverría C, Huber A, Taberlet F (2007) Estudio comparativo de los componentes del balance hídrico en un bosque nativo y una pradera en el sur de Chile. In: Bosque. http://www.redalyc.org/resumen.oa?id=173113292013. Accessed 19 Jun 2014

Elizalde R (1970) La sobrevivencia de Chile: la conservación de sus recursos naturales renovables. Ministerio de Agricultura, Servicio Agrícola y Ganadero

Gallopin GC (2006) Linkages between vulnerability, resilience, and adaptive capacity. Glob Environ Change-Human Policy Dimens 16:293–303. https://doi.org/10.1016/j.gloenvcha.2006.02.004

Heilmayr R, Echeverría C, Fuentes R, Lambin EF (2016) A plantation-dominated forest transition in Chile. Appl Geogr 75:71–82

Holling CS (2001) Understanding the complexity of economic, ecological, and social systems. Ecosystems 4:390–405

Huber A, Iroumé A, Mohr C, Frêne C (2010) Efecto de plantaciones de Pinus radiata y Eucalyptus globulus sobre el recurso agua en la Cordillera de la Costa de la región del Biobío, Chile. Bosque (Valdivia) 31:219–230. https://doi.org/10.4067/S0717-92002010000300006

INE (2017a) Compendio estadístico 2016: region de La Aracucanía. Instituto Nacional de Estadisticas, Temuco

INE (2017b) Resultados Censo 2017 por Región, Provincia y Comuna. Santiago de Chile

INFOR (2016) Anuario Forestal 2016. Instiuto forestal, Sanitago de Chile

Jara JC, Palma P, Pantoja R (2006) El bosque ya no es matorral: mujeres rurales revalorizando el bosque a través de la avellana. In: Bosques y comunidades del sur de Chile. Editorial Universitaria

Jullian C, Nahuelhual L, Mazzorana B, Aguayo M (2018) Evaluación del servicio ecosistémico de regulación hídrica ante escenarios de conservación de vegetación nativa y expansión de plantaciones forestales en el centro-sur de Chile. Bosque 39:277–289

Kates RW, Clark WC, Corell R et al (2001) Sustainability science. Science 292:641–642. https://doi.org/10.1126/science.1059386

Lara A, Solari ME, Prieto MDR, Peña MP (2012) Reconstrucción de la cobertura de la vegetación y uso del suelo hacia 1550 y sus cambios a 2007 en la ecorregión de los bosques valdivianos lluviosos de Chile (35o - 43o 30´ S). Bosque (Valdivia) 33:13–23. https://doi.org/10.4067/S0717-92002012000100002

Library of the National Congress of Chile (ed) (2008) Historia de la Ley No 20.283. In: Ley sobre recuperación del bosque nativo y fomento forestal

Little C, Lara A, McPhee J, Urrutia R (2009) Revealing the impact of forest exotic plantations on water yield in large scale watersheds in south-Central Chile. J Hydrol 374:162–170

Manuschevich D (2014) Linking politics to changes in ecosystem services: a social-ecological approach to land use change in Chile. Ph.D. thesis, State University of New York College of Environmental Science and Forestry

Manuschevich D, Beier CM (2016) Simulating land use changes under alternative policy scenarios for conservation of native forests in south-Central Chile. Land Use Policy 51:350–362. https://doi.org/10.1016/j.landusepol.2015.08.032

Manuschevich D, Sarricolea P, Galleguillos M (2019) Integrating socio-ecological dynamics into land use policy outcomes: a spatial scenario approach for native forest conservation in south-Central Chile. Land Use Policy 84:31–42. https://doi.org/10.1016/j.landusepol.2019.01.042

Max-Neef MA (2005) Foundations of transdisciplinarity. Ecol Econ 53:5–16

McWethy DB, Pauchard A, García RA et al (2018) Landscape drivers of recent fire activity (2001-2017) in south-Central Chile. PLoS One 13:e0201195. https://doi.org/10.1371/journal.pone.0201195

Meyfroidt P, Rudel TK, Lambin EF (2010) Forest transitions, trade, and the global displacement of land use. Proc Natl Acad Sci 107:20917–20922

Miranda A, Altamirano A, Cayuela L et al (2015) Different times, same story: native forest loss and landscape homogenization in three physiographical areas of south-central of Chile. Appl Geogr 60:20–28. https://doi.org/10.1016/j.apgeog.2015.02.016

Miranda A, Altamirano A, Cayuela L et al (2016) Native forest loss in the Chilean biodiversity hotspot: revealing the evidence. Reg Environ Chang. https://doi.org/10.1007/s10113-016-1010-7

Moguillasky G, Silva V (2001) Estrategias empresariales y políticas públicas: el futuro del complejo forestal en Chile. In: Mas alla del bosque. LOM ediciones, Santiago, pp 107–144

Myers N, Mittermeier RA, Mittermeier CG et al (2000) Biodiversity hotspots for conservation priorities. Nature 403:853–858

Niklitschek ME (2007) Trade liberalization and land use changes: explaining the expansion of afforested land in Chile. For Sci 53:385–394

Norgaard RB, Baer P (2005) Collectively seeing complex systems: the nature of the problem. Bioscience 55:953–960. https://doi.org/10.1641/0006-3568(2005)055[0953:CSCSTN]2.0.CO;2

Oyarzún C, Aracena C, Rutherford P et al (2007) Effects of land use conversion from native forests to exotic plantations on nitrogen and phosphorus retention in catchments of southern Chile. Water Air Soil Pollut 179:341–350. https://doi.org/10.1007/s11270-006-9237-4

Oyarzún CE, Frêne C, Lacrampe G et al (2011) Soil hydrological properties and sediment transport in two headwater catchments with different vegetative cover at the Coastal Mountain range in southern Chile. Bosque 32:10–19

Pizarro R, Zolezzi C (2003) Opinión sobre la Ley de Bosque Nativo: Aspectos Económicos

Pliscoff P (2015) Aplicación de los criterios de la Unión Internacional para la Conservación de la Naturaleza (IUCN) para la evaluación de riesgo de los ecosistemas terrestres de Chile. MMA, Sanitago

Reyes R, Nelson H (2014) A tale of two forests: why forests and forest conflicts are both growing in Chile. Int For Rev 16:379–388. https://doi.org/10.1505/146554814813484121

Rocheleau D, Ross L (1995) Trees as tools, trees as text: struggles over resources in Zambrana-Chacuey, Dominican Republic. Antipode 27:407–428. https://doi.org/10.1111/j.1467-8330.1995.tb00287.x

Ruhoff AL, Souza BSP, Giotto E, Pereira RS (2006) Avaliação dos processos erosivos através da equação universal de perdas de solos, implementada com algoritmos em legal. Geomática 1:12–22

Sarricolea P, Herrera-Ossandon M, Meseguer-Ruiz Ó (2017) Climatic regionalisation of continental Chile. J Maps 13:66–73. https://doi.org/10.1080/17445647.2016.1259592

Sedjo R (1983) Comparative economics of plantation forestry a global assessment. Resources for the future, Washington, DC

Soto L, Galleguillos M, Seguel O et al (2019) Assessment of soil physical properties' statuses under different land covers within a landscape dominated by exotic industrial tree plantations in south-Central Chile. J Soil Water Conserv 74:12–23. https://doi.org/10.2489/jswc.74.1.12

Stehr A, Aguayo M, Link O et al (2010) Modelling the hydrologic response of a mesoscale Andean watershed to changes in land use patterns for environmental planning. Hydrol Earth Syst Sci 14:1963–1977. https://doi.org/10.5194/hess-14-1963-2010
Swyngedouw E (1999) Modernity and hybridity: nature, regeneracionismo, and the production of the Spanish waterscape, 1890–1930. Ann Assoc Am Geogr 89:443–465
Turner BL, Robbins P (2008) Land-change science and political ecology: similarities, differences, and implications for sustainability science. Annu Rev Environ Resour 33:295–316. https://doi.org/10.1146/annurev.environ.33.022207.104943
Turner BL, Lambin EF, Reenberg A (2007) The emergence of land change science for global environmental change and sustainability. Proc Natl Acad Sci U S A 104:20666–20671. https://doi.org/10.1073/pnas.0704119104
Ulianova O, Estenssoro F (2010) The Chilean environmentalism: emergency and international insertion. Si Somos Americanos Revista de Estudios Transfronterizos 12:183–214
Verburg PH, Overmars KP (2009) Combining top-down and bottom-up dynamics in land use modeling: exploring the future of abandoned farmlands in Europe with the Dyna-CLUE model. Landsc Ecol 24:1167–1181. https://doi.org/10.1007/s10980-009-9355-7
Verburg R, Filho SR, Lindoso D et al (2014) The impact of commodity price and conservation policy scenarios on deforestation and agricultural land use in a frontier area within the Amazon. Land Use Policy 37:14–26. https://doi.org/10.1016/j.landusepol.2012.10.003
Watts MJ, Bohle HG (1993) The space of vulnerability: the causal structure of hunger and famine. Prog Hum Geogr 17:43–67. https://doi.org/10.1177/030913259301700103
Zamorano-Elgueta C, Rey Benayas JM, Cayuela L et al (2015) Native forest replacement by exotic plantations in southern Chile (1985–2011) and partial compensation by natural regeneration. For Ecol Manag 345:10–20. https://doi.org/10.1016/j.foreco.2015.02.025
Zhao Y, Feng D, Yu L et al (2016) Detailed dynamic land cover mapping of Chile: accuracy improvement by integrating multi-temporal data. Remote Sens Environ 183:170–185. https://doi.org/10.1016/j.rse.2016.05.016

Chapter 6
Between Extractivism and Conservation: Tree Plantations, Forest Reserves, and Peasant Territorialities in Los Ríos, Chile

Alejandro Mora-Motta, Till Stellmacher, Guillermo Pacheco Habert, and Christian Henríquez Zúñiga

6.1 Introduction

Throughout the world, economic growth remains dependant on the material expansion of economic processes, which are based on the growing extraction of natural resources in rural areas (Krausmann et al. 2018). Particularly in Latin America, such expansion has created continuous waves of large amounts of resource extraction often aligned with national development goals. This process has been referred as extractivism[1] (Brand et al. 2016; Gudynas 2015). Concurrently nature conservation often follows global and national sustainability goals. Both, extractivism and conservation strategies are mainly territorialized in rural areas, where local – often traditional and smallholder based – modes of living base their livelihoods and territorialities (Escobar 2014).

[1] In the definition of Acosta (2013), extractivism consists in "those activities which remove large quantities of natural resources that are not processed (or processed only to a limited degree), especially for export. Extractivism is not limited to minerals or oil. Extractivism is also present in farming, forestry and even fishing" (2013: 62).

A. Mora-Motta (✉) · T. Stellmacher
Center for Development Research, University of Bonn, The Right Livelihood College Campus Bonn, Bonn, Nordrhein-Westfalen, Germany
e-mail: amotta@rlc-bonn.de

G. P. Habert
Centro Transdisciplinario de Estudios Ambientales y Desarrollo Humano Sostenible (CEAM—NIAP), Universidad Austral de Chile, Valdivia, Región de Los Ríos, Chile

C. H. Zúñiga
Institute of Economía, University Austral de Chile, Valdivia, Chile

F. Fuders and P. J. Donoso (eds.), *Ecological Economic and Socio Ecological Strategies for Forest Conservation*, https://doi.org/10.1007/978-3-030-35379-7_6

Recently, in view of the socio-environmental crisis in Chile, a 'green growth' economy has been instituted as a sustainable way forward (Gobierno de Chile 2013), positioning what might be called a neoliberal thinking applied on sustainability[2] (Brand and Lang 2015; Lander 2011). In this context 'green growth' also masks a framework of institutions and processes that facilitates the extraction of natural resources in specific territories. Tree plantations in Chile show a extractivist structure, since they are mainly based on the expansion of pine and eucalyptus monocultures with the aim to produce wooden raw materials for global markets (Brand et al. 2016; Gudynas 2015). While on the one hand, this goes along with rather short-term private profits for multinational companies, on the other hand, this produces long term societal and environmental costs, locally, nationally, and globally (Clapp, 1995; Donoso et al. 2015, Millaman et al. 2016, Reyes and Nelson 2014). This 'productive' approach on sustainability is nowadays complemented by a system of protected areas in Chile, which is a mixture of public and private conservation initiatives seeking to preserve nature, particularly forests (Pauchard and Villarroel 2002). It certainly serves to limit environmental destruction and protects biodiversity but can also be coined as a neoliberal way of grabbing nature, or 'green grabbing' (Holmes 2015).

These concerns have triggered discussions about conflicting modes of nature appropriation, which became manifest in different interpretations around 'sustainability'. At the centre of the debate we find questions like: sustainability of what? – but less frequently – sustainability for whom? This last question is the one that politicizes nature and forces a political ecology of analysing divergent forms of perceiving, valuing, and using nature, i.e., forms of nature appropriation. Nature materializes in concrete 'local' places with particular territorial dynamics. Thus, from a theoretical point of view, we approach nature and territory from a relational and ontological angle (Escobar 2014; Leff 2014; Max-Neef 2016), i.e., under the premise that social, economic, cultural, and biophysical aspects are co-founded creating and recreating spaces, but often through tensions and conflicts. We understand territoriality as a way in which specific human groups construct a geographic territory under a complex network of relationships. Territory is thus a political entity, which is disputed by different territorialities that set different ways of appropriating nature (Porto-Gonçalves and Leff 2015). From this viewpoint, sustainability is politicized and constituted through specific and diverse cultural forms of nature appropriation. Nature is disputed among regional, national, or even international actors who seek the extraction of resources for economic reasons or its conservation, and local actors who look for ways of living that allow them to reproduce their particular territorialities.

This chapter aims to question the de-politicized idea of 'sustainability' and its materialization in a concrete place, also seeing 'sustainability' as a new veil to legitimize the bitter side of economic growth and the territorial transformations caused by it. For that, we analyse one case in which three territorialities overlap with

[2]For a definition of sustainability see Azkarraga et al. (2011).

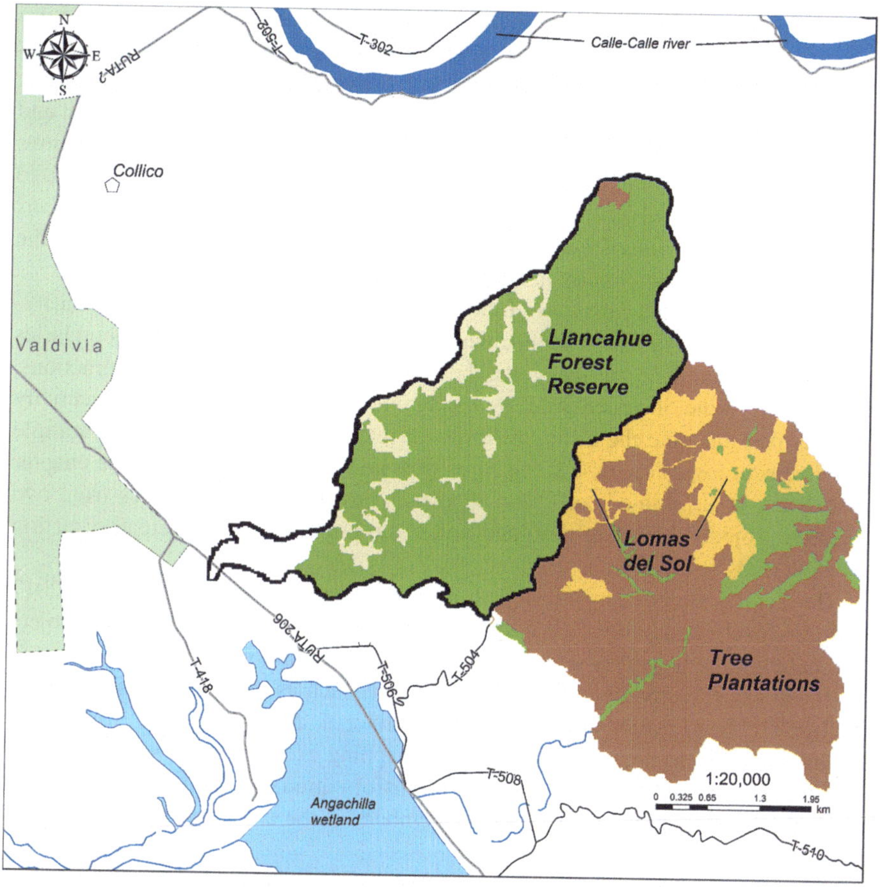

Fig. 6.1 Schematic map of the study area. (Source: own elaboration based on information from UACh 2017)

divergent interests, perceptions, values, uses, and strategies for nature appropriation. We centre our attention on the territoriality of a social peasant group[3] who inhabit the sector[4] Lomas del Sol – LdS –, Municipality of Valdivia, Province of Valdivia, Region of Los Ríos, south-central Chile (see Fig. 6.1). The group consists of

[3]The use of the concept of 'community', as highlighted by Roberto Morales (personal comment to an early version of the paper), can be problematic in this case. We use the broader notion of 'social group', which share a common history, productive practices and common way to appropriate nature. Therefore, we opted for the concept of territoriality, which captures the strategies and actions of these shared practices.

[4]'Sector' corresponds to a political unit in Chile under the level of the municipality ('comuna' in Chilean nomenclature). At this level, local associativity institutions are legally recognized, such as peasant or indigenous associations.

about 24 families, some of them members of the peasant association "*Comité pro Adelanto Lomas del Sol*". The inhabitants of LdS are one of the many groups throughout south-central Chile whose way of living is trapped between nature extractivism and nature conservation. LdS is, on one side, a biodiversity hotspot and hydrologically important for providing potable water to the city of Valdivia (Donoso et al. 2014). On the other side, LdS lies adjacent to vast pine and eucalyptus tree plantations established by three companies, namely: *Hancock Chilean Plantations (former Forestal Tornagaleones* of the consortium MASISA), *Arauco Sur*, and the plantations of the *Fried* family in a medium-size property.

We approached this case study following the idea of 'strong transdisciplinarity' (Max-Neef 2005; Nicolescu 1996), which seeks to dilute disciplinary boundaries and the Cartesian divide to recognize the complexity of human-nature interactions, the dialectic of the 'included third' by taking into consideration that pure objectivity is never possible, and the different 'levels of reality' which recall us that multiple and divergent logics co-exist at the same time and place place (on this see chapter by Erlwein et al. in this book). The chapter is based on empirical work from two research projects that were designed under the principles of participatory action research (Fals-Borda 2015).

The first of the projects developed a series of actions between 2014 and 2017 with the local community (see Pacheco and Henríquez 2016a). The other project was based on participatory action research in the area and was conducted between 2016 and 2017, including ethnographic methods such as participatory observation (Guber 2011) and participatory mapping (Sletto et al. 2013; Rister and Ares 2013). We particularly use the information of the mapping exercise of the second project as a pivotal input to understand territorial transformation and complemented it with the multiple interactions with different members of the community, listed in both projects.

After this introduction, the second section of this chapter briefly presents the history of tree plantations in Chile. The third section sketches nature conservation through protected areas in Chile with a particular focus on the Llancahue forest reserve. The fourth section briefs the plantation boom in Valdivia. The fifth section presents the territorial transformation from the viewpoint of the inhabitants of LdS. The sixth section discusses the politicized sustainability embedded in this particular territorial conflict. Finally, the last section concludes that while the sustainability approach of Llancahue clashes with that of the inhabitants of LdS, the underlying reason for the conflict is the expansion of tree plantations.

6.2 Historical Context of the Chilean Tree Plantation Model

The Chilean tree plantation model carries over one century of expansion processes. It consolidated from a state development strategy, through a neoliberal development strategy with a primary role of the private companies and a subsidiary state, mostly through the promotion of large private pine and eucalyptus plantations. The following is a summary of these processes.

6.2.1 *Before the 1970s*

The process of nature appropriation in south-central Chile traces back to the Spanish colony and the enclave economy they built in the attempt to control the Mapuche territory.[5] For over three centuries, the Spanish were never able to fully control these territories. After the so-called 'pacification of the Araucanía' in the XIX century,[6] in which the newly independent Chilean state consolidated its military control over the Mapuche territory, the first Forest Law was introduced in 1872. It sought to regulate the practices of clearing and burning of primary forest, and to encourage the planting of introduced species. However, this law did not have significant real effects. With the beginning of the twentieth century, the model of scientific forestry was introduced in Chile, the first forest management service was created, and a law project to implement incentives for the promotion of tree plantations was presented to congress in 1911, being rejected (Klubock 2014).

In 1931, the second Forest Law came into force, including the spirit of the project rejected decades before. During this time, logging of primary forests and large-scale planting of introduced species increased. In 1920 the *Compañía Manufacturera de Papeles y Cartones* [Papers and Paperboards Manufacturing Company] – CMPC–, the first private industrial forestry company, was established. In 1937, with the creation of the *Corporación de Fomento* [The Chilean Economic Development Agency] – CORFO – Chile articulated an industrialization programme with the forestry industry as a key economic sector. With the agrarian reform of the 1960s, the forestry development model was restructured, and small-scale plantations were promoted through cooperatives (Clapp 1995; Donoso et al. 2015; Klubock 2014).

6.2.2 *Implementation of the Neoliberal Model*

The military dictatorship (1973–1990) set a partial agrarian counter-reform. The new policies favoured private entrepreneurship, the transfer of public property to private hands, and its concentration in a few economic groups (Gómez 2014). In addition, barriers for international trade were removed (Clapp 1995). Thanks to the

[5]Group of indigenous people in Chile and Argentina. Mapuche in the original language means "people of the earth".

[6]We propose to call this process "Usurpation of Araucanía". The process starts with a series of negotiations and finalized with a military campaign covering most of the nineteenth century. Military control of the frontier was gained by the Chilean state in the 1870s.

Decree Law 701 of 1974 – DL 701 – which foresaw subsidies for tree plantations and guaranteed no expropriation – there was a massive expansion of tree plantations in south-central Chile (Clapp 1995; Donoso et al. 2015). Also, other policies, such as the Water Code of 1981, privatized resources and rights. Besides, the period was characterized by systematic violations of human rights, particularly of peasants and indigenous people. In this period, Chilean forestry plantation companies, now in private hands, became transnationals, expanding mainly to other Latin American countries (Gómez 2014).

6.2.3 Return to Democracy

Despite the establishment of social and environmental institutions and the recognition of indigenous communities shortly after the end of the dictatorship, several public organisms created in the period of the dictatorship remained intact. In democracy the tree plantation model, through the before mentioned DL 701, underwent modifications that tried to encourage small-scale plantations, with some achievements, but without substantially changing the dominance of large-scale monoculture plantations.

The government of Chile sought to become part of the Organization for Economic Co-operation and Development – OECD –. For that purpose, a joint study was performed by OECD and the Economic Commission for Latin America and the Caribbean – ECLAC – (OECD & ECLAC, 2005) assessing the economic, social and environmental state of Chile in order to become a member country of the OECD. Based on the reports' recommendations, the Chilean government implemented a series of changes towards a 'green economy'. As part of the restructuration, a Ministry of Environment was created in 2010 and other institutional bodies, such as Environmental Courts, where introduced to promote, evaluate, and monitor the conservation and sustainable use of the environment as well as to impart environmental justice and arbitrate conflicts.

A focus was given to primary forests and tree plantations. For example, the Law on the Promotion of Native Plantations (Native Forest Law), which came into force in 2008 after 16 years of discussion, was introduced with the aim to create a subsidy for sustainable management and recovery (restoration) of native forests, but at the end it became a mechanism to promote monoculture plantations with single native species, with limited public funds. In concert with the global trend, private actors also increasingly engaged in: (i) eco-labelling initiatives such as the Forest Stewardship Council[7] – FSC –, which is widely used by companies such as

---

[7]FSC is a global organization dedicated to promoting responsible forest management. In short, their activity engages stakeholders in the process of verification of responsible practices of companies. Companies which decide to comply to the protocols pf FSC have to demonstrate certain standards, so their products get an eco-label. With this eco-label, consumers can verify that the companies are following the standards (Heilmayr and Lambin 2016; Millaman et al. 2016).

ARAUCO and MASISA (Heilmayr and Lambin 2016; Millaman et al. 2016); (ii) private conservation reserves, such as Oncol private reserve in Valdivia, owned by ARAUCO (Holmes 2015); and (iii) different approaches to eco-tourism (Klubock 2014).

6.2.4 *The Recent Tree Plantation Model*

The large-scale pine and eucalyptus monocultures stopped being supported by DL 701 in 2012, meaning that plantations established after this year could not apply for the subsidy for a refund of the 75% of the establishment investment (extensive to 90% during some years). The law period came to an end and a new one was not enforced by the Senate of Chile. However, in the meantime, Chile became one of the top 10 countries in the world in terms of land allocated to tree plantations (CONAF 2014). In 2019 the Senate is still discussing to follow with DL 701 for 20 years more, and at present this subsidy is still being given to private companies.

At the end of 2016, tree plantations covered more than 2.4 million hectares in Chile. The main species used were radiata pine (57.6%) and eucalyptus (globulus and nitens) (35.6%) (INFOR 2018). Most plantations in Chile are found in the south-central region mainly in the regions of Biobío (38.5%), Araucanía (20.4%), Maule (17.8%), and Los Ríos (7.7%), i.e. between 37 and 41°S Lat. In Chile, there are three large size companies, 11 medium size companies, 714 medium size owners, and 22,747 small size owners[8] (INFOR 2018).

Tree plantations are logged to produce mainly sawn wood and pulp. The total production capacity for sawn wood was more than 4.5 ton (eight million m^3) and more than 5 ton of pulped wood (Fig. 6.2). The main actors are the three large size companies, which are multinational consortiums with a presence in several Latin American countries (Gómez 2014): ARAUCO S.A., CMPC, and MASISA S.A. Figure 6.2 illustrates the expansion of tree plantations in Chile between 1978 and 2016. The figure also displays the production of sawn wood from 1930 to 2016, and the production of pulp wood from 1975 to 2016.

In this long-term process, as shown in Fig. 6.2, tree plantations started spreading steadily, especially since the dictatorship. Consequentially, the production of wooden raw materials increased. The democratic turn, instead of curbing the expansion of tree plantations even fuelled it. The only major decline (2006–2010) can be explained by decreasing demand due to the global financial crises. In sum, the process left a consolidated model of nature extractivism based on the extensive use of land and tree plantations throughout the country, but most strongly in south-central Chile.

---

[8]Planted land ownership (x): (i) Large-size company: x >30,000 ha; (ii) Medium-size company: 5000 ha < x < 30,000 ha; (iii) Medium-size owner: 200 ha < x < 5000 ha; and (iv) Small-size owner: x <200 ha

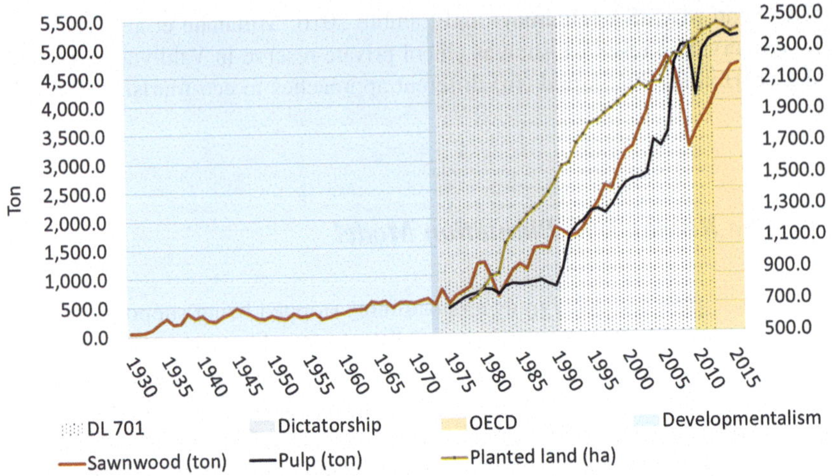

Fig. 6.2 Expansion of tree plantations and forestry industry in Chile. (Source: Own elaboration based on statistics of INFOR (2018). (Note: Growth of national production of sawn wood (ton*) and pulp (ton) are displayed to the left axis. Land used for plantations (ha) in the right axis). Important historic periods are coloured. *Statistics for sawn wood are originally presented in m^3. We used the conversion factor of the source: 1 m^3 = 0.55 t)

6.3 Conservation Model of Llancahue: A Turn toward Sustainability?

In Chile, the first protected areas emerged in the early XXth century. They basically had two objectives: to protect the remaining primary forests from rampant deforestation and to exercise control of the political borders shared with Argentina (Klubock 2014; Otero 2006). The first 'Forest Reserves' where thought as reservoirs of exploitable woods. In the 1970s and 1980s, a second wave of protected areas emerged based on the concept of 'wilderness'. In 1984, the National Public System of Protected Areas (SNASPE, acronym in Spanish) was established with four categories: (i) Virgin Region Reserves, (ii) National Parks, (iii) Natural Monuments, and (iv) National Reserves (Pauchard and Villarroel 2002). Since 2003, properties owned by the fiscal governmental unit, the Ministry of National Property [*Ministerio de Bienes Nacionales*], can enter the SNASPE.

The forest reserve of Llancahue is a 'fiscal property' and gained the status of 'Protected National Good' in 2005 (Ministry of National Property 2006). It is a public property of 1270 ha, owned by the government since 1929, located 2 km south-east to the city of Valdivia in the Region of Los Ríos, south-central Chile. The purpose of the reserve is to protect the hydrological basin that provides water for the city of Valdivia, to conserve biodiversity, and to prevent deforestation and forest degradation (Moorman et al. 2013a). The hydrographic basin of this forest ecosystem supplies nearly 80% of the water for the city of Valdivia. Besides, the Ministry of the Environment selected Llancahue forest as one of the 40 high priority areas for the conservation of natural ecosystems within the regions of Los Ríos and Los Lagos (Farías et al. 2004).

The Llancahue forest reserve contains one of the last reserves of primary forest (approximately 700 ha of old-growth stands) plus secondary forests of different types (Donoso et al. 2014). The geomorphology of the Llancahue forest reserve consists of moderate hills between 181 and 424 m a.s.l. with gentle slopes (<30%) except near the river basins. According to the Koeppen classification, the climate is temperate rainy warm with Mediterranean influence. The annual average rainfall is 2357 mm, with July being the rainiest month and February the driest (Núñez et al. 2006). The average temperature is 12.0 °C, with January being the warmest month with 17.0 °C and July being the coldest with 7.6 °C (Donoso et al. 2014).

In 2008, the Universidad Austral de Chile – UACh – in Valdivia received a concession for the Llancahue forest reserve free of charge by the Ministry of National Property (2008) for a period of 20 years to implement a sustainable forest and watershed management project together with local communities (UACh 2017). The goals of the project included forest management especially in some areas covered with secondary forests, "to promote old-growth conditions through the use of ecological or restoration thinnings while working with the local community" (Moorman et al. 2013a). Since then, the UACh initiated a process with the aim of conserving the primary forest and ensuring a balanced water supply while meeting the needs of people living in the surroundings (Donoso et al. 2014; Moorman et al. 2013a, b).

The Llancahue forest reserve is located next to a small scattered hamlet of 24 peasant families in the sector of Lomas del Sol. These families' livelihoods depend on tree logging in their own small forest plots and in the forest reserve, which they have used for decades without regulations and after 2008 under the rules of the management plan of the Llancahue project. The wood is used for traditional charcoal production, which provides the main income to the community. Besides, their livelihoods depend on smallholder farming and, since recent times, on tourism (Pacheco and Henríquez 2016a). A co-management sustainability plan was designed with the participation of members of LdS. Other organizations such as the NGO 'Agrupación de Ingenieros Forestales por el Bosque Nativo' – AIFBN – and several public agencies were involved in a committee for the Management of Llancahue.

The plan of UACh proposed a sustainable use of the Llancahue forest reserve based on the concept of co-management, an approach founded on a rather technical vision of sustainability. The initial co-management conservation model employed people from LdS as loggers to do the thinnings in the 317 ha of secondary forests within the reserve. Each worker received a wage plus a share of the extracted timber, following estimations on the willingness to accept (Moorman et al. 2013a). The estimations suggested loggers from LdS would accept a salary of $346/month USD ($ 200,000 CLP). The authors also performed an evaluation which signalled that people in LdS expected mainly employment as a benefit from the Llancahue project.

However, the study also estimated that the project could pay only a maximum of nine workers for 3 months a year. Between 2009 and 2016 the co-management plan was applied for the Llancahue forest reserve. Table 6.1 shows some statistics. The

Table 6.1 Features the Llancahue forestry co-management plan

	2009	2010	2011	2012	2013	2014	2015	2016
Area under management (ha)	11	10	9	12	7	12	14	9
Wood extraction (mt)	1700	1200	270	1500	920	2400	1430	1320
Wood for UACh	1600	1100	180	1400	800	40	n.d.	610
Wood for loggers of LdS	100	100	90	100	120	n.d.	300	710
Local workers (number)	6	6	6	6	6	9	9	9
Wages (CLP/mt)	4000	4500	5000	5000	6000	6000	7000	6000
Total annual wages [II*VI] (1000 CLP)	6800	5400	1350	7500	5520	14,400	10,010	7920
Average annual wage per worker (1000 CLP)	1133	900	225	1250	920	1600	1112	880

Source: Pablo Donoso (in Pacheco and Henríquez 2016a)
Units: *CLP* Chilean Peso, *mt* meter

area under management reached its maximum in 2015, when 14 ha were managed. The average annual wage per worker oscillated between \$225 CPL and \$1600 CPL, and the amount of wood per worker per year was between 15 and 79 m^3 of firewood (1 m^3 of firewood approximately 0.66 m^3 of solid wood).

The co-management plan was not without criticism. The initial approach sought to present the Llancahue forest reserve "as a demonstration conservation project where public primary forests are managed sustainably and consistent with the needs of local communities" (Moorman et al. 2013a). However, while the programme was successful in reaching the first goal, i.e. to manage the native forest without degrading it, the second goal was not met: the model is not consistent with the needs of the local community. The main reason behind this is that the project cannot deliver what people in LdS expected from it, mainly employment. Highly seasonal working opportunities for 9 individuals in 24 households can certainly not deemed to be sufficient. The second issue is that other needs, such as education, were not taken into account. Finally, a more fundamental problem is that the notion of 'participation' did not reach decision-making, therefore, is not 'co-management' but rather participatory management.

Although the plan recognizes that "a major challenge in community based approaches is determining how to engage local community participation", the people of LdS is considered a threat for conservation per se: "they [protected areas] are threatened by illegal uses by neighbouring communities such as illegal timber harvesting and grazing of animals... The adaptive co-management model in Llancahue has recognized the need to work with the campesinos (...) in the neighbouring community who were illegally logging the forests in the watershed" (Donoso et al. 2014).

How to use and manage the forest are in fact an interplay of bundles of rights. UACh is the manager and not the owner of the Llancahue forest reserve. There is no binding obligation for UACh as the acting authority to include LdS needs into the decision-making, beyond an advisory role. However, this stands in contrast with the needs and rights to autonomously determine a mode of living of the people of LdS, jointly with the historical marginalization of basic elements of well-being such as health or education (Pacheco and Henríquez 2016a). In this sense, the obligation scales up to the state, which through the Ministry of National Property provided the concession of Llancahue expecting its conservation but without any consideration for the interplay with the surrounding uses of property, neither the community nor the plantations.

UACh recognized these major shortcomings and widened the cooperation approach with the social group since 2014. The pure work-based strategy was complemented with community based eco-tourism (Pacheco and Henríquez 2016a) that was inspired by participatory action to improve the local conditions by means of constituting strategies of eco-socio-development (Henríquez et al. 2010, 2018; Sampaio 2005). However, the pivotal role of the sustainability approach remains on the forestry management.

We see the main shortcomings in the limited understanding of the relations between people living around the reserve, their interests, their rights, and their relationship with the forest and the adjacent tree plantations. Instead of pointing at LdS group as 'the threat' for Llancahue's forest, a more in-depth analysis of the territorial processes must be done to understand local needs, dependencies, and forest uses in relation to tree plantations. We find three types of actors in conflict about the management of this territory: Forest companies, local community of LdS (some of the families are dedicated to illegal logging, or in activities related to forest production), and the University as a manager of Llancahue Reserve.

6.4 The Plantations Boom in Valdivia

In 2016, the region of Los Ríos had 185.108 ha of eucalyptus and pine plantations, and the municipality of Valdivia alone had 22.933,6 ha (INFOR 2018). An early industry in Valdivia, mainly the CMPC cellulose plant and INFODEMA, a local plywood company, had operations since 1942. MASISA, one of the three biggest forestry companies in Chile, was funded in Valdivia in 1960, opening the plant in the city in 1965 and creating its logging filial, *Forestal Tornagaleones*, in 1967. However, the plantation boom in Los Ríos and in particular in Valdivia, started in the 1970s and accelerated since the 1990s. The dynamism of the industry in Valdivia received another push with the inauguration of the cellulose plant in San José de la Mariquina in 2004.

The first tree plantations in the area of Llancahue appeared in the 1980s. Until 2013, they heavily expanded conforming an extractivist matrix (Fig. 6.3c). Figure 6.3 displays the land use in the area of Llancahue for the years 1960, 1980, and 2013. Between 1960 and 1980, while native forests increased within the borders of the Llancahue forest reserve, they decreased outside the reserve. Between 1980 and

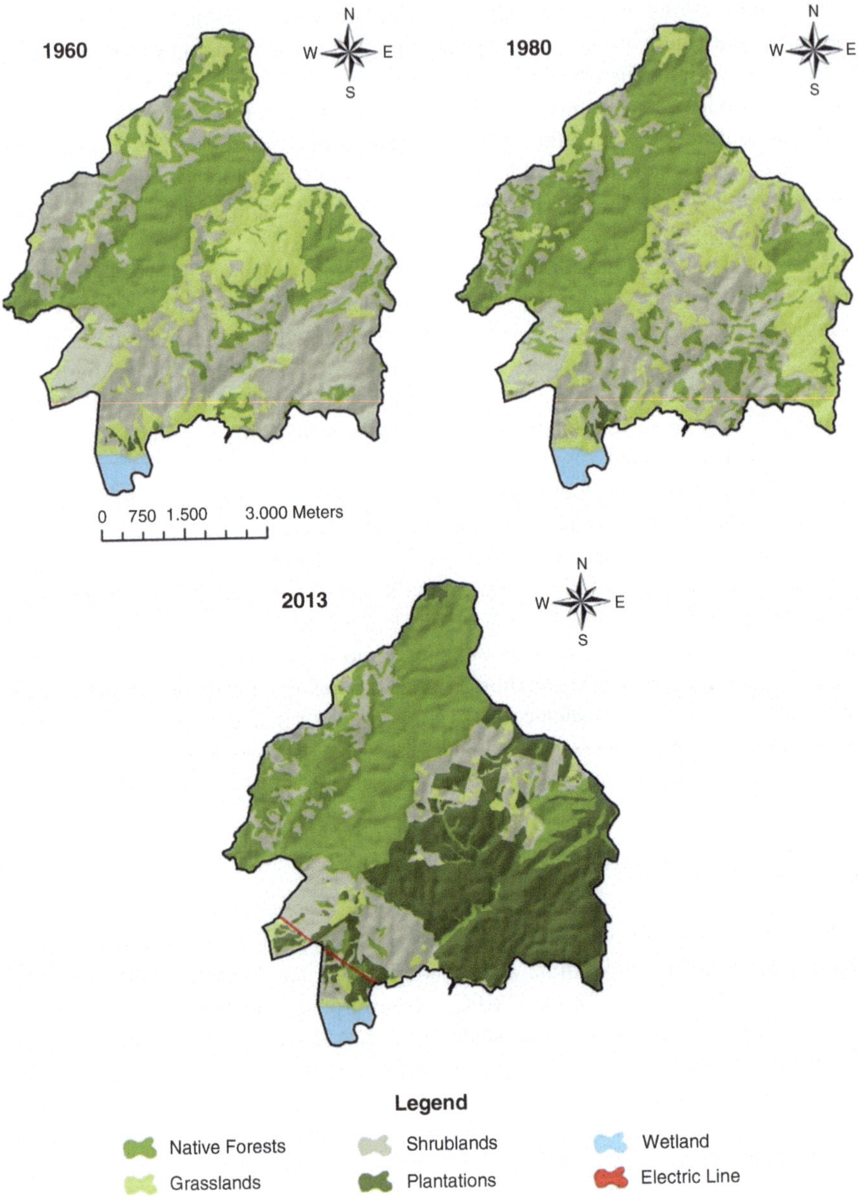

Fig. 6.3 Changes in land use in Llancahue area between 1960 and 2013. (Source: Medel (2013) in UACh 2017)

2013, there was a growth of native forests within the reserve, while tree plantations replaced both native forests and private agricultural land outside the margins of Llancahue.

6.5 Living Between the Llancahue Reserve and Tree Plantations

The peasant group in LdS consists of 24 families, some of which are grouped under the peasant association without lucrative purpose *Comité pro Adelanto Lomas del Sol*, established in 1997 and legally recognized as a decentralized local social organization by the municipality of Valdivia. LdS is located in the buffer zone of the Llancahue forest reserve, but also besides blocks of tree plantations (see Fig. 6.1). This makes it an interesting case to reflect the territorial transformation of a social group that is trapped in between extractivism and conservation. The main source of income is firewood and coal production, sold mainly in Valdivia, complemented by small-scale farming. People have orchards and greenhouses, and raise livestock such as pigs, sheep, and cattle. The formal education is mostly low, not exceeding 8 years of schooling. Basic infrastructure is lacking or scarce. There is no health service and no potable water, which is ironic as the area provides the water for the whole city of Valdivia. Generally, there are no job opportunities in LdS, especially for women and young people, or for men who wish to work outside forestry or agriculture. Some families migrated to urban centres, mainly Valdivia (Pacheco and Henríquez 2016b).

The management of forest conservation controlled the illegal logging, but on the other side, it reduced the working opportunities for local families dedicated to produce charcoal and firewood. The tree plantations expansion made pressure over property of local families, resulting in some of them are selling their properties to forest companies. LdS is located between the conservation and the forest extractivism. Even if LdS social group could make illegal practice and harmful for conservation, the effects of this practice are less harmful than the impacts of the industrial model of plantations in all water basins of Llancahue.

To understand the loss of the territoriality of the peasant group of LdS in the interplay with the territorialities of the conservation scheme and the tree plantation blocks these two, it is necessary to start from the local notion of territory. In the following we show our empirical findings gathered with a mixed method approach involving focus group discussions and a participatory mapping, conducted together with residents of LdS and complemented by observations in several local activities. Focus groups were used to reconstruct the preponderant territorial dynamics in three historical periods: (i) 1970–1990, (ii) 1990–2010, and (iii) 2010–2016. However, the exercise included questions to contextualize the time frame before 1970. The group division was made by age. The exercise revolved around the construction (and later analysis) of maps. The exercises provided a synthesis of the main elements of territorial transformation, which ultimately inform on the local (externalized) costs of tree plantations in a qualitative way.

6.5.1 Territorial Context Before 1970

The first Chilean settlers, the families with surnames Hernández and Noches, arrived in the Llancahue area between 1915 and 1930, in a process of late agrarian frontier expansion. At first, only male settlers arrived and began the land occupation as squatters, 'opening' land for agriculture. Later on, they brought their wives and children. These first families were followed by others, starting a local economy based on the extraction of wood from native species to establish their homesteads, as firewood, and mainly as the raw material for charcoal production. Charcoal was produced with artisanal coal furnaces, a technology used all over south-central Chile. The coal was transported with carts and yokes of oxen to Valdivia and Collico for sale, often constituting the only source of monetary income. The trip took more than a day with a one-way transportation time of around 5 h. In the 1960s, the population of the sector *LdS* had grown enough to justify the construction of a school, which operated between 1960 and 1972 and constituted the first permanent state agency in LdS. In 1974, as a *comunero*[9] mentions, there were enough inhabitants to "play football".

The physical accessibility to LdS was difficult, since the bridle paths up the hill were precarious from the surrounding towns. Even the road from Valdivia to LdS was rudimentary until it was paved in 2009, when a gavel carpet was added to the road. In total, until that year, LdS was quite isolated, despite the very close proximity to the city of Valdivia.

Land property of peasants in LdS was not formally titled until a private property law recognized them as owners in 1974. Until this point, the families lived as squatters between the public property of the forest reserve and large private properties.[10] Until the late 1970s there were no pine or eucalyptus plantations in the area, and, as stressed by the older interviewees, water was abundant.

6.5.2 Period 1970–1990

During the 1970s, the orchards in LdS were still mainly for subsistence and forests products were used for day to ,day life. Multiple species of fungi from the forest,

[9] An individual of a community.

[10] However, as reported by Donoso et al. (2014), even in the present several of the families do not have formal titles.

such as 'Loyo' and 'Digüeñe',[11] for example, were common in diet. In agriculture, as one of the focus group participants commented, "each household needed to sow in order to sustain itself", and "there was not much potato left for sale", pointing to the lack of yield surplus. Income was mainly derived from the production of charcoal using timber from the surrounding primary forests.

Although in 1974 land property of peasants in LdS was formally titled, peasants were not allowed to sell their land until 1979. In 1979, forestry companies began to buy land from peasants in LdS. The participants of our focus group discussions did not recall the year in which the first forest plantation was established in the area. But they know that logging companies purchased the first land in 1979. This was followed by several deals at the beginning of the 1980s.

Large-scale tree plantations grew during the 1980s, while several peasants in LdS established small–scale tree plantations on their plots to sustain their firewood availability. The reasons behind were the degradation of the primary forests on their own plots and the clearcutting of the primary forests on the holdings of the plantation companies. The participants of our focus group discussions associated the land purchases by plantation companies to migration of peasants from LdS to cities, mainly to Valdivia. They remembered a purchase of 42 ha of land in 1989 for a price of $CLP 2.5 million, meaning approximately $ CLP 59,000 per ha. Other reasons included the lack of job opportunities and difficulties to access basic education, as the primary school was closed in 1972, forcing children to assist to a boarding school in Huellelhue, a town 6 km away from the sector, 2 h walking through the tree plantations.

Figure 6.3 shows that the degradation of the native primary forest in the area of the Llancahue forest reserve was greater in 1960 than in 1980 when the plantations were beginning to arrive. This suggests that the families in LdS had slowly overused the forest up to this point (1960). However, the combination of the advance of the tree plantations on the one side, and the establishment of stronger norms of forest conservation on the other, gradually led to a crisis of the way of life of the LdS community and ultimately to the loss of their livelihoods.

Focus group participants point out a migration from 'below', the sector of Piedras Blancas, 'upwards' to the sector of LdS (see magenta arrow in the middle of Fig. 6.4). In addition, a block of native forest outside the Llancahue forest reserve is clearly identified (green area), which can also be seen in the land cover maps (Fig. 6.3). The map also displays coal kilns in almost every household, the abundance of charcoal, and Valdivia as the main coal destination (purple arrows). Two more elements were revealed by the exercise. First, there was a church in LdS. Second, the symbol used for flora and fauna is not only placed at the native forest which corresponds to the Llancahue reserve, but also at their individually owned land, signalling that there were native forests on their plots. One of the main elements of the territorial transformation during 1970 and 1990 is the first tree plantation block, located between LdS and the river Calle-Calle (brown box

[11] For a description of Chilean fungus see Furci (2008).

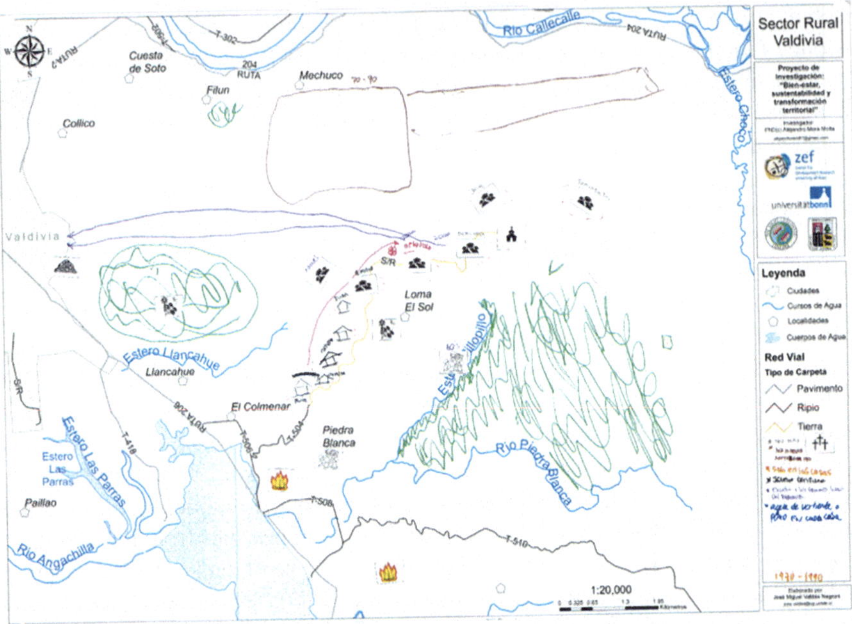

Fig. 6.4 Elements of territorial transformation during 1970–1990. (Source: first author, participatory mapping workshop, Lomas del Sol, Valdivia, December 2016)

and rectangle). This is the tree plantation of the *Fundo Fried*, a land owned by the individual of the same surname, which is still present today.

One interesting event came out in one of the focus group discussions. The abundant native forest near the Pillpillo creek was being occupied by members of the community. They remember they and their parents used to extract timber and other forest products such as local fungi form this forest. In the 1980s, an official of the state visited the place started a legal process calming that it was a public land. Finally, the land was recognized as public, so the families had to move away. The land, however, was sold to MASISA and the forest, they remember, was transformed into plantations.

6.5.3 *Period 1990–2010*

Between 1990 and 2010, tree plantations spread on the hills next to LdS. Plantation companies introduced socio-ecologically doubtful agronomic practices, such as aerial application of agrochemicals, which were dispersed by air currents falling over the orchards in LdS and the surrounding primary forests. As one of the participants reported "In the eighties and nineties . . . we lived near the Pillopillo creek, and the forestry company used [aerial] fumigation. Near my house we had a small orchard ... and all the chickpeas were ruined, all was burned"; the reason was that in that time "plantations were just starting to grow, and they had to fumigate due to a small animal that the pine had". "Yes, fumigations where due to a worm supposedly", complemented another participant. Practices were banned later on.[12]

Plantation companies often highlight their important role in creating jobs. However, the operations around LdS did not offer much working opportunities for local people. The jobs that were created required a level of qualification which most community members did not have. In addition, the number of newly created jobs was very limited, as the processes were highly technical, with intensive use of machinery.

Between 1990 and 2010, the importance of charcoal production increased while subsistence agriculture decreased in LdS. This was associated with an intensified migration, in which between seven and ten families left the sector. However, some of the young members got qualified jobs, mainly in Valdivia, and even positions within the forestry industry but away from the sector. This was the case of the son of one participant, who gained a position as a truck driver for ARAUCO. His job was to transport logs from all logging sites through Los Ríos and Los Lagos to the cellulose plant in San José de la Mariquina (50 km north of Valdivia).

Our interviewees perceived an important change during this period. They started noticing that water was becoming scarce. Water is crucial for agriculture and forestry, and especially given the fact that LdS was not connected to a pipe water system, the water shortages started creating massive problems, mainly during summer. One of the participants declared: "there has been moderate water scarcity, almost at each slope. Before we had rain 15 days on a raw during winter, now it rains only for two days" as a result, the dependence on income from jobs outside agriculture and forestry grew.

Since 2005, new infrastructure came to LdS in the form of brick kilns and a local water system (not connected to the aqueduct which provides water for Valdivia). This is associated with the arrival of two new actors, namely: (i) the NGO AIFBN, which developed a program related to the establishment of the water system, and (ii) UACh, which took the management concession of the forest reserve.

Figure 6.5 shows the territorial transformations between 1990 and 2010. The most important element highlighted by the informants is that the area got surrounded by tree plantations (brown area). The plantations of the company *Forestal Tornagu-leones* (now Chilean Hancock Plantations) now joined *Forestal Fried* and created a continuous block of tree plantations around LdS. The place where plantations are shown in Fig. 6.5 was still occupied with native forests in 1990 (Fig. 6.4). When we compare this with Fig. 6.3, it becomes evident that the advancement of the tree plantations led to the deforestation in that area. However, our informants perceived the deforestation during that time as being larger as it actually was, as it is evident when contrasting the land use changes of Fig. 6.3 and the change in the perceived

[12]These practices were only banned in the municipality of Osorno (Ministry of Health 2000). However, it might have been a corporate decision which changed practices.

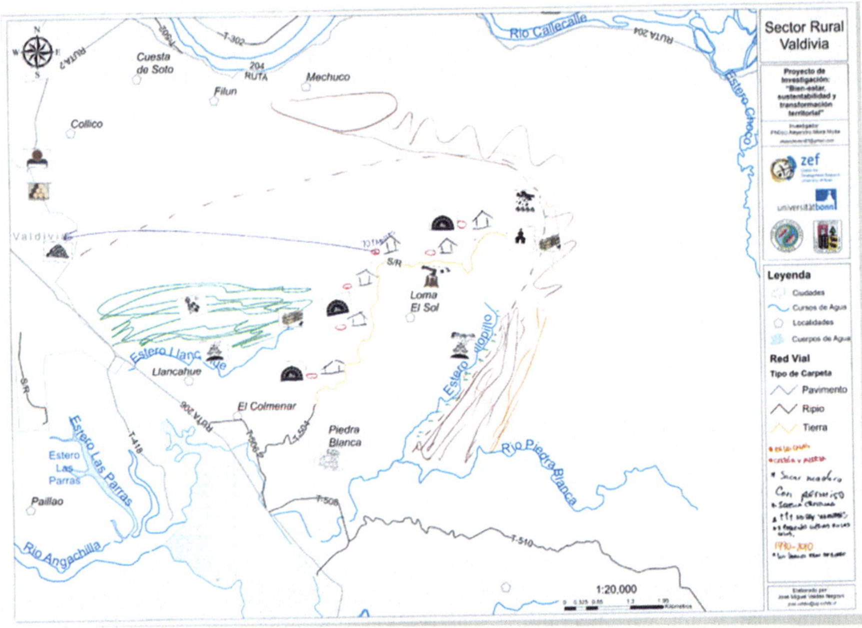

Fig. 6.5 Elements of territorial transformation during 1990–2010. (Source: first author, participatory mapping workshop, Lomas del Sol, Valdivia, December 2016)

native forest beside the Pillopillo creek (Estero Pillopillo) in Figs. 6.4 and 6.5. One possible explanation might be the high economic and cultural value the participants attached to the native forest that was replaced by plantations.

6.5.4 *Period 2010–2016*

The year 2010 brought a fundamental change for the inhabitants of LdS: the dirt road became a gravel road, and parts of it were asphalted shortening the transport time to Valdivia to 30 min by car. Some families bought pick-ups in order to transport firewood or charcoal to Valdivia, now being able to do two or even tree journeys per day. The improved road also spurred a community-based ecotourism project, started by UACh. The project introduced a change in perspective regarding the ‘participation’ of people of LdS. While previously their involvement was mainly needed as loggers in few selected areas of the reserve, with the community-based eco-tourism[13]the strategy changed by implementing a co-design process for the

[13]Community-based tourism (CBT) is a tool for communities to achieve defend their territories from threats such as real estate speculation and cultural de-characterization. With this tourism it

selection of desired economic activities and actions pointing towards them. Ecotourism trails were built up within the Llancahue forest reserve and signals with local stories and biological history facts were introduced. UACh staff trained five inhabitants of LdS as ecotourism tour guides. Also, other projects were encouraged. One of these was a pilot project for photovoltaic solar energy designed by Debus et al. (2017), which included the installation of solar panels at the LdS community meeting house in a series of collaborative workshops. This project shows to one of the problems of LdS. LdS is close to the regional capital city of Valdivia but is not connected to the electric network. Families mainly rely on fuel-based power generators.

However, at this point, as is evident in Fig. 6.6, the tree plantations already surrounded the entire sector of LdS, spatially trapping the peasant community between the Llancahue forest reserve and the tree plantations. According to our informants, the Chilean forestry consortium MASISA, one of Latin America's largest wood panel manufacturers, is the owner of most of the tree plantations around LdS (although recently MASISA sold these lands to Hancock Chilean Plantations). By then, the community seemed to have a conflict with MASISA

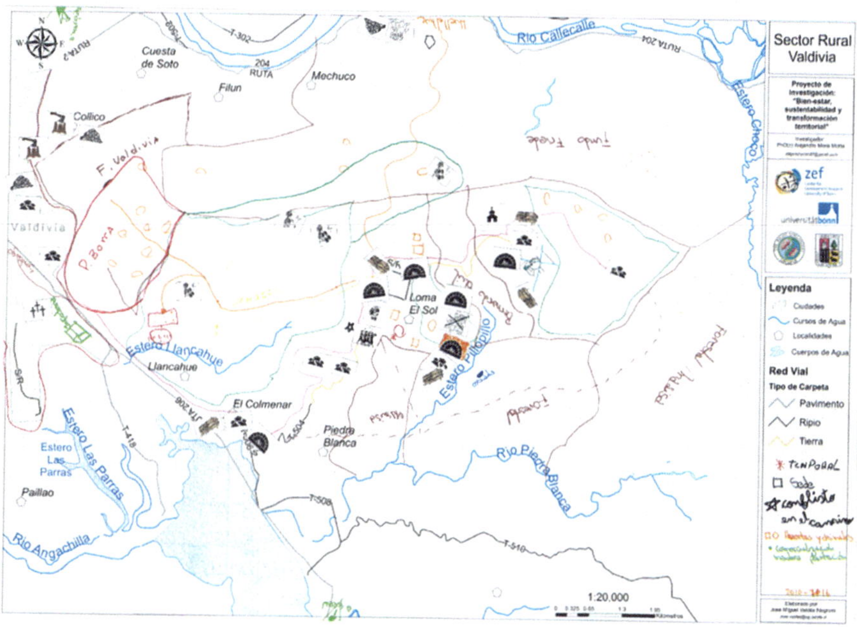

Fig. 6.6 Elements of territorial transformations during 2010–2016. (Source: first author, participatory mapping workshop, Lomas del Sol, Valdivia, December 2016)

---

has been possible, the generation of work and income, protect biodiversity and cultural identity, conserving ways of life of rural and indigenous communities (Pacheco and Henríquez 2016a).

authorities about the use of the public road. People complained that the company, which normally uses its own roads within its properties, started using the only communal road to transport the logs. Families in LdS were not only affected by the noise and the dust, but also because the road was gradually destroyed by the heavy trucks. One day they decided to block the road. Once they did it, MASISA stopped sending its truck through the public road. According to Vásquez (2014), the NGO AIFBN made a formal complaint to Forest Stewardship Council – FSC –, since MASISA possessed the eco-label certification of this global organization. Due to the conflict with the community they deemed it to be no more justified.

With the advancement of tree plantations, the shortage of water increased. As one of the members of LdS mentioned: "Before [1970] in the creeks the stream flow had more water than now, that's the difference, the wells were completely filled".

During the time of the research, the coal prices were relatively low, and the efforts in time and work involved in the process of coal making almost did not offer added value. For these reasons, the use and dependence on firewood increased. Peasants tend to cut down the last remnants of native forest on their plots. According to one of the young participants, the weekly production of coal of the own plot of his father is around 15 sacks of coal, but the effort of doing the coal is so high and badly paid that they were considering leaving only the firewood. Two other participants mention that, on average, they produce 40 sacks of charcoal, a lower estimate. The price at which they sell in Valdivia was between $ CLP 4000 and $ CLP 5000 per sack, but the commercial price in the city for final consumers is between $ CLP 8000 and $ CLP 9000. Also, the participant reported that there is no real differential prize for the type of wood between pine or native species, since the consumers do not have the knowledge to differentiate. When selling the firewood in Valdivia, the prices for the 1 m^3 of firewood are around $ CLP 22,000 for eucalyptus and between $ CLP 26,000 and $ CLP 30,000 for native species and pine. The main difference in costs is that the coal requires one-week production time at the kilns, thus making the firewood more appealing.

In reflect which are the most fundamental changes from 1970 to the present (2016) the participants of the focus group highlighted that for the community, the infrastructure, including the road and the new kilns make a difference. However, the most important change perceived was the process in which people of the sector decided to leave and sold their land to the logging companies. One of the participants recognized that it was not a forced process, but rather "because due to their needs, people did not value their assets". Regarding the prices, the participant reflected that while in the 1980s 1 ha of land was worth $ CLP 40,000 to $ CLP 50,000, one charcoal sack was worth $ CLP 300, and one lunch in Valdivia was $ CLP 500, in 2016 1 ha of land was worth $ CLP ten million, one charcoal sack was worth $ CLP 5000, and one lunch in Valdivia was $ CLP 4000. This reflection offers a rough estimate of how land valorisation has taken place, having very low prises when the land deals from the smallholders to the companies happened.

6.6 Towards a Social Model for Sustainability?

We explored three forms of nature appropriation and their interactions in our particular case. First, the Llancahue forest reserve is a public conservation area of 1270 ha managed by UACh under a 'sustainable co-management' plan (Donoso et al. 2014; Moorman et al. 2013a, b) that seeks to protect the watershed hydrological cycle and biodiversity while involving the community of LdS. Second, large monoculture tree plantations owned by logging companies (big and small ones), that use the land to generate their (short-term) benefits while contributing to 'green growth', ascribing to the 'sustainability' seal offered by self-imposed eco-labelling companies, mainly FSC (Heilmayr and Lambin 2016; Millaman et al. 2016; Reyes and Nelson 2014). Finally, the territoriality of the peasant community of LdS, which appropriates nature in a far less extractive manner than the plantation companies but depends on resources from the native forests. Viewed in perspective, the interplay of these three forms of nature appropriation shows a process in which tree plantations slowly cornered the LdS community against the Llancahue forest reserve.

With the expansion of tree plantations, starting from 1979, the peasant community in LdS gradually lost its cohesion and became more and more dependent on timber extraction from the Llancahue reserve. Companies, in general, do not allow external individuals to harvest their plantations. However, when people have accessed few trees, the companies have not raised concerns. Yet, issues with logging companies are avoided, as one of the LdS members stated: "we don't want to have any problem with the forestry companies". However, another participant stated that the logging companies sometimes give a formal letter allowing them to use the remaining after harvest residuals. Besides, several members of the community introduced pine or eucalyptus in their smallholdings.

The population of Valdivia, a city with over 150,000 inhabitants, depends on water supply from the Llancahue reserve. Scaling up, national and international interests see the Valdivian forests as biodiversity hotspots, being Llancahue one among the few relatively well conserved forests. However, for the local community, native forests represent a traditional source of timber for firewood and artisanal charcoal production, which constitutes its main source of income. In 2008, UACh received a concession for Llancahue reserve for a period of 20 years to implement a sustainable forest and watershed co-management plan. With this plan, UACh assumed the role of authority and imposed restrictions on the uses of Llancahue forest in order to conserve the old-growth stands and the hydrological cycle. However, the forest was previously perceived by the peasant community in LdS as a 'common' good, a free-access space where no withdrawal rights. Fuelled by a lack of job and income opportunities, this triggered tensions between the territoriality of the community versus the co-management plan introduced by UACh.

The main problem lies in that model of 'sustainability' in which co-management has not been able to adequately assume the historical territorial transformation processes in which the community gradually became more and more dependent on a forest reserve, even though it does legally not have the right for wood extraction (while the adjacent tree plantation owners and timber companies do have it). Therefore, when prohibiting or regulating the extraction of wood from within the forest reserve, the unwanted outcome is to limit the main source of income for the local community. This would not be so problematic if peasants had alternative means to cover their daily expenses. The orchard system, for example, which was traditionally important for subsistence, has however also been weakened since plots were sold due to the plantation expansion.

Despite investments derived from projects (e.g. infrastructure) the sustainability approach offered by UACh has limitations: it can guarantee the protection of the watershed and forest biodiversity, but not the relationship that the community of LdS held with that same nature. In fact, it negatively effects the way of life of the LdS peasant community in general, although there are increasing job opportunities for some. Nevertheless, the underlying cause is often overlooked or ignored: the large-scale expansion of tree plantations into native forests and smallholdings, with full support of the Chilean state.

Locally, the recent economic problems of the inhabitants of LdS seem to be more related to the conservation model than to the plantation model. However, looking at the broader territorial and historical picture we have presented, it becomes clear that the progressive spreading of tree plantations gradually led the community to depend almost exclusively on the exploitation of the native forest in a reserve since many other forests were clear-cut for plantation development. This was precisely the case of the native forest surrounding the Pillo Pillo Creek, from which families used to extract wood and other forest products. The satisfaction of needs became more and more dependent on monetary income, particularly from the extraction of wood from the reserve. Thus, it is inevitable that conflicts arise when restrictions are established and implemented on the use of Llancahue forest reserve, because LdS group did not develop economic association, productive cooperation, or other strategies to diversify the local economy. Their isolation and low educational levels played a negative role for local development, unlike in other rural territories of southern Chile.

The 'sustainability' idea embedded in a 'green growth' concept to which logging companies tend to refer is a euphemism at the local scene. Logging companies do not necessarily engage in illegal activities, but rather in unethical ones. The community has mainly interacted with *Forestal Tornagaleones*, of the MASISA consortium (now HCP). While the company has 'collaborated' with the donation of 0.5 ha to the project to create a space for environmental education (Vásquez 2014), they have used their economic and bargaining power to green wash their activities by complying to FSC. The problem that emerges with logging companies, is that the only way to modify their actions, in a sensitive scale, is by changing the legal framework they rely upon. Within their property, they could deforest native forests and replace them with plantations in the past, only mediated by the management plan, which required an official permission. This was the case with the native forest present outside the reserve, acquired by logging companies and transformed into plantations.

A 'social sustainability' approach must start by understanding the relationships between fundamental human needs (Max-Neef et al. 1991) of local people and 'their' territory. At stake is the reproduction of the peasant territoriality in LdS. The notion of sustainability proposed by the co-management approach of UACh at Llancahue involves a wider preoccupation by the local peasant group but is not yet able to construct a 'sustainability' for that group. Then, how is it possible to build a social sustainability that considers the satisfaction of the fundamental human needs of the peasant families of LdS? It seems impossible to reverse the territorial transformation that was triggered, ultimately by the tree plantation expansion. However, alternatives that allow new territorialities where communities satisfy their fundamental human needs are possible.

The sustainability problem for the people at LdS, and social groups alike, starts by a process of regenerating meaningful opportunities which allow to build territorial processes for the satisfaction of needs. It cannot be solved with a single policy implementation. As the sustainability preoccupations are multi-scale, measures to create such meaningful opportunities should also include multiscale changes. We presented the problem of tree plantations as an extractivist phenomenon, which is focused on the extraction of raw materials for exports. In this regard, the problem relies on the obsession of an ever-growing economy and the set of regulations which support such extraction. Therefore, measures which actually can generate changes at the national level rely on changes on the indicators of progress (Azkarraga et al. 2011). However, this the type of policies suggested in this direction are contrary to those of green growth.

This type of measure must be echoed with some sort of territorial action which balances the negative effects of monocultures. There is an interesting approach of territorial use based on the PROT, and taken as... What we perceive is that the state is still subsidiary to the companies' extractive decisions. It is important that the state engage in a more direct regulation, using tools such as local taxes, which would allow the municipality income out of the land dedicated to plantations. However, while taxes may regulate decisions, stronger restrictions should apply in the neighbouring areas to forest reserves.

Finally, to benefit local actors who have been systematically excluded from the state, as has been historically the case of LdS community, requires a stronger state engagement. Solving structural inequalities is not an exclusive responsibility of a conservation management programme, but also of the state action. While the state has been present in the case of LdS funding the construction of the road, their engagement needs to be enhanced in talking multiple inequalities at the same time.

6.7 Conclusions

In this chapter, we presented a case in which differing perceptions, values, and uses of nature structured tensions and often conflicts around its appropriation. The interplay of the interests that set the territorialisation of these differing views of nature fostered a territorial transformation. Three territorialities interacted in such process: (i) The tree plantation model under which nature is appropriated in an extractivist manner, legitimized with the 'green growth' discourse and policies; (ii) the Llancahue forest reserve under which nature is protected, with a focus on watershed and biodiversity conservation; and (iii) the territoriality of a peasant community which uses nature mainly for firewood and charcoal production and small-scale agriculture. The peasant community is geographically located right between the tree plantations and the Llancahue forest reserve. Nevertheless, the local problems are complex, since some local residents and external agents have been involved in illegal native wood extraction from Llancahue and other near properties.

Logging companies started the massive expansion of tree plantations in the area in the 1980s, promoted by the Chilean state. In LdS, and many other parts of Chile, plots from peasant families were sold to logging companies, which established plantations with fast growing pine and eucalyptus to produce timber primarily for sawn wood and cellulose production, mainly for export. This export-focused activity is at the centre of the extractivist processes that are present in all Latin America (Gudynas 2015).

The Llancahue reserve, north to LdS, is a public property that became a forest reserve in 2005 mainly for watershed and biodiversity conservation. The LdS peasant community had however perceived it as a free-access common and used the forest for timber, firewood, and coal production without permission for decades. In 2008 UACh, was given a concession for Llancahue reserve for 20 years. Since then, the university implemented a sustainable management plan, with the attempt to build a co-management plan with LdS people but without participation beyond seasonally working relations in logging. The plan limited the use of the woods from Llancahue reserve. In 2014, the plan was revised and involved other actions, such as the development of a community-based ecotourism programme and a photovoltaic energy project.

The peasant group of LdS that exercised a territoriality based on smallholder agriculture and the use of resources from the surrounding native forests, ultimately got trapped in between these two territorialized models of nature appropriation. On one side, the group has not enough bargaining power to confront the advance of the logging companies, which are backed by state regulations and policies. Plantations advanced progressively, clearcutting the relicts of native forest, reducing the space for (the now illegal) timber extraction on plots of forests of peasant's own plots, nature that was highly valued for its economic functions for the peasant mode of living. As a consequence, peasants relied more and more on the woods from the forest reserve.

The process and extent of the territorial transformation allowed us to argue that, even if there are conflicts between the UACh forest conservation approach and the peasant group of LdS, the strongest factor – in impact and size – triggering the conflict arena are the tree plantations, covered by the sustainability discourse of 'green growth'. It becomes clear that the problem is limited by the regime of appropriation of nature on which extractivism is based, that is, a set of regulations including laws and policies that interact with the local institutions. Therefore, the dichotomy of nature 'conservation' versus 'human needs' requires to be revisited since extractivism, which plays a silent role in the local conflict arena, becomes the trigger when looking to the broader picture.

As an alternative, we see that actions towards a 'social sustainability' that allows for the reproduction of the peasant territoriality of LdS should point towards the legal recognition of more organic ways to appropriate nature (Max-Neef 2016), which maintain cultural practices while satisfying fundamental human needs (Max-Neef et al. 1991). With a binding recognition of territorial rights, new peasant territorialities could emerge by restructuring a way of life based on a respectful appropriation of nature. But this is not easy in the practice, because in the territory a climate of distrust predominates between peasants.

However, both the territorial models and the resource regimes that support extractivism require to be transformed to give space to other forms of exercising territoriality, but such structural transformation is not an easy task. Therefore, in the short term, it is necessary to 'manage' the tensions and conflicts by negotiating alternatives for the co-construction of sustainability, which must respect the ecosystem limits and understand that sustainability is a social construct and problem. Nevertheless, this is so difficult in the territory because the peasants are disconnected with the sustainable concept in their local knowledge, and in their productive practices based in timber and charcoal production.

Acknowledgments AMM is grateful to Laura Fúquene, for her assistance, Viola Debus for her collaboration, and José Valdés Negroni for the elaboration of the mapping tool. Also, to the German Academic Exchange Service (DAAD) which provided the financial support for this research through the RLC Campus Bonn. We are especially grateful to the community members of LdS who shared their experiences and knowledge with us. Besides, we are grateful to the line of Economics and Politics of the TESES research centre for discussion, comments, and criticism on an early draft of the document.

References

Acosta A (2013) Extractivism and neoextractivism: two sides of the same curse. En M. Lang & D. Mokrani (Eds.), Beyond development: alternative visions from Latin America (1. transl. ed), pp. 61–86. Amsterdam: Transnational Institute – Rosa Luxemburg Stiftung

Azkarraga J, Max-Neef M, Fuders F, Altuna L (2011) La Evolución Sostenible II – Apuntes para una salida razonable, 117 páginas. Lanki, Eskoriatza

Brand U, Lang M (2015) Green economy. In: Pattberg PH, Zelli F (eds) Encyclopedia of global environmental governance and politics. Edward Elgar Publishing, Cheltenham, pp 461–469

Brand U, Dietz K, Lang M (2016) Neo-Extractivism in Latin America – one side of a new phase of global capitalist dynamics. Ciencia Política 11(21):125. https://doi.org/10.15446/cp.v11n21.57551

Clapp RA (1995) Creating competitive advantage: forest policy as industrial policy in Chile. Econ Geogr 71(3):273–296. https://doi.org/10.2307/144312

Corporación Nacional Forestal [CONAF] (2014) Catastro de los recursos vegetacionales nativos de Chile. Monitoreo de cambios y actualizaciones alanho 2013. CONAF, Departamento Monitoreo de Eco-sistemas Forestales, Santiago. 35 p

Debus V, Henríquez C, Fuders F (2017) La tecnología apropiada frente al cambio climático: Barreras del mercado eléctrico nacional y posibilidades para Paneles Solares Caseros a escala humana. Master thesis by Viola Debus, Facultad de Ciencias Económicas y Administrativas. Universidad Austral de Chile
Donoso PJ, Frêne C, Flores M et al (2014) Balancing water supply and old-growth forest conservation in the lowlands of south-Central Chile through adaptive co-management. Landsc Ecol 29(2):245–260. https://doi.org/10.1007/s10980-013-9969-7
Donoso PJ, Romero J, Reyes R, Mujica R (2015) Precedentes y efectos del neoliberalismo en el sector forestal chileno, y transición hacia un nuevo modelo. In: Pinol A (ed) Democracia versus Neoliberalismo: 25 años de neoliberalismo en Chile. Fundación Rosa Luxemburgo – ICAL – CLACSO, Santiago, pp 210–233
Escobar A (2014) Sentipensar con la tierra: nuevas lecturas sobre desarrollo, territorio y diferencia (1st edición). Medellín: Ediciones Unaula
Fals-Borda O (2015) In: Moncayo VM (ed) Una sociología sentipensante para América Latina. Siglo XXI Editores – CLACSO, México
Farías A, Tecklin D, Pliscoff P (2004) Análisis del avance hasta la fecha en la definición de las áreas prioritarias para la conservación de la biodiversidad en la región de Los Lagos. WWF Report, for the consultive system of SNASPE, Region of Los Lagos. WWF, Valdivia, 19 p. Retrieved from http://d2ouvy59p0dg6k.cloudfront.net/downloads/areas_prioritarias_para_la_conservacion_de_la_biodiversidad_region_los_lagos.pdf
Furci G (2008) Hongos. En CONAMA, Biodiversidad de Chile, Patrimonio y Desafíos (pp. 366–375). Santiago: Ocho Libros Editores
Gobierno de Chile (2013) Estrategia Nacional de Crecimiento Verde. Gobierno de Chile – Ministerio de Ambiente – Ministerio de Hacienda, Santiago, p 96
Gómez S (2014) El caso de Chile. In: Capitalismo: Tierra y poder en América latina (1982–2012): Argentina, Brasil, Chile, Paraguay, Uruguay (Vol. I). Universidad Autónoma Metropolitana – Consejo Latinoamericano de Ciencias Sociales – Ediciones Continente, México, pp 137–171
Guber R (2011) La etnografía: método, campo y reflexividad. Siglo Ventiuno Editores, Buenos Aires
Gudynas E (2015) Extractivismos: ecología, economía y política de un modo de entender el desarrollo y la naturaleza (1st edición). Cochabamba: CEDIB, Centro de Documentación e Información Bolivia
Heilmayr R, Lambin EF (2016) Impacts of nonstate, market-driven governance on Chilean forests. Proc Natl Acad Sci, 113(11), 2910–2915. doi.org/https://doi.org/10.1073/pnas.1600394113
Henríquez C, Zechner TC, Sampaio CAC (2010) Turismo y sus Interacciones en las transformaciones del espacio rural. Revista Ciencias Sociales 18:21–31
Holmes G (2015) Markets, nature, neoliberalism, and conservation through private protected areas in southern Chile. Environ Plan A 47(4):850–866. https://doi.org/10.1068/a140194p
Instituto Forestal [INFOR] (2018) Chilean statistical yearbook of forestry 2018. Statistical Bulletin N° 163. INFOR, Santiago
Klubock TM (2014) La Frontera: forests and ecological conflict in Chile's frontier territory. Duke University Press, Durham
Krausmann F, Lauk C, Haas W et al (2018) From resource extraction to outflows of wastes and emissions: the socioeconomic metabolism of the global economy, 1900–2015. Glob Environ Chang 52:131–140. https://doi.org/10.1016/j.gloenvcha.2018.07.003
Lander E (2011) La Economía Verde: el lobo se viste con piel de cordero (Justicia Agraria y Ambiental). Transnational Institute, Amsterdam, p 10. Recuperado de https://www.tni.org/files/download/green-economy_es.pdf
Leff E (2014) La apuesta es por la vida: Imaginación sociológica e imaginarios sociales en los territorios ambientales del sur. Siglo Ventiuno Editores, México
Max-Neef MA (2005) Foundations of transdisciplinarity. Ecol Econ 53(1):5–16. https://doi.org/10.1016/j.ecolecon.2005.01.014
Max-Neef M (2016) Philosophy of ecological Economics. Int J Econ Manag Sci:1–5. https://doi.org/10.4172/2162-6359.1000366

Max-Neef M, Elizalde A, Hopenhayn M (1991) Human scale development: conception, application and further reflections. The Apex Press, New York
Millaman R, Hale C, Aylwin J et al (2016) Chile's forestry industry, FSC certification and Mapuche communities. Independent Report 205 p. Chile. Retrieved from: https://ga2017.fsc.org/wp-content/uploads/2017/10/Chiles-Forestry-Industry-FSC-Certification-and-Mapuche-Communities-FINAL.pdf
Ministry of National Property (2006) Decreto 634: Por el cual se destina al Ministerio de Bienes Nacionales predio denominado Fundo Llancahue, X Región de los Lagos. Ministry of National Property, Government of Chile, Santiago. Retrieved from: https://www.leychile.cl/N?i=249078&f=2006-04-22&p
Ministry of National Property (2008) Decreto 160: Por el cual se otorga concesión gratuita de inmueble fiscal en la Región de los Ríos a la Universidad Austral de Chile. Ministry of National Property, Government of Chile, Santiago. Retrieved from: https://www.leychile.cl/N?i=271532&f=2008-05-28&p
Moorman M, Donoso PJ, Moore S et al (2013a) Sustainable protected area management: the case of Llancahue, a highly valued Periurban forest in Chile. J Sustain For 32(8):783–805. https://doi.org/10.1080/10549811.2013.803916
Moorman MC, Peterson N, Moore SE et al (2013b) Stakeholder perspectives on prospects for co-management of an old-growth forest watershed near Valdivia, Chile. Soc Nat Resour 26(9):1022–1036. https://doi.org/10.1080/08941920.2012.739676
Nicolescu B (1996) La transdiscilplinariedad. Manifiesto. Ediciones Du Rocher. Traducción al español
Núñez D, Nahuelhual L, Oyarzún C (2006) Forests and water: the value of native temperate forests in supplying water for human consumption. Ecol Econ 58(3):606–616. https://doi.org/10.1016/j.ecolecon.2005.08.010
Otero L (2006) La huella del fuego: historia de los bosques nativos; poblamiento y cambios en el paisaje del sur de Chile. Pehuén Ed, Santiago
Pacheco G, Henríquez C (2016a) El turismo de base comunitaria y los procesos de gobernanza en la comuna de Panguipulli, sur de Chile. Gestión Turística, 25, enero-junio, pp 42–62
Pacheco G, Henríquez C (2016b) Plan de Gestión Territorial de Llancahue. Un camino hacia el ecodesarrollo a través de la diversificación productiva. Informe oficial, Núcleo Teses, UACh. UACh, Valdivia, p 32
Pauchard A, Villarroel P (2002) Protected areas in Chile: history, current status, and challenges. Nat Areas J 22(4):318–330
Porto-Gonçalves CW, Leff E (2015) Political ecology in Latin America: the social re-appropriation of nature, the reinvention of territories and the construction of an environmental rationality. Desenvolvimento E Meio Ambiente 35:65–88. https://doi.org/10.5380/dma.v35i0.43543
Reyes R, Nelson H (2014) A tale of two forests: why forests and forest conflicts are both growing in Chile. Int For Rev 16(4):379–388. https://doi.org/10.1505/146554814813484121
Rister J, Ares P (2013) Manual de mapeo colectivo: recursos cartográficos críticos para procesos territoriales de creación colaborativa. Tínta Limón, Buenos Aires. Retrieved from https://issuu.com/iconoclasistas/docs/manual_de_mapeo_2013?reader3=1
Sampaio C (2005) Turismo como Fenómeno Humano, princípios para se pensar a ecossocioeconomía. EDUNISC, Santa Cruz do Sul
Sletto B, Bryan J, Torrado M et al (2013) Territorialidad, mapeo participativo y política sobre los recursos naturales: la experiencia de América Latina. Cuadernos de Geografía 22(2):193–209
Universidad Austral de Chile (UACh) (2017) Informe de gestión Reserva Llancahue 2008–2016.. Internal Report, p. 33. UACh, Valdivia
Vásquez P (2014) Caracterización social sector Llancahue y Lomas del Sol. Independent report. Temuco. 27 p. Retrieved from: https://www.google.com/url?sa=t&rct=j&q=&esrc=s&source=web&cd=1&cad=rja&uact=8&ved=2ahUKEwj8qK-tz9beAhVPzBoKHQk1BoMQFjAAegQICBAC&url=http%3A%2F%2Fwww.teses.cl%2Fteses%2Fwp-content%2Fuploads%2F2016%2F05%2FVasquez-P-2014-Caracterizacion-Social-Llancahue-y-LS.pdf&usg=AOvVaw2wnCsnbqUXO6RtYJ5mQ15R

Chapter 7
Land Tenure Insecurity and Forest Conservation in Chile: The Case of the Mapuche Huilliche Indigenous Communities in the Coastal Range Rainforests of Mapu Lahual

Manuel von der Mühlen, José Aylwin, Teodoro Kausel, and Felix Fuders

7.1 Introduction

Today, many indigenous communities live in a situation of land tenure insecurity (FAO 2009). Without secure land rights they are exposed to involuntary land loss and can simultaneously become victims to the deterioration of their natural habitat through large-scale resource extraction or development projects (Agyepong 2013; DFID 2009). Customary land rights and community conservation areas could protect indigenous peoples and local communities from potential land grabbing and protect the natural environment (Almeida 2015). This research argues that lack of recognition of customary rights to land and limited recognition of territories conserved by indigenous peoples and local communities, known under their acronym 'ICCA', is one of the main causes of land tenure insecurity.[1] Figure 7.1 displays the results of the problem analysis in the form of a problem tree. The results are based on the case study of the research, which took place in the territory of the Mapuche Huilliche indigenous communities of Mapu Lahual, Osorno Province, Los Lagos Region in southerns Chile. These communities live in a situation of acute land tenure insecurity because many people lack formal land titles (McAlpin 2004). However, since the late 1990s, several development interventions have taken place

[1] ICCA: Indigenous community conservation areas.

M. von der Mühlen (✉)
Programme Department, Plan International Switzerland, Zürich, Switzerland

J. Aylwin
Observatorio Ciudadano, Temuco, Araucania, Chile

T. Kausel · F. Fuders
Economics Institute, Faculty of Economics and Administration, Universidad Austral de Chile, Valdivia, Chile

F. Fuders and P. J. Donoso (eds.), *Ecological Economic and Socio Ecological Strategies for Forest Conservation*, https://doi.org/10.1007/978-3-030-35379-7_7

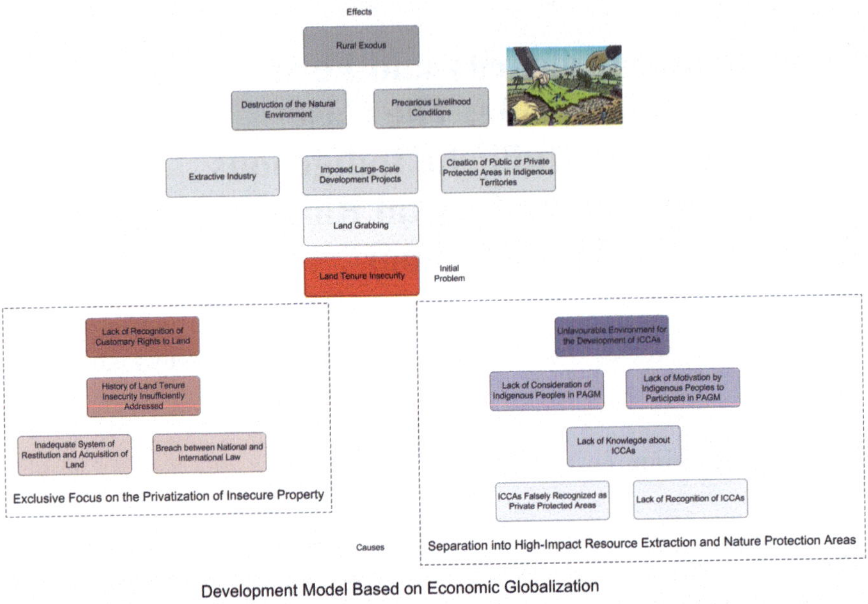

Fig. 7.1 Problem tree. (Source: own elaboration)

to strengthen their land tenure security and facilitate the creation of community conservation areas (McAlpin 2004). The case study region covers an area of approximately 60,000 hectares (ha). The communities are located at the coastal edge of the three communes San Juan de la Costa, Río Negro, and Purranque (Fig. 7.2; Maggi 2012).

7.2 Material and Methods

The research uses a mixed-methods approach combining two qualitative research methods: literature review and 13 expert interviews. The interview partners were selected based on non-probability purposive sampling. An interview guide was developed to facilitate the flow of the interview as proposed by Henning et al. (2011). The questions of the follow-up interview were always refined based on the answers obtained in the previous interview. This is a method that allows the researcher to draw inductive inferences (Henning et al. 2011). The interviews lasted 45 min on average. All interviews were recorded and transcribed. Interpretation of the findings was based on content analysis[2] and treated confidentially. While several

[2]*Content analysis* is used to capture the meaning contained within text and talk, see Denscombe (2012).

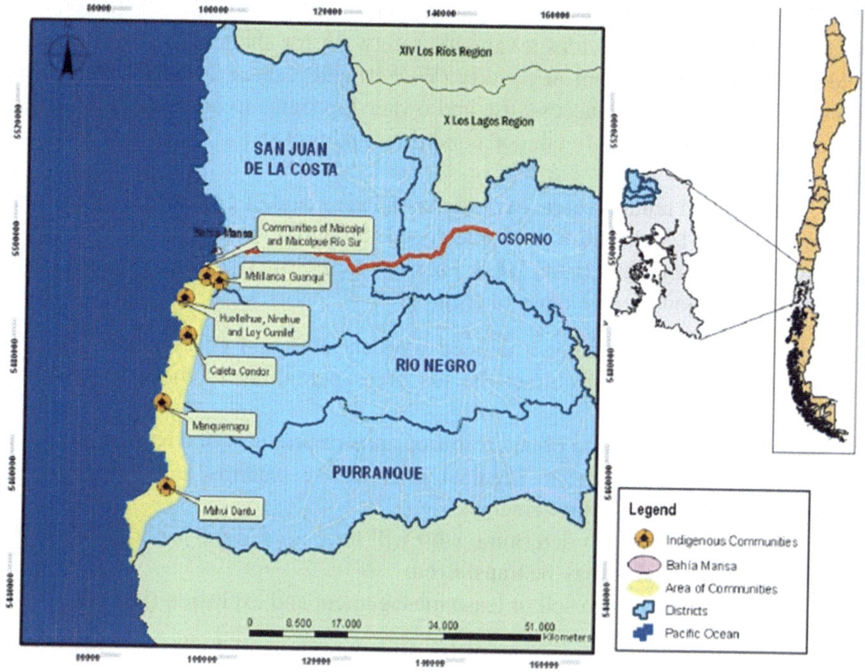

Fig. 7.2 Geographic location of Mapu Lahual. (Source: Maggi 2012)

studies have addressed the relationship between land titles, indigenous peoples, and protected area governance, few studies have analysed this relationship in detail (Mathiesen 1998; Ovideo (n.d.); Secretariat of the CBD 2012). This study addresses these issues.

The main research question was: What are the underlying reasons and possible solutions to land tenure insecurity of indigenous peoples and local communities? In this context, the following sub-questions were answered: (1) What are the consequences of land tenure insecurity? (2) What are the implications for other indigenous peoples and local communities in southern Chile?

7.3 Theoretical Framework

7.3.1 Insecure Land Tenure

In Chile, approximately 3% of the national territory is legally recognised to indigenous peoples (RRI 2015). However, this is not Chilean-specific. Around one quarter of the world population does not have secure access to land (Wickeri and Kalhan 2010). Insecure access to land and natural resources often is associated with extreme hunger, poverty, and increased inequality (FAO 2002, 2012). In contrast, secure access to land provides a valuable safety net for shelter, food and income, and can help lift people out of poverty (Wickeri and Kalhan 2010). Providing the poor and vulnerable with access, use and ownership rights to land and resources is key to achieve sustainable livelihood conditions and protect the natural environment (FAO 2012).

The term 'land tenure' refers to property rights to land of groups and individuals (Kuhnen 1982). Land tenure is characterised through a 'bundle of rights', rather than a single right to land (FAO 2002; Feeny et al. 1990). The literature commonly identifies the following five types of tenure[3]:

1. *Access:* The right to enter a defined physical area and enjoy non-subtractive benefits. The right to subtract the resources pertaining to these lands is not guaranteed;
2. *Withdrawal:* The right to obtain resource units or products of a resource system;
3. *Management:* The right to regulate internal use patterns and transform the resource by making improvements;
4. *Exclusion:* The right to determine who will have access and withdrawal rights, and how those rights may be transferred;
5. *Alienation:* The right to sell or lease management and exclusion rights.

Land rights are held under one of the following four globally recognised land tenure regimes[4]:

1. *State property regime:* The property rights are held by some authority in the public sector, but can partly be transferred to individuals (e.g., by leaseholds or concessions);
2. *Individual property regime:* The property rights are held by an individual or legal body, but can partly be restricted by the state;
3. *Common property regime:* The property rights are held by the community. Members can use the commons within a demarcated territory independently, based on strict rules and procedures. Non-members are excluded;
4. *Open access:* Property rights are not assigned. Access is unregulated. Common property regimes are sometimes mistakenly treated as open access, if the former is not adequately recognised in the national legal framework.

Insecure property rights to land can have devastating impacts on the lives of indigenous peoples and local communities. In the worst case, they can lead to land grabbing and its various negatively associated consequences (Wehrmann 2008, 2011), including uncontrolled deforestation and a lack of care for the surrounding environment. Land grabbing is defined as *'governments and private investors [. . .] securing large tracts of agricultural land [. . .] by means of long-term lease or purchase agreements'* (Foljanty and Wagner 2009). Land is under pressure for

[3] *Source:* Ostrom and Hess (2007).

[4] *Source:* Davy (2009), GTZ (1998), Heller (1997), Wehrmann (2011).

a variety of reasons. These include, but are not limited to tourism development, conservation, and food and energy security (FAO 2012). In theory, land grabbing takes place on lands of open access. In practice, however, there is hardly any pure open access left (Wehrmann 2011). Nevertheless, land treated as open access can be taken away from indigenous communities in situations where law enforcement mechanisms are weak (ILC 2011b, 2013).

7.3.2 Local Community Conservation Areas

There are around 6000 indigenous peoples comprising 370 million individuals in over 90 countries worldwide (FAO 2009; ILC 2013). Although indigenous peoples comprise just 5% of the world population, they represent 15% of its global poor (FAO 2009). Many indigenous peoples live in remote areas, often on unproductive soil and isolated from the rest of the country (FAO 2009). Indigenous peoples have a "totally distinct" relationship to land than non-indigenous peoples (Interview Partner 5, 2016).[5] According to Martinez Cobo, former UN Special Rapporteur on the Rights of Indigenous People:

> It is essential to know and understand the deeply spiritual special relationship between indigenous peoples and their ancestral land as basic to their existence as such and to all their beliefs, customs, traditions and culture [. . .]. For such people, the land is not merely a possession and a means of production [. . .]. Their land is not a commodity which can be acquired, but a material element to be enjoyed freely (Cobo in: Aylwin 1999: 4).

It is noteworthy that indigenous peoples' entitlement to land is widely recognised in international law. The two most well-known and important legal documents are:

1. Convention 169 Concerning Indigenous and Tribal Peoples in Independent Countries (ILO Convention 169, 1989); it is legally binding on those member states that have signed it.
2. The UN Declaration on the Rights of Indigenous Peoples (UNDRIP 2007); it is of non-binding character, but nevertheless has become an important reference document.

The International Union for Conservation of Nature (IUCN) defines ICCAs as "natural and/or modified ecosystems, containing significant biodiversity values, ecological benefits and cultural values, voluntarily conserved by indigenous peoples and local communities, through customary laws or other effective means" (Secretariat of the CBD 2012). The following three characteristics distinguish ICCAs from other forms of conservation (Almeida 2015; Jonas et al. 2012; Secretariat of the CBD 2012):

---

[5] *Original words:* "Es una relación cultural totalmente distinta con la naturaleza".

1. The people or community is closely connected to a well-defined territory, area, or species' habitat;
2. The people or community is the major-decision-maker and implementer of the site and/or species. This implies that community institutions are entitled with the capacity to develop and enforce regulations. This role can be *de jure* or de facto;
3. The management decisions by the people or community lead to the conservation of the territory, area or species' habitat and associated cultural values, even if the primary objective of management is not conservation per se.

The idea behind ICCAs is to provide "ownership and use with little human impact" (Interview Partner 4A, 2016).[6] The core idea is for these territories to re-gain the status of an autonomously governed place. Therefore, ICCAs are never only a conservation initiative, but instead always must be seen in the bigger picture of territorial autonomy (Holmes 2013; McAlpin 2004). ICCAs need to be recognised simultaneously to customary rights to land (Almeida 2015; Stevens 2010). The recognition of ICCAs can facilitate the regularisation of indigenous lands, while the legalisation of customary rights to land can facilitate the creation of ICCAs (Almeida 2015; Jonas et al. 2012).

ICCAs should not be over-romanticised as territories of communities that function in perfect harmony (Assies 2009). There is a general tendency to overlook internal conflicts taking place within and between communities. Even if these territories become officially recognised, there needs to be some form of regulation to decrease internal conflicts (Assies 2009). Not all indigenous communities are in favour of community conservation. Some communities prefer to exploit their resources in a short-term profitable way rather than preserve them sustainably over the long term (Interview Partner 2, 2016). However, research suggests that where customary rights are recognised and the degree of autonomy is high, community conservation functions well (Almeida 2015; Benda-Beckmann 2001).[7]

7.4 Results

7.4.1 The Underlying Reasons of Land Tenure Insecurity

7.4.1.1 History of the Legal Recognition of Land Titles

The Mapuche constitute by far the largest group among the indigenous peoples in Chile. Today, they represent an economically vulnerable group of society, which is also the result of discrimination and land loss (Aylwin 2006; Aylwin et al. 2013a). '*Mapu*' means 'land' and '*che*' means 'people'. Hence, '*Mapuche*' translates into

[6] *Original words:* "[D]e poca escala, de poco impacto".

[7] However, private ownership does not necessarily guarantee the protection or a sustainable use of natural resources, see Chap. 2 of this book.

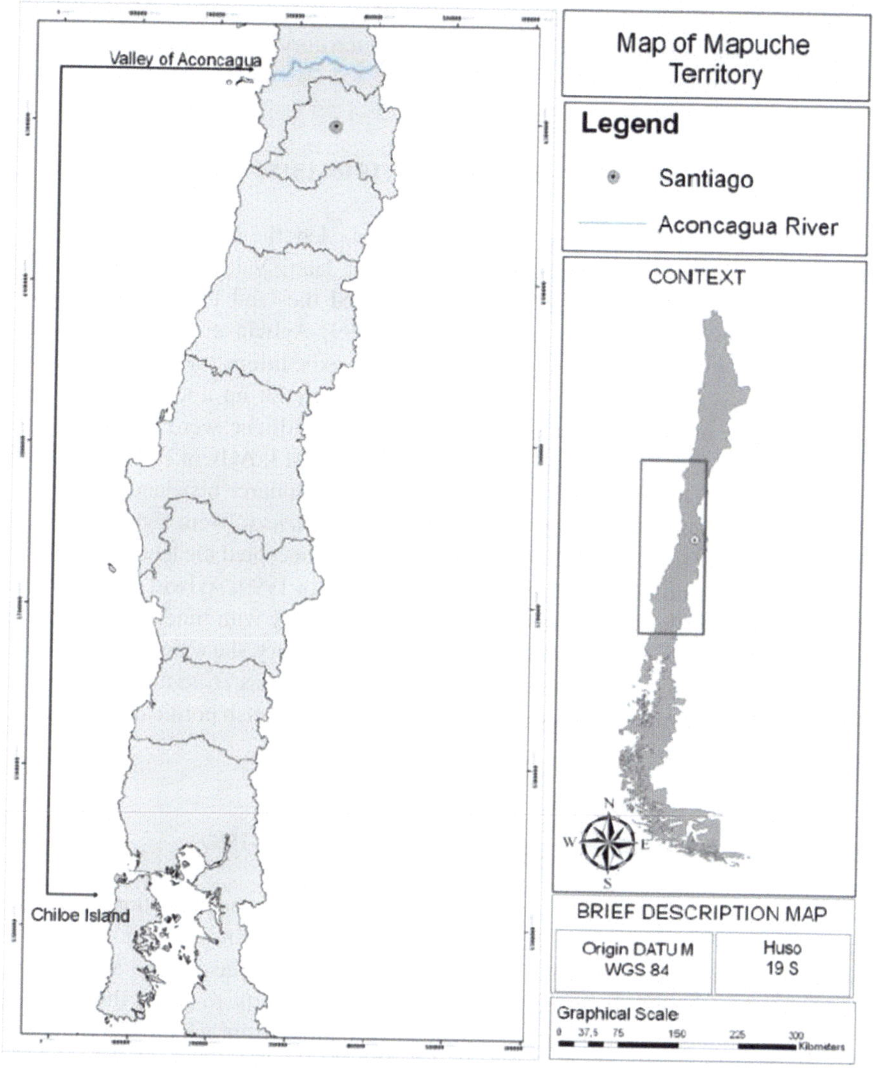

Fig. 7.3 The Ancestral Mapuche Territory. (Source: Maggi 2012)

'people of the land' (Llancaqueo 2005). The Mapuche are composed of several different groups. 'Huilliche' is not an ethnic, but a geographic determination (CVHNT 2008). The *Huilliche* are known as the 'people from the south' (Azócar et al. 2005; Llancaqueo 2005; Sallés et al. 2012). Figure 7.3 shows the ancestral Mapuche territory. Before the arrival of the Spanish, the territory stretched from the valley of Aconcagua above Santiago de Chile all the way to the archipelago of Chiloe. The Huilliche territory (marked by the pin) is located in the south and known as *Futahuillimapu*, which means 'great territory of the south' (CVHNT 2008; Mathiesen 1998).

7.4.1.2 The Colonial Land Tenure Regime (1540–1818)

In 1540, the Spanish explorer Pedro de Valdivia led the first expedition on a quest for gold to what is today Chile. One year later, Valdivia established the City of Santiago. The Spanish colonisers declared the land to be *terra nullius*, which is Latin for 'nobody's land' (Aylwin 1999; Aylwin et al. 2013a). With the dual goal to compensate its soldiers for their expeditions and to establish an economic base for its new colony, the Spanish crown set up a system of forced labour, known as *encomienda* (Aylwin 2006). The Huilliche were displaced into different encomiendas across the country (Agyepong 2013; Aylwin 1999; Aylwin et al. 2013a). However, the Spanish were encountered with fierce resistance for nearly 300 years. Several peace agreements, known as *parlamentos*, were signed between the Mapuche and the Spanish crown. The parlamentos accepted the land south of the BíoBío river as autonomous Mapuche territory (Aylwin 1999; Aylwin et al. 2013a). These agreements had the status of an international treaty with binding obligations (Aylwin 1999; Aylwin et al. 2013). In the Huilliche territory, the ultimate defeat was eventually besieged with an agreement that became known as *Tratado de Paz de Las Canoas*. This agreement subjected the Huilliche under Spanish political and judicial control (CVHNT 2008).

7.4.1.3 The Post-Colonial Land Tenure Regime (1818–1989)

In 1818, Chile became officially recognised as an independent nation. Strongly influenced by the liberal notion of the French and American Revolution, the first president of the young Chilean Republic, Bernardo O'Higgins, issued a Decree in 1819 that proclaimed the principles of freedom and liberty to guide the nation's policies. Although it was originally intended as a policy from which everyone was to benefit, this Decree established the basis for land laws and policies in the years to come, which had negative impacts on indigenous peoples (Aylwin 1999; Aylwin et al. 2013a). Because indigenous land was no longer treated differently, the Decree of 1819 was used by non-Mapuche to purchase Mapuche land in the Huilliche territory through contracts that the latter did not properly understand (CVHNT 2008). Instead of becoming free and equal citizens, as predicted by O'Higgins, the Huilliche became the victims of scrupulous farmers and soldiers of the settler nations (Aylwin 1999; Aylwin et al. 2013a). In order to limit the steady increase in unregulated access of private individuals into Huilliche territory, the "young Chilean Republic handed out titles of communal property to the Huilliche called

Títulos de Comisario" (Interview Partner 8, 2016).[8] The titles were handed out in the period between 1824 and 1848 adding legal ownership to material possession for the Huilliche (CVHNT 2008; Egaña Rodríguez 2008).

About a decade later, in the Araucania territory, the only independent Mapuche territory left, under the euphemist name 'Pacification of Araucania'[9] (1861–1883) the Chilean army annexed the territory with the intention to assimilate the Mapuche into mainstream society (Aylwin 2006; Hager 2013). The Mapuche were moved to so-called *reducciones,* land parcels provided by the Chilean government. Land titles were handed over to the head of Mapuche families (Maggi 2012; Miller et al. 2010). These land titles were called *Títulos de Merced* or 'titles of mercy.' 2.918 land titles were handed over to Mapuche families in total, displacing almost 83,000 individuals in the period between 1884 and 1929 (Aylwin et al., 2013b; Interview Partner 12, 2016). The land allocated to the Mapuche through the 'reducciones' was reduced to approximately 6% of its original size (approximately 510,386 ha) (Aylwin 1999; Aylwin et al. 2013a).

During the government of Salvador Allende (1970–1973), an agrarian reform process was initiated. Between 1964 and 1973, the Mapuche belonged to those groups of society who benefited from large-scale redistribution of farms larger than 80,000 ha. In the territory of the Huilliche, approximately 5.394 ha of indigenous lands were recovered through the process of the agrarian reform (CVHNT 2008). The period of expropriation ended with the *coup d'état* by the military regime on September 11th, 1973.

General Augusto Pinochet (1973–1989) introduced an agrarian counter reform. The previously expropriated lands were either returned to private landholders or transferred to the state and made available for purchase and investment (CVHNT 2008). In this sense, "the process of agrarian reform never existed in Chile because [during the counter reform] the advances that had been developed [by the Allende administration] were taken away" (Interview Partner 12, 2016).[10] The 'reducción' system was ended and any person still occupying a 'reducción', whether Mapuche or not, could from now on subdivide his or her land. The 'reducciones' were divided into a total of 72,068 individual plots, equalling approximately 463,409 ha (Aylwin 2006). The counter reform was part of a process privatizing the national economy, deregulating the market, and facilitating access to lands by foreign investors (Hager 2013; Holmes 2014; Murray 2002). Based on this model of economic globalization, a process of legal and policy reforms was initiated (Murray 2002). By the end of the 1970s, Chile had become the most open free-market economy in the world (Murray 2002).

[8] *Original words:* "[L]a joven república va a entregar títulos de propiedad comunitaria a los indígenas Huilliche en este caso que se van a llamar 'Títulos de Comisario.'"

[9] Spanish: *Pacificación de la Araucanía.*

[10] *Original words:* "[E]ntonces, básicamente el proceso de reforma agraria en Chile nunca existió porque se robó los avances que se desarrollaban en este momento."

7.4.1.4 The National Legal Framework Governing Indigenous Peoples Today

Since the return to democracy in 1990 the situation of indigenous peoples has improved, although a lot remains to be done.[11] The Constitution guarantees fundamental human rights. However, as it was implemented during the military dictatorship (1973–1990), indigenous peoples are omitted from its content as a specific group. Law No. 19.253 of 1993 on the 'encouragement, protection and development of indigenous peoples', better known as 'Indigenous Law' as well as Law No. 20.249 of 2008, which 'creates the marine and coastal spaces of native peoples', better known as the 'Lafquenche Law',[12] are the two principal ordinary laws governing indigenous peoples' rights to terrestrial and marine and coastal space in Chile today. These laws apply to all indigenous peoples of Chile.

As specified by Article 5 (2) of the Constitution (1980), "it is the duty of the organs of the state to respect and promote the rights guaranteed by the Constitution, and by the international conventions which have been ratified by Chile and which are in force".[13] Hence, international conventions, from their moment of ratification, assume constitutional hierarchy in the national legal framework (Núñez 2015).

7.4.2 The Nature and Scope of Indigenous Peoples' Rights to Land

This section puts indigenous peoples' rights to land in Chile into separate categories. However, the categories sometimes overlap, the same law and article mentioned in one category may re-appear in another.

7.4.2.1 Category I: The Concept of Indigenous Peoples' Rights to Land, Territories and Natural Resources

The Indigenous Law (1993) recognises the special relationship of indigenous peoples to their ancestral lands. It recognises that "land is the principal fundament of their existence and culture" (Article 1). The Indigenous Law recognises indigenous lands as those that indigenous peoples "currently occupy in ownership or posses-

[11] *Source:* Agyepong (2013), Gómez (2010), Moeckli et al. (2013).

[12] This law is known as 'Lafquenche Law' because it was greatly influenced by this people. It supports the protection of indigenous peoples' marine ecosystems "[f]or the exploitation and conservation of marine and coastal resources" Interview Partner 4A (2016).

[13] *Original text:* "El ejercicio de la soberanía reconoce como limitación el respeto a los derechos esenciales que emanan de la naturaleza humana. Es deber de los órganos del Estado respetar y promover tales derechos, garantizados por esta Constitución, así como por los tratados internacionales ratificados por Chile y que se encuentren vigentes".

sion" stemming from titles recognised by the state, those they have historically occupied and possessed, and which are recognised in the land registry, those that in the future will be declared by the courts as belonging to them, and those bestowed upon them by mechanisms established by the state (Article 12). All the lands referred to in Article 12 are transcribed in a land registry that has been created to administer indigenous lands and that is operated and maintained by the state agency for indigenous matters, *Corporación Nacional de Desarrollo Indígena* or CONADI (Article 15). The land registry provides legal recognition to these lands.

7.4.2.2 Category II: The Right to Ownership and Possession of Land

Indigenous lands recognised by the Indigenous Law (1993) (Article 12) and inscribed in the land registry (Article 15) are those to which indigenous peoples have the right to ownership. To actively promote this right, a Land and Water Fund, administrated by CONADI, has been established with the following objectives: (i) To grant subsidies for the acquisition of lands by indigenous peoples where the surface of their lands is considered insufficient, (ii) to finance mechanisms that contribute to the solution of land conflicts, and (iii) to finance the regulation and purchase of water rights, or projects that are aimed at obtaining access to water (Article 20).

A shortcoming is the separate recognition of land and the resources pertaining to these lands. As Aylwin (1999) states, "the acknowledgment and protection of indigenous rights over the natural resources existing within their lands" has been very complex. With the exception of water rights granted to the Andean peoples, the Chilean state has a monopoly on subsurface resources (Agyepong 2013). In contrast to the Chilean Constitution (1980), the ILO Convention 169 (Article 15) does recognise the use and administration of natural resources to indigenous peoples' lands.

7.4.2.3 Category III: The Right to Protection of Land and Land Rights under Customary Tenure Systems

Customary rights are guaranteed in Article 54 of the Indigenous Law (1993), "as long as it is not incompatible with the Constitution of the Republic".[14] Furthermore, as stated in Article 18 of the same law, "the succession of indigenous lands for individuals is subject to customary law, with the limitations set forth in this law, and

[14]*Original text:* '[L]a costumbre hecha valer en juicio entre indígenas pertenecientes a una misma etnia, constituirá derecho, siempre que no sea incompatible con la Constitución Política de la República. En lo penal se la considerará cuando ello pudiere servir.'

by custom itself".[15] The recognition of customary law further contains provisions on specific groups of indigenous peoples in Chile. Customary rights to land are political rights recognised to indigenous peoples by the state of Chile today (Aylwin et al. 2013a). However, the application of this law is limited. The state's policies are sometimes contradictory in matters related to indigenous peoples. In environmental conflicts over Mapuche lands, sectoral laws are often given preference (Aylwin 2008; Skjævestad 2010). As a result, Chile pursues a state policy with the dual objective to protect indigenous peoples' rights on the one hand, while pursuing export-oriented economic development on the other hand (Aylwin 2006, 2008).

7.4.2.4 Category IV: The Right to Protection Against the Loss of Land (I.E. Division, Transfer, Land Rent, Restitution and Acquisition)

The Indigenous Law (1993) was implemented with the objective to end the division of indigenous land and to prevent further land loss (Aylwin 1999; Aylwin et al. 2013a). To put this objective into action, the law establishes that the remaining indigenous lands, both individual and collective properties, be exempted from tax payments. Furthermore, the law establishes that due to 'national interest', lands belonging to indigenous communities may not be sold or leased to non-indigenous peoples and, in the case of individual plots, that lease be restricted to a period of 5 years (Article 13). Furthermore, lands may only be exchanged between indigenous and non-indigenous peoples, if these lands are of similar value and authorized by CONADI, as inscribed in the land registry (Article 15).

The rights laid out in Article 13 protect indigenous peoples from displacement. However, the Constitution (1980) mandates the state to expropriate lands in the name of public interest. The market-oriented land policy has benefited several indigenous families in the past. Between 1995 and 2014, approximately 722.167 ha have been distributed to indigenous peoples, the vast majority of them Mapuche (INDH 2014). However, most of the lands purchased by CONADI were those that had in the past been acknowledged to indigenous peoples before they had been expropriated by non-indigenous peoples. Therefore, these purchases have so far mainly addressed restitution, and not acquisition (Agyepong 2013).

The policy to restitute 'lost' and acquire 'new' lands is based on a marked oriented approach to development (Baranyi et al. 2004). It is beneficial to export-oriented, large-scale farms and has led to the development of an enormous forestry industry. However, it can also exacerbate the situation of indigenous communities (Aylwin 2006; Holmes 2014). Although some advances have been made concerning the restitution of indigenous territories, all expert interviewees who contributed

[15] *Original text:* '[L]a sucesión de las tierras indígenas individuales se sujetará a las normas del derecho común, con las limitaciones establecidas en esta ley, y la de las tierras indígenas comunitarias a la costumbre que cada etnia tenga en materia de herencia, y en subsidio por la ley común.'

to this research, agreed that the current system has become very expensive and inefficient. "CONADI has many individuals and communities, who apply for land, and it has very few resources" (Interview Partner 7, 2016).[16] This fostered speculation and created a market of private property holders on paper (Interview Partner 7, 04.04.2016), who wait for the highest price until they sell 'their' land (Interview Partner 12, 2016; Interview Partner 7, 2016; Interview Partner 9, 2016).

7.4.3 Governance of Protected Areas in Chile

Chile is home to a rich level of biodiversity, natural resources, and endemism thanks to its unique flora and fauna. This is a result of the country's geography, which stretches more than 4000 km from north to south, and is home to deserts, tropical rainforests, lakes and rivers, as well as the Arctic ice (Arce and Aylwin 2012a, b). The concept of the Yellowstone National Park in the USA has inspired the creation of public protected areas throughout the twentieth century in Chile (Arce and Aylwin 2012a, b). There are different estimations about the extent to which protected areas created by the state overlap with indigenous lands; estimates are as high as 90% (Arce and Aylwin 2012b). It is further estimated that out of the current 100 protected areas in Chile, 21 are established on lands where indigenous peoples are present; excluding those lands to which they may claim rights (Arce et al. 2016). There is a lack of awareness about community conservation areas (Pauchard and Villarroel, 2002). Although in recent years an increased level of awareness about indigenous conservation practices can be notified, "it is still not a highly discussed topic" (Interview Partner 1, 2016).[17]

7.4.4 The Consequences of Land Tenure Insecurity: Uncontrolled Resource Extraction and Social Costs

7.4.4.1 Resource Extraction

In Chile, large tracts of land are secured for the extraction of resources (Aylwin 2008; ILC 2011a). While mining is geographically localizable to northern and central Chile, in southern Chile the extractive industry mainly consists of forest plantations and salmon farms (Hager 2013; ILC 2011a). Approximately 15,637,233 ha of the country's terrestrial surface is covered by forests, which corresponds to an area of 21.7% of the national territory (Aylwin et al. 2013b).

[16] *Original words:* "[H]ay muchas comunidades que solicitan tierra a CONADI. CONADI tiene muchas comunidades que solicitan tierra y tiene poco recursos."

[17] *Original words:* "[N]o es un tema muy discutido todavía en Chile."

Approximately 13,430,603 ha (85.9%) of this number is native forest, while the remaining approximately 2.7 million ha are planted forests. These plantations have primarily been set up in traditional Mapuche lands (Aylwin 2008; Aylwin et al. 2013b). The expansion of the forest industry began in the 1970s with the enactment of Decree 701 (1974) enabling large-scale forestry throughout the country (Aylwin 2006; see also Chap. 5). The salmon industry began its work as early as in the 1920s and has since then become an important export-oriented economic branch of the country. Today, Chile is the second largest salmon exporter in the world behind Norway (Hager 2013). There are also many examples of large-scale development projects extending into Mapuche territory in southern Chile. Today, the biggest so-called 'megaproject' in the traditional Huilliche territory is the installation of hydroelectric dams and hydropower stations, and, to a lesser extent, highway extensions (Interview Partner 12, 2016; Susskind et al. 2014).

7.4.4.2 Land Acquisition of Forest Areas

There are several examples of land acquisitions for nature conservation in Chile. Over the last 150 years –facilitated through policies that have enabled an environment favourable to private investments– powerful private actors have acquired lands for conservation. Today, the intensity of these types of land purchases is increasing. There is a tendency to regard private protected areas as a counter movement to protect the natural environment. However, some see them as part of the incorporation of nature (Holmes 2013, 2014); and within them a hidden form of land grabbing. While land acquisitions for nature conservation are not detrimental to the natural environment, they nevertheless undermine local sovereignty. Therefore, land grabs for the conservation of nature, also known as 'green grabs', can potentially lead to the displacement of local communities. For instance, the US entrepreneur Douglas Tompkins (1943–2015) has become a controversial figure in Chilean conservation efforts. On the one hand, he has contributed to the conservation of large forest areas. On the other hand, he has also been accused of displacing smallholder farmers and indigenous peoples for the creation of the Pumalin National Park (Holmes 2013, 2014). Private conservation areas can lead to loss of livelihood opportunities and, in response, intensified out-migration (rural exodus) (Aylwin 2008; Interview Partner 12, 2016).

7.4.4.3 Social Costs

The social costs attributed to insecure land tenure in southern Chile are high. There are many conflicts over different types of resources involving the public and private sector and local communities. One type of conflict concerns the advance of state or private-owned forest companies on the one hand and indigenous communities on the other hand (Aylwin and Cuadra 2011; Carruthers and Rodriguez 2009). Another type of conflict concerns the installation of the salmon industry in traditional Huilliche lands. Local communities claim that the excessive use of antibiotics by this industry has reduced the wildlife fish stock and put their main food source at risk (Hager 2013). Yet another type of conflict concerns large-scale hydropower dams (Aylwin 2006; Hager 2013; Susskind et al. 2014). Opponents accuse companies of lack of adequate consultation, lack of participation, and lack of compensation for the damages caused (Susskind et al. 2014). An example of a hydropower conflict that has created a substantial amount of controversy is the Ralco hydroelectric dam, which has led to the displacement of 675 people, including 500 Mapuche Pehuenche (Susskind et al. 2014).

7.4.5 *Possible Solutions to Land Tenure Insecurity: The Case of Mapu Lahual*

7.4.5.1 Stakeholders Involved in Mapu Lahual

Several stakeholders have become involved in development interventions in Mapu Lahual. They have worked together with the communities to implement several projects. The stakeholders have been divided into four groups: (i) public institutions, (ii) indigenous communities' representatives, (iii) private sector, and (iv) external organisations. The results are based on the respective literature on the case study region, complemented by information obtained from the expert interviews.[18] Altogether, 20 different stakeholders could be identified from the public and private sector, including external organisations, and the communities. Table 7.1 provides detailed information about each individual stakeholder.

Based on the information obtained through the stakeholder matrix, Fig. 7.4 was developed. It visualizes the relationships among the various stakeholders. Relationships are described 'among' the four groups and 'within' them. Stakeholders relate to one another through dependence (or influence), cooperation or potential cooperation, and conflict (see legend). Two stakeholders can share more than one form of relationship. The relations are exemplified through the example of CONAF, one of the most important stakeholders.

[18] *Source:* CONAF 2001; Correa et al. 2002; Pauchard and Villarroel 2002; McAlpin 2004; Mapu Lahual 2006; Molina et al. 2006; Guala and Szmulewicz 2007; Arce 2010, 2011; Maggi 2012; Montenegro and Aldo 2012; Holmes 2013; Ancapan et al. 2014; Interview Partner 1 (2016); Interview Partner 5 (2016); Interview Partner 7 (2016); Interview Partner 8 (2016); Interview Partner 9 (2016); Interview Partner 10 (2016); Interview Partner 11 (2016); Interview Partner 12 (2016).

Table 7.1 Stakeholder Matrix for Mapu Lahual

	Stakeholder	Characteristics	Interests	Challenges	Planning implications
Public institutions	Corporación Nacional Forestal (CONAF	National forestry agency; First institution in Mapu Lahual	Forest conservation; Compliance with laws	Assure stable alerce resources; Compliance with Decree 490 (1976) in the absence of land titles	Authorized to deliver work and forest management plans; Authorized to fine in case of non-compliance; Technical support
	Corporación Nacional de Desarrollo Indígena (CONADI); Subsecretaria de Pesca (SUBPESCA)	CONADI: Public agency in charge of indigenous matters SUBPESCA: Public agency in charge of marine and coastal protected areas	CONADI: Land acquisitions SUBPESCA: Administration of marine and coastal protected areas	CONADI: Many claims, few resources SUBPESCA: Application of Lafquenche Law (2008)	CONADI: Land tenure security SUBPESCA: Supervision of marine and coastal protected areas
	Gobierno regional (GORE)	Regional government of Los Lagos	Integrated and balanced regional development	Revenue generation; Competing development regions	Potential source of funding
	Ministerio de Bienes Nacionales (MBN); Comisión Nacional del Medio Ambiente (CONAMA); Ministerio del Medio Ambiente (MMA)	Public agencies in charge of environmental protection	Environmental protection; Creation of a national system of protected areas (GEF-SNAP); Creation of a regional system of protected areas (GEF-SIRAP)	Land tenure insecurity	Decree 2.695 (MBN, 1979); GEF-SNAP; GEF-SIRAP
	Servicio Nacional del Turismo (SERNATUR) (potential partner); Ministerio de Obras Publicas (MOP)	Public agency in charge of tourism development; Public agency in charge of physical infrastructure	Tourism development inside protected areas; Highway extension connecting northern and southern Chile	Lack of free, prior and informed consent (FPIC) for concessions of protected areas; Highway extension through Mapu Lahual	Concessions for tourism development; Physical infrastructure

Community	Individual indigenous communities	Inhabitants of Mapu Lahual	Secure land titles; Exploitation of alerce and alternative income-generating activities	Absence of land titles	Participatory planning
	Asociación Indígena Mapu Lahual (AIML)	Executive organ of indigenous parks' network	Representative organ of the indigenous communities of Mapu Lahual	Equal representation; Internal division; Controlling mechanisms over funds	Mapu Lahual Master Plan (2006)
	Cooperativa (inactive)	Cooperative to promote community-based tourism	Development of community-based tourism	Lack of coordination and cooperation	Participatory planning

(continued)

Table 7.1 (continued)

	Stakeholder	Characteristics	Interests	Challenges	Planning implications
Private sector	WWF Chile	One of the world's biggest conservation organisation	Protection of the Valdivian ecoregion	Potentially prioritising conservation over livelihoods; Working with all communities together	Planning instruments; Handbooks and maps; Potential source of funding
	Observatorio Ciudadano	Chilean human rights organisation	Promote indigenous peoples' rights	Awareness and interest in indigenous peoples' rights	Legal council; Information hub
	ICCA consortium	Worldwide coalition of conservation organisations	Protection of the natural environment through community conservation areas	Awareness and interest in ICCAs	Information hub
	Extractive industry	Private corporations	Economic profit; National economic development	Exploitation of natural resources	Potential threat to conservation and livelihood objectives; Important for national economic development and job creation
	Individual landholders	Legal landholders, who are physically absent	Speculation in land prices; Investments in land	Potential conflicts over territory	Potential threat to conservation and livelihood objectives; Potential role as job creators

External organisations	New Zealand Aid (NZAID) (past involvement)	Development agency of New Zealand	Development of indigenous peoples (history of Māori support)	Limited financial and technical support	Experience of past projects; Potential future project partner
	Global Environmental Fund (GEF)	International fund for environmental projects	Environmental protection; Creation of a national system of protected areas (GEF-SNAP); Creation of a regional system of protected areas (GEF-SIRAP); Creation of marine and coastal protected areas (GEF-Marino)	Concerns regarding practical application (top-down, lack of cultural and social assessments)	GEF-SNAP; GEF-SIRAP; GEF Marino
	UNDP	Development agency of the United Nations (UN)	Creation of a national system of protected areas (GEF-SNAP); Creation of a regional system of protected areas (GEF-SIRAP)	Limited contextual understanding	GEF-SNAP; GEF-SIRAP

Source: own elaboration

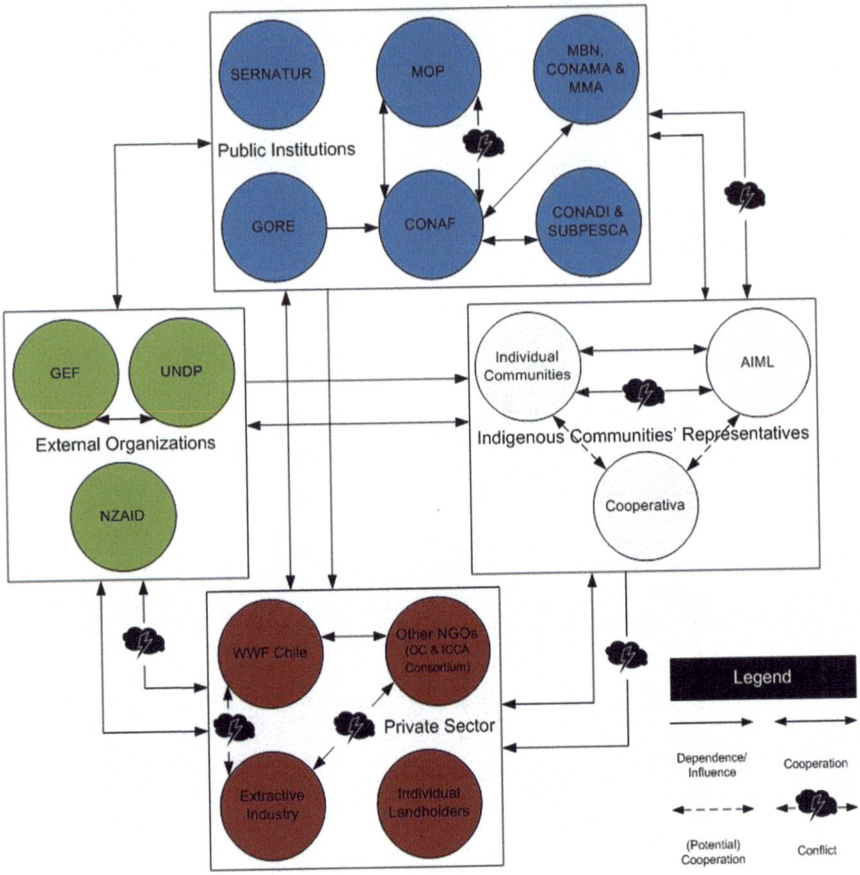

Fig. 7.4 Stakeholder map. (Source: own elaboration)

7.4.5.2 Relationships 'Among' the Stakeholder Groups

The group 'public institutions' acts in cooperation with the 'indigenous communities' representatives in form of the AIML. CONAF, CONADI, SUBPESCA, MBN, and GORE are all directly or indirectly in contact with community representatives. SERNATUR, as a potential but not yet fully integrated stakeholder, is left aside for now. There is conflict between MOP and the Huilliche communities, confirmed through the protests against the planned highway construction in the past (Correa et al. 2002). No conflict can be observed between public institutions and the private sector. The relationship is characterized through cooperation between CONAF and WWF Chile. The restitution or acquisition through CONADI (Land and Water Fund) or by means of MBN Decree 2.695 (1979) shows the engagement between the public and the private sector. The closest type of relationship is in the form of 'dependence' since CONADI is dependent on the price that the private landholders can demand according to the market value (Interview Partner 12, 2016). Public institutions cooperate with external organisations. NZAID worked closely together with CONAF in the past; and GEF and UNDP are implementing the GEF-SNAP and GEF-SIRAP projects in cooperation with the MMA today.

7.4.5.3 Relationships 'Within' the Stakeholder Groups

CONAF is in cooperation with the MBN (Decree 2.695 of 1979) and CONADI (Land and Water Fund). CONAF is financially dependent on GORE for its operations in Mapu Lahual. There is no relation with SERNATUR and only minor relations with the other public institutions now. CONAF and the AIML closely cooperate in the construction of work plans. CONAF also works closely together with WWF Chile and, recently, with GEF. All these relationships are based on close cooperation. In the same way, the relationships of the other 'among', 'within', and 'cross-sectoral' relationships can be analyzed.

7.4.5.4 The Development Interventions in Mapu Lahual

In the end, the main objective sought by these communities, with the support of some of the stakeholders involved, is to be recognised as an autonomously governed conservation area. Today, the communities have demarcated their area of influence. They have also increased the size of the territory legally recognised to them. Furthermore, parts of the marine and coastal areas of Mapu Lahual are under protection of the Lafquenche Law (2008). That means that these areas are governed by customary law of the respective communities. The communities of Mapu Lahual have gained a relatively high level of societal acceptance. It has become a *priority development zone* of the regional government (CONAF 2001; Interview Partner 10, 2016). Several successful interventions have already taken place. The individual outputs can be categorized as follows: (i) capacity building for local communities, (ii) diversified income opportunities, (iii) improved communication means, (iv) the creation of an indigenous parks network, (v) improved infrastructure, and (vi) secure land tenure. Importantly, these six categories are not isolated, but interconnected.

Category I: Capacity Building for Local Communities

An important strategy employed by the stakeholders involved in Mapu Lahual is organisational capacity building. CONAF and WWF Chile have systematically increased the governance capacity of the indigenous communities in the region (McAlpin 2004). Technical assistance was provided by CONAF through work plans, which included the delimitation of community borders. CONAF further introduced social encounters, which strengthened communication among the communities. Community members have met on a yearly basis since the first social encounter in the late 1990s (CONAF 2001; Interview Partner 10, 2016). The following is a selection of planning instruments that have been developed to train the communities.

- Alerce work plans, including delimitation of the area of influence;
- Forest management plans;
- Parks management plans;
- Community plans;
- Ecotourism business plans;
- Mapu Lahual master plan.

Three of the main stakeholders –CONAF, WWF Chile, and the AIML– have been involved in the creation of these plans, which is also a result of the consistent participatory planning processes. Under institutional assistance, the AIML created a master plan, in which it envisions the future development of the region. The master plan is guided by four pillars: (i) territory and nature, (ii) sustainable economy, (iii) culture and education, and (iv) politics and organisation (Arce 2010; Mapu Lahual 2006). Each of these pillars is either directly or indirectly linked to the creation of community conservation areas. Land title clearance is a clearly defined objective of the plan.

Category II: Diversified Income Opportunities

To strike a balance between community and conservation needs, the various initiatives that have taken place in Mapu Lahual aimed at diversifying the income of the communities to lessen the dependence on alerce of its residents (Arce, 2010; Interview Partner 5, 2016). An important form of income diversification is community-based tourism (CBT) (Guala and Szmulewicz 2007). The idea behind CBT is to restrict the number of tourists, attract environmentally responsible tourists, and implement activities that do not harm the environment (Montenegro and Aldo 2012; Rivera and Pavez 2012; Skewes et al. 2015). CBT has specifically been promoted as an alternative form of development in Mapu Lahual by WWF Chile and CONAF (Interview Partner 7, 2016; Molina and Pavez 2012). Both organisations regard CBT as an alternative form of local economic development with low environmental impacts (Ancapan et al. 2014; McAlpin 2004). CBT contributes to regional development, too. If well executed, it strikes a balance at the challenge to combine economic development and environmental protection (Interview Partner 6, 2016; Montenegro and Aldo 2012).

Category III: Improved Communication Means

Improvements in telecommunications and road connections have been pursued by NZAID, amongst others (Arce 2010, 2011; Interview Partner 5, 2016). The goal was (and remains) to facilitate access to the region and communication within and among communities as well as with places outside Mapu Lahual. Improvements in communication both through new road connections and new means of telecommunication have positive impacts on tourism development and the organisational capacity, as it improves reachability to and from the region and facilitates coordination.

Category IV: Creation of an Indigenous Parks Network

Through the foundation of the AIML, governance and management responsibility were transferred to the local level. Another major achievement of territorial development in Mapu Lahual is the indigenous parks network (RED).[19] This network comprises six interconnected parks inside the Mapu Lahual territory, which had been declared 'indigenous conservation areas' by the Mesa Hueyelhue working group (Interview Partner 10, 2016). The AIML oversees these parks. Although not legally recognised as protected areas, these parks have achieved a substantial level of social recognition (Interview Partner 10, 2016). Their existence has (i) raised awareness about the unique environmental value of Mapu Lahual, (ii) raised awareness about the existence of conservation areas under the control of indigenous communities, independent of the legal status of these areas, and (iii) attracted a significant number of tourists to the region (Interview Partner 5, 2016; Interview Partner 7, 2016; Interview Partner 10, 2016).

Category V: Improved Infrastructure

Physical planning has taken place in the past and continues to take place in Mapu Lahual today. NZAID has developed trails and paths, assisted in the construction of community houses and invested in material construction in general. These constructions have directly benefited other strategies. For example, the paths created have made it easier for tourists to reach the territory and for the local community to connect these paths to attractive tourist locations (Interview Partner 5, 2016; Interview Partner 7, 2016).

Category VI: Secure Land Tenure

Arguably the most important development strategy pursued is the clearance of land titles by means of the Land and Water Fund (CONADI) and Decree 2.695 of the MBN (1979). As an example, one can look at the community of Maicolpi. Maicolpi

[19] RED: *Red de Parques Indígenas Mapu Lahual*

theoretically has land titles of approximately 14,000 ha resulting from the Títulos de Comisario (Antriao 2009). Maicolpi was the last place where the Títulos de Comisario had been handed out (Interview Partner 11, 2016; Interview Partner 8, 2016). The Títulos de Comisario lost their legal validity, while the law "*Ley de la Propiedad Austral*" facilitated the inscription of lands to private landholders. The community of Maicolpi demands restitution of its original 14,000 ha. In 2001, CONADI bought a total of 1298 ha for the community by means of the Land and Water Fund. In 2003, the community intended to regain 2500 ha by means of Decree 2.695 (1979) (Antriao 2009).

7.4.6 *Challenges and Recommendations of the Interventions*

Despite noticeable improvements in Mapu Lahual compared to the situation two decades ago many challenges remain.

7.4.6.1 Challenge I: No Collective Action

As stated by Interview Partner 7, "there is no collective action"[20] among the communities (04.04.2016). Many communities work separately because they do not have an integral territorial vision, as presented in the region's own master plan (AIML). The lack of unity and vision among the communities can be traced back to the isolation in which they still live (Interview Partner 5, 2016). *Recommendation*: Stronger promotion of territorial development through AIML and stronger capacitation of AIML to represent community interests through the vision defined in its own regional master plan.

7.4.6.2 Challenge II: Lack of Community Representation

Some communities do not feel that their interests are adequately represented in the AIML. This is partly a legal problem because indigenous associations cannot claim political representativeness. However, it is also an internal problem. Many Huilliche feel that some families are overly represented, while others are completely left aside (Interview Partner 10, 2016; McAlpin 2004). *Recommendation*: The AIML should clearly communicate its purpose to all communities, and all communities should be equally represented.

[20]*Original words:* [No existe una acción colectiva."

7.4.6.3 Challenge III: Little Regulation of Tourism Activities

Tourism at the local level requires regulation to avoid environmental harm (Interview Partner 2, 2016; Interview Partner 6, 2016; Pauchard and Villarroel 2002). At present, some communities face the difficulty of organising a still predominantly disorganised tourism environment. Many tourists leave their garbage behind causing environmental damage (Interview Partner 10, 2016; McAlpin 2004). This counteracts the central idea of CBT. *Recommendation*: Follow guidelines on CBT and develop regulations for a well-coordinated and well-functioning CBT environment. This can be done through a capacitated AIML.

7.4.6.4 Challenge IV: Land Tenure Insecurity

The situation of land tenure security has certainly improved (more communities have legal titles to their lands than before). Nevertheless, still less than one fifth of the whole territory belongs to the community in contrast to more than four fifth, which remain in public or private hands (Arce and Aylwin 2012a, b; McAlpin 2004). Furthermore, because of the lack of territorial vision, the land that has been restituted is not guided by a communal sense of the territory 'Mapu Lahual'. Land is being subdivided by the individual communities as a result of the lack of this integral, territorial vision, which forecloses a view on the benefits of customary land rights and ICCAs (Interview Partner 8, 2016). *Recommendation*: Communicate the advantages of customary land rights and ICCAs (i.e. indigenous territory in favour of *all* communities of Mapu Lahual) and the disadvantages of land subdivision (i.e. the slow, but steady decrease in the total land size).

7.4.6.5 Challenge V: Highway Extension

The continued plans to make vast parts of the Mapu Lahual territory available for an extension of the north-south highway ('*Ruta Costera Sur*') is even more reason for development practitioners and community members to continue the ongoing process of territorial planning. *Recommendation*: In the words of Interview Partner 10: "The other person, who still has not resolved the problems, does not receive the same level of assistance. This is what we must fix. That the process continues until the last community has achieved a secured future and its territorial rights recognised".[21]

---

[21] *Original words:* "[E]l otro, que todavía no soluciona, va a seguir arreglando, pero no cuenta con toda mi fuerza del resto para poder seguir apoyando. Eso es lo que uno tiene que hacer. En que el proceso termina hasta que la última comunidad ha logrado en los futuros común que el reconocimiento de su derecho territorial."

7.5 Discussion and Conclusions

7.5.1 The Social and Environmental Consequences of Land Tenure Insecurity

The Mapuche Huilliche in Chile have experienced a well-documented history of land tenure insecurity. The Chilean government after achieving independence disregarded the peace agreements that existed between Spanish colonisers and the Mapuche, which had had the character of an international treaty at the time. The failure to respect and integrate customary land rights into the national legal framework, combined with a large-scale expansion by settler communities into indigenous territories, resulted in the loss of land for indigenous communities across the country.

Today, indigenous peoples' rights to land are firmly established by national and international law. The remaining indigenous territories are protected, and there is even a mechanism in place for indigenous peoples to reclaim some of their ancestral land. However, the non-recognition of Articles 1 and 12 of the Indigenous Law (1993) in the Constitution and the lack of application of the former still impedes land claims of indigenous communities. The ratification of ILO Convention 169 (1989) opens new possibilities to legally recognise and protect lands that have traditionally been occupied by indigenous peoples. The concept of public protected areas originated with the adoption of the Yellowstone model. The model is based on the idea to remove permanent human settlements from protected areas. In contrast, community conservation areas do not regard the two as mutually exclusive. This creates an unfavourable environment for the participation of indigenous peoples and local communities in forest conservation and the establishment of locally-managed conservation areas.

Without secure land rights, indigenous peoples and local communities live in precarious livelihood conditions. Losing access to land and resources can lead to uncontrolled resource extraction and forest degradation with negative consequences for local livelihoods and the natural environment. Finding solutions to strengthen land rights of indigenous peoples is, thus, key to protect the environment and secure livelihoods.

7.5.2 Possible Solutions to Land Tenure Insecurity

Through the adoption of the Indigenous Law (1993) and the Lafquenche Law (2008), the Chilean government has already taken steps to strengthen the land rights of indigenous peoples. Recognising customary land rights in their entirety requires recognising not only that communities use the land, but also recognising the communal authorities who define the rules and regulations to manage it. Local communities should be regarded as independent actors and complementary partners for development. To further strengthen customary land rights, ILO Convention 169 (1989) and the UNDRIP (2007) should be applied in addition to national law in cases regarding land restitution and acquisition.

Development interventions should look at the entire territory. They should be targeted to include all communities who belong to that territory, rather than at individual communities. The interventions that have taken place in Mapu Lahual, including the indigenous parks network, provide an example of such an intervention that could be replicated elsewhere in Chile and Latin America. The internationally accepted principle of *Free Prior Informed Consent* (FPIC) aims to establish consultation and participation of indigenous peoples prior to the implementation of any proposed development project in territories that are home to indigenous peoples and local communities. This principle should be upheld.

7.5.3 *Limitations*

The findings outlined in this chapter may not be generalizable to other communities in Chile. What works relatively well in Mapu Lahual may not work equally well in other places, and vice versa. Furthermore, Mapu Lahual receives a relatively large amount of public and private financial aid and technical support because it is part of the second largest region of coastal rainforests in the world and as such considered a priority conservation area. Other regions lacking this characteristic may not receive the same level of attention as Mapu Lahual.

7.6 Outlook

This study has presented a general overview of the territorial development interventions in Mapu Lahual from a legal-historical and regional planning perspective. Future research could look in more detail at individual components of this study, for instance, community-based tourism (see also Chaps. 6 and 12 in this book). Future research could also evaluate the success of the interventions that have taken place so far and conduct a comparative analysis between Mapu Lahual and other regions in Chile and Latin America. This could provide further implications about the generalisability of the findings and the replicability of the proposed development interventions in another context.

Finally, it is worth mentioning that the described land tenure conflicts would most probably be substantially attenuated if we understood that all land, as well as other natural resources, exist for everyone alike independent of their ethnic background or who occupied them first as, for example, stated by Henry George (1935). As was outlined in Chap. 3 all pure nature-made produce to which humans did not add any value –as is the case with land– should therefore equally belong to everyone; not only because of fairness reasons but also to improve allocative efficiency. In other words, no one should be able to claim private ownership to land (e.g., George 1935; Gesell 1949). A solution could be that all land is considered state land, communal land, or common land. A similar situation existed for centuries when all land belonged to the respective king and maybe a restrictive net of nobles (see Chap. 3), while buildings and agricultural or forestry products planted on it were privately owned. Following this proposal, anyone who wishes to use the land for forestry, agriculture or other motives has to lease it from the government or respective local community. Houses of natural persons and small-scale agricultural activities in their surroundings could be expected from the leasing rents.

References

Agyepong E (2013) Securing land tenure in ethnically-diverse societies: a comparative study of the laws and land policies of Chile and Ghana. Spring Res Ser 58

Almeida F (2015) Collective land tenure and community conservation: exploring the linkages between collective tenure rights and the existence and effectiveness of territories and areas conserved by indigenous peoples and local communities (ICCAs). Companion Doc Policy Brief 2:3–32

Ancapan J, Paillamanque G and Barrientos M (2014) Mapu Lahual: Territorio Indígena para la Conservación del Pueblo Originario en el Sur de Chile. Revista Parques 3:1–8

Antriao B (2009) Información sobre la Situación de las Tierras en el Territorio Huilliche. Informe Aplicación Ley Indígena, Osorno

Arce L (2010) Governance of indigenous communities conserved areas: the case of Mapu Lahual in the rainforest of Chile

Arce L (2011) Connectivity and accessibility of isolated territories: the case of Mapu Lahual in Southern Chile. Master's Thesis, Universidad Austral de Chile. Valdivia

Arce L, Aylwin J (2012a) Análisis de Derecho Internacional, Legislación Nacional, Fallos, e Instituciones al Interrelacionarse con Territorios y Áreas de Conservación de los Pueblos Indígenas y Comunidades Locales: Report No. 9. Natural Justice, Chile

Arce L, Aylwin J (2012b) Recognition and support of ICCAs in Chile. CBD Secretariat Tech Ser 64:3–25

Arce L, Aylwin J, Guerra F, et al (2016) El Estado Chileno y la Conservación de la Naturaleza Frenta a una Decisión Histórica: ¿Cómo Reconocer a los Territorios y Áreas Conservadas por Pueblos Indígenas y Comunidades Locales?

Assies W (2009) Land tenure, land law and development: some thoughts on recent debates. J Peasant Stud 36(3):573–589

Aylwin J (1999) Indigenous Peoples' rights in Chile and Canada: a comparative study. Master's Thesis, University of British Columbia. Vancouver

Aylwin J (2006) Land policy and indigenous peoples in Chile: progress and contradictions in a context of economic globalization. Dev Cooperation (ICCO) 12:14

Aylwin J (2008) Globalization and indigenous people's rights: an analysis from a Latin American perspective. Cah Dialog 1:1–37

Aylwin J and Cuadra X (2011) Los Desafíos de la Conservación en los Territorios Indígenas en Chile. Observatorio Ciudadano 14:9–78

Aylwin J, Meza-Lopehandía M, Yáñez N (2013a) Los Pueblos Indígenas y el Derecho. LOM, Santiago de Chile

Aylwin J, Yáñez N and Sánchez R (2013b) Pueblo Mapuche y Recursos Forestales en Chile: Devastación y Conservación en un Contexto de Globalización Económica. Observatorio Ciudadano

Azócar G, Sanhueza R, Aguayo M et al (2005) Conflicts for control of Mapuche-Pehuenche land and natural resources in the Biobío highlands, Chile. J Lat Am Geogr 4(2):57–76
Baranyi S, Diana C, Morales M (2004) Tierra y Desarrollo en América Latina: Perspectivas para la Investigación sobre Políticas. National University of Canada, Ottawa
Carruthers D, Rodriguez P (2009) Mapuche protest, environmental conflict and social movement linkage in Chile. Third World Q 30(4):743–760
CONAF (2001) Alerce y Huilliche: "Estado y Perspectiva"
Correa M, Catalán R and Paillamanque M (2002) Percepción de las Comunidades Huilliches sobre el Proyecto Ruta Costera Sur. Ambiente y Desarollo 18
CVHNT (2008) Informe de la Comisión Verdad Histórica y Nuevo Trato con los Pueblos Indígenas. Santiago de Chile
Davy B (2009) The poor and the land: poverty, property, planning. Town Plan Rev 80(3):227–265
Denscombe M (2012) The good research guide for small-scale social research projects. Open University Press, Berkshire
DFID (2009) Poverty and environment. Available at: https://ec.europa.eu/europeaid/sites/devco/files/methodology-dfid-guide-to-environmental-screening-200306_en_2.pdf
Egaña Rodríguez G (2008) Identidades Territoriales como Estrategias de Adaptación Cultural a la Ecología del Estuario de Choroy-Traiguén, Provincia de Osorno. Master's Thesis, Universidad de Chile. Santiago de Chile
FAO (2002) Land tenure and rural development. Land Tenure Stud 3:1–42
FAO (2009) Indigenous and tribal peoples: building on biological and cultural diversity for food and livelihood security. Available at: http://www.fao.org/3/a-i0838e.pdf
FAO (2012) On the voluntary guidelines on the responsible governance of tenure. Land Tenure J 1:1–10
Feeny D, Berkes F, McCay B et al (1990) The tragedy of the commons: twenty-two years later. Hum Ecol 18(1):1–19
Foljanty K and Wagner J (2009) Development policy stance on the topic of land-grabbing: the purchase and leasing of large areas of land in developing countries. BMZ Discourse Ser
George H (1935) Progress and poverty – an inquiry into the cause of industrial depressions and of increase of want with increase of wealth – the remedy. 50th anniversary. New York: Robert Schalkenbach Foundation
Gesell S (1949) Die natürliche Wirtschaftsordnung durch Freiland und Freigeld. Rudolf Zitzmann, Lauf
Gómez S (2010) Update: essential issues of the Chilean legal system. Available at: http://www.nyulawglobal.org/globalex/Chile1.html
GTZ (1998) Land tenure in development cooperation: guiding principles. Available at: https://www.mpl.ird.fr/crea/taller-colombia/FAO/AGLL/pdfdocs/englisch.pdf
Guala C, Szmulewicz P (2007) Evaluación de Buenas Prácticas en Servicios de Ecoturismo Comunitario en la Eco-Región Valdiviana, Chile. Gestión Turística 8:9–24
Hager K (2013) Die Mapuche in Chile: Zwischen staatlicher repression und Widerstand. Available at: https://www.gfbv.de/fileadmin/redaktion/Reporte_Memoranden/2013/Mapuche_Memorandum__November_2013_HP.pdf. Accessed 2 Dec 2015
Heller M (1997) The tragedy of the anticommons: property in the transition from Marx to markets. William Davidson Working Paper 111:1–84
Henning M, Hutter I, Bailey A (2011) Qualitative research methods. SAGE Publications, London
Holmes G (2013) Private protected areas and land grabbing in southern Chile. Available at: http://povertyandconservation.info/sites/default/files/Holmes%20-%20Private%20protected%20areas%20and%20land%20grabbing%20in%20Southern%20Chile_0.pdf
Holmes G (2014) What is a land grab? Exploring green grabs, conservation, and private protected areas in southern Chile. J Peasant Stud 41(4):547–567
ILC (2011a) The concentration of land ownership in Latin America: an approach to current problems. Available at: http://www.landcoalition.org/sites/default/files/documents/resources/LA_Regional_ENG_web_11.03.11.pdf

ILC (2011b) The tragedy of public lands: the fate of the commons under global commercial pressure. Knowl Chang

ILC (2013) Indigenous peoples' rights to land, territories and natural resources. Available at: http://www.landcoalition.org/sites/default/files/documents/resources/IndigenousPeoplesRightsLandTerritoriesResources.pdf

INDH (2014) Territorios y Derechos Humanos. Available at: http://www.indh.cl/wp-content/uploads/2014/12/Territorios-y-derechos-humanos-INDH-2014.pdf. Accessed June 2016

Jonas H, Makagon E, Booker S, et al (2012) An analysis of International Law, National Legislation, Judgements, and Institutions as they interrelate with territories and areas conserved by indigenous peoples and local communities: report no. 1: International Law and Jurisprudence. Available at: http://www.iccaconsortium.org/wp-content/uploads/images/stories/Database/legalreviewspdfs/kenya_lr.pdf

Kuhnen F (1982) Man and land: an introduction into the problems of agrarian structure and agrarian reform. Available at: http://www.professor-frithjof-kuhnen.de/publications/man-and-land/1.htm

Llancaqueo V (2005) Pueblo Mapuche Derechos Colectivos y Territorio: Desafíos para la Sustentabilidad Democrática. LOM Ed, Santiago de Chile

Maggi D (2012) Implementation of a marine-protected area in Chile: consequences of neglecting socio-cultural factors. PhD Thesis, University of Otago, Otago

Mapu Lahual (2006) Rewe Lafquen: plan Maestro (Master Plan)

Mathiesen M (1998) Una Mirada a la Identidad de los Grupos Huilliche de San Juan de la Costa. Universidad ARCIS, Centro de Investigaciones sociales

McAlpin C (2004) Una Evaluación de la Red de Parques Indígenas Mapu Lahual. PhD Thesis, University of Colorado. Colorado

Miller R, Lesage L, Escarcena S (2010) The international law of discovery, indigenous peoples, and Chile. Nebraska Law Rev 89:819

Moeckli D, Shah S, Sivakumaran S et al (2013) International human rights law. Oxford University Press, New York City

Molina J, Pavez C (2012) Territorios Indígenas de Conservación: Aprendizajes desde la práctica en el Sur de Chile. Available at: https://d2ouvy59p0dg6k.cloudfront.net/downloads/guia_de_territorios_indigenas_de_conservacion__aprendizajes_desde_la_practica_en_el_sur_.pdf

Molina R, Correa M, Smith-Ramirez C et al (2006) Alerceros Huilliches de la Cordillera de la Costa de Osorno. ANDROS Impresores, Santiago de Chile

Montenegro I, Aldo F (2012) Ordenamiento Territorial en el Sur de Chile: Experiencia de WWF con Pueblos Indígenas y Comunidades Locales. Available at: https://d2ouvy59p0dg6k.cloudfront.net/downloads/lineamientos_de_ordenamiento_territorial_en_el_sur_de_chile__experiencia_de_wwf_con_pueb.pdf

Murray W (2002) The neoliberal inheritance: agrarian policy and rural differentiation in democratic Chile. Bull Lat Am Res 21(3):425–441

Núñez C (2015) Bloque de Constitucionalidad y Control de Convencionalidad en Chile: Avances Jurisprudenciales. Anuario de Derechos Humanos 11:157–169

Ostrom E, Hess C (2007) Private and common property rights. Indiana Univ Bloomington: Sch Public Environ Aff Res Paper 11:1–116

Ovideo G (n.d.) Lessons learned in the establishment and management of protected areas by indigenous and local communities. Available at: http://cmsdata.iucn.org/downloads/cca_goviedo.pdf

Pauchard A, Villarroel P (2002) Protected areas in Chile: history, current status, and challenges. Nat Areas J 22(4):318–330

Rivera F, Pavez C (2012) Planificación y Gestión del Eco-Turismo Comunitario con Comunidades Indígenas. Available at: http://d2ouvy59p0dg6k.cloudfront.net/downloads/guia_de_planificacion_y_gestion_del_ecoturismo_comunitario_con_comunidades_indigenas.pdf

RRI (2015) Who owns the world's land? A global baseline of formally recognized indigenous and community land rights. Available at: http://www.rightsandresources.org/wp-content/uploads/GlobalBaseline_web.pdf

Sallés A, Pizarro N, Luna C (2012) Mapuche: Lengua y Cultura. Pehuén, Santiago de Chile

Secretariat of the CBD (2012) Recognising and supporting territories and areas conserved by indigenous peoples and local communities: global overview and national case studies. CBD Tech Res Ser 64:11–91

Skewes JC, Zuñiga CH, Vera MP (2015) Turismo Comunitario o de Base Comunitaria: Una Experiencia Alternativa de Hospitalidad Vivida en el Mundo Mapuche. CULTUR: Revista de Cultura e Turismo 6(2):73–85

Skjævestad A (2010) The Mapuche people's battle for indigenous land: litigation as a strategy to defend indigenous land rights. Cultures of legality: judicialization and political activism in Latin America 207

Stevens S (2010) Implementing the UNDRIP and international human rights law through the recognition of ICCAs. Policy Matters 17:181–194

Susskind L, Kausel T, Aylwin J et al (2014) The future of hydropower in Chile. J Energy Nat Resour Law 32(4):425–481

von Benda-Beckmann F (2001) Legal pluralism and social justice in economic and political development. IDS Bull 32(1):46–56

Wehrmann B (2008) Land conflicts: a practical guide to dealing with land disputes. Gesellschaft für Internationale Zusammenarbeit GmbH (GIZ), Eschborn

Wehrmann B (2011) Land use planning: concepts, tools and guidelines. Gesellschaft für Internationale Zusammenarbeit GmbH (GIZ), Eschborn

Wickeri E, Kalhan A (2010) Land rights issues in international human rights law. Malays J Hum Rights 4(10):1–10

Chapter 8
Towards a New Forest Model for Chile: Managing Forest Ecosystems to Increase Their Social, Ecological and Economic Benefits

Pablo J. Donoso and Jennifer E. Romero

8.1 Introduction

Although Chile is a quite small country located in the southern tip of the world, its forest sector has become an interesting case of study, especially due to the contrasts between the so-called two sub-sectors (Salas et al. 2016): the short-rotation industrial plantations (2.8 million ha), and the threatened and mismanaged native forests (14 million ha). Native forests grow from 33 to 56°S, while plantations (68% *Pinus radiata* and 23% *Eucalyptus* species) are grown mostly from 35°to 41°S. This geographical region where plantations grow best and (25 m^3/ha/year for *P. radiata* and 35 m^3/ha/year for *Eucalyptus*; Cubbage et al. 2007) overlaps with the area of most biodiverse and productive native forests, including many endemic species of flora and fauna (Bannister et al. 2012; Armesto et al. 1998). This region concentrates 30% of the Chilean population (5.3 M people, in a country where 40% of the population lives in Santiago). Also, it is within these latitudes that most Chilean industrial agriculture develops, especially in the Central Valley between the Coastal and the Andean ranges, including wine, cattle for milk and meat, and fruits and vegetables. In addition, this is the region where most industrial plantations have been established, especially in the foothills of both the Andean and Coastal ranges. Therefore, in a region where most of the native forests located in the lowlands have disappeared, also in the more mountainous lands they have been subjected to losses due to replacement by tree exotic plantations, in addition to continuous degradation and fragmentation (Echeverría et al. 2006; Fig. 8.1).

P. J. Donoso (✉)
Institute of Forests and Society, Universidad Austral de Chile, Valdivia, Chile
e-mail: pdonoso@uach.cl

J. E. Romero
Agrupación de Ingenieros Forestales por el Bosque Nativo, Valdivia, Chile

F. Fuders and P. J. Donoso (eds.), *Ecological Economic and Socio Ecological Strategies for Forest Conservation*, https://doi.org/10.1007/978-3-030-35379-7_8

Fig. 8.1 The four main forest land uses in south-central Chile. From the top left, in a clockwise direction, pictures show old-growth forests, secondary forests, high-graded forests and finally a landscape dominated by clear-cut harvested and unharvested even-aged plantations of exotic species. (Photos by P. Donoso, except the one of high-graded forest taken by Daniel Uteau)

The above scenario has led to considering south-central Chile as a Biodiversity Hotspot (Myers et al. 2000), i.e. a biogeographic region that is both a significant reservoir of biodiversity and is threatened with destruction. While diversity and productivity of native forests, as well as human population, decline to the south, still degradation of native forests is the norm in this region (Zamorano-Elgueta et al. 2014). Overall, while industrial plantations allow the development of the forestry industrial sector that annually exports more than five billion dollars, native forest mostly provide firewood harvested through forest high-grading. Under this scenario, we raise the question *Can the forest sector in Chile increase the provision of ecosystem goods and services?* We predict that this goal is achievable, and that the target has to be to develop more resilient forest stands and landscapes.

8.2 A Brief Review of the Recent Chilean Forest History

The first proposals to regulate the use of forests in Chile date from the late XIX century, which is triggered by the concerns on millions of hectares of soils being eroded through agriculture in Central Chile. In 1872 the "Ley de Bosques" (Forest Law) was enacted. This law regulated the use of fire and promoted afforestation with introduced tree species. This law was not very effective due to institutional limitations, but it served as the basis to enact an improved version in 1931 (Camus 2004). However, deforestation continued hand in hand with the colonization of south-central Chile, with agriculture, and the increasing demand of timber and firewood (Ramírez 2003). Between 1931 and 1973 the state also created forest industries, established more than 300,000 ha of forest plantations of exotic species, and developed the public forest institutions that until today are responsible to enforce the law, administer state-owned conservation areas, and conduct research (Donoso and Otero 2005). In the 1950s one public and one private forestry school were created, and were the only ones until the 1980s, when a new law for universities was enacted during the military dictatorship, which jointly with temporary market signals resulted in 17 forestry schools for a country with a population around 15 million people by then (Donoso and Otero 2005). Many of these schools closed when the oversupply of foresters became evident (Donoso 2012), and currently there are six forestry schools in Chile which annually graduate close to 100 foresters.

Following the military coup of 1973, the government enacts the Decree Law 701 (1974), which provided highly attractive subsidies to plantations. This and other measures (see Niklitschek 2007) became a great incentive for private companies to establish plantations, which occurred on eroded soils but also replaced thousands of hectares of native forests (Lara and Veblen 1992). On the other hand, and after 16 years of discussion in the National congress, only in 2008 the so-called "Law of Native Forests" (N°20,283) was enacted, but is has been very ineffective (Manuschevich and Beier 2016). The overall result of this history is that the region of Chile that allows the development of temperate forests, from Mediterranean forests in the north to Magellanic rainforests in the extreme south, has been diminished through alternative land uses (especially agriculture and industrial forest plantations). These forests are today confined mostly in the two mountain ranges (and especially in the Andes), but their conservation status in the south-central part of the Coastal range is dramatic since this region is mostly covered by industrial plantations (Fig. 8.2).

8.3 Native Forests and Its Current and Potential Provision of Ecosystem Goods and Services

Native forests include the Mediterranean forests (33–38°) and the temperate rainforests, the latter divided into the Valdivian Rainforests (37°45′ to 43°30'S lat.), the North Patagonian Rainforests (43°20′ to 47°30′S lat.) and the Magellanic Rainforests (south of 47°20′ lat.) (Veblen et al. 1983; Veblen and Alaback 1995). In addition, these native forests are divided into 12 forest types, of which four are dominated by conifers (including the emblematic Araucaria (*Araucaria araucana*) and Alerce (*Fitzroya cuppressoides*)) and are mostly allocated for preservation. Native forests in the northern portion of the country's forestland (south-central

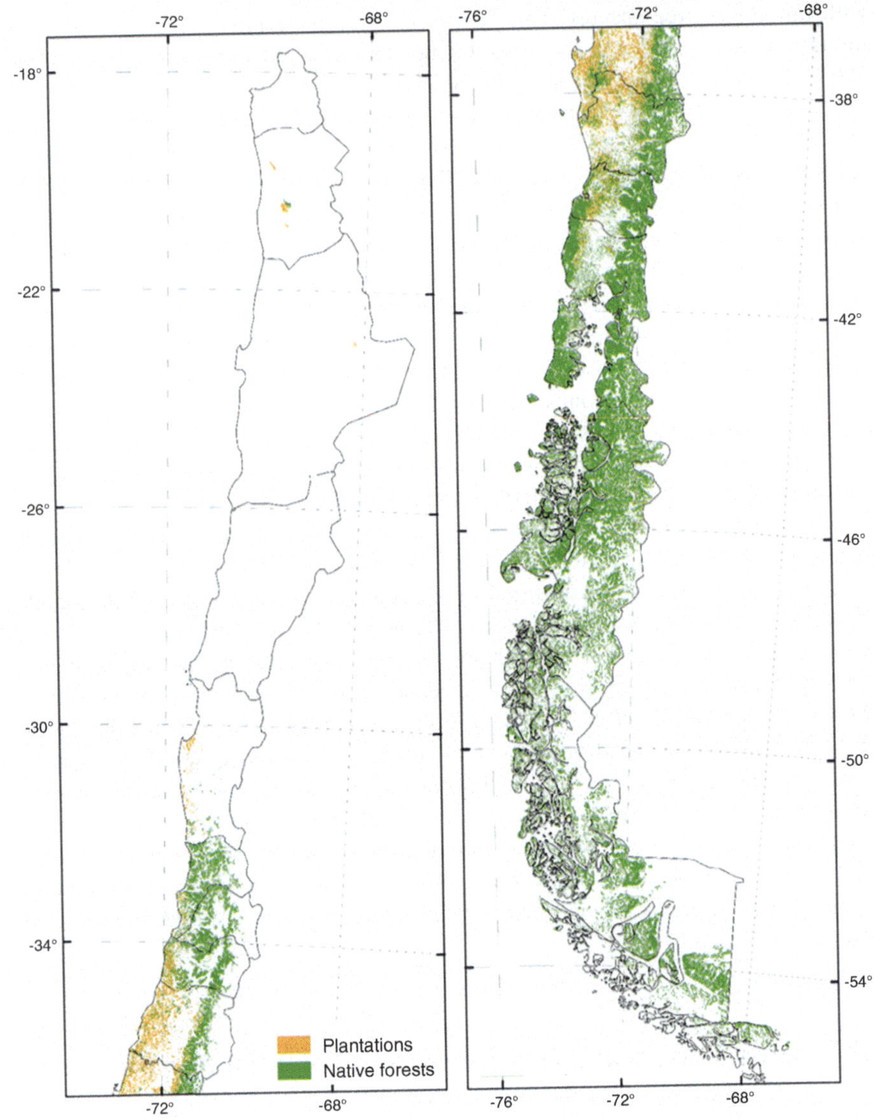

Fig. 8.2 Area of native forests and plantations in Chile. This map illustrates the concentration of native forests to the south and their diminished lands in the Central Depression and Coastal range in central Chile due to land use changes

Chile) have a high degree of degradation, whereas in the southernmost part (Chilean Patagonia) they are in better shape due to the lower human population density and other ecological factors that facilitate recovery of these forests. Overall, Mediterranean forests and Valdivian Temperate Rainforests are the ones in a poorest and most fragile state of conservation due to the high human population density and conflicting land-use demands on productive sites.

Valdivian Temperate Rainforests are highly productive mixed-species forests (Loguercio et al. 2018). Mixed-species forests are not only in general more productive than pure forests, but they also provide more ecosystem services, are more resilient and therefore have a greater adaptive capacity (e.g., Morin et al. 2011; Vila et al. 2013). Therefore, these forests have a great potential to provide goods and services of local and global importance. The challenges ahead are to properly manage native forests especially where they have a good potential to provide goods and services (accessible, good growth rates, etc.), and to recover (restore or rehabilitate) degraded forests.

There has been a great improvement in silvicultural options for these forests. Most experiments and some operational management have been conducted in secondary forests dominated by *Nothofagus* species or by *Drimys winteri*, the latter a fast-growing species well adapted to growing in poorly-drained soils (Loguercio et al. 2018; Donoso et al. 2018). Thinnings in these forests increase the growth of selected trees but generally also tree regeneration and the growth of companion species. Therefore, it is of major importance when managing these forests to define from the onset whether the forest is to be managed through even-aged systems to continue the promotion of one or few dominant species, or it will be managed for future uneven-aged forests with the aim of developing productive mixed forests. Actually, there are also some experiments with uneven-aged silviculture of mature and old-growth forests (Donoso 2013), which if not properly managed are likely to be subjected to high-grading, and experiments with restoration and variable-density thinnings aimed to convert even-aged forests into uneven-aged forests (Donoso et al. 2018). Managed secondary forests can attain from 10 to 20 m^3/ha/year, while managed old-growth forests can reach from 5 to10 m^3/ha/year (Salas et al. 2018; Donoso et al. 2018, Donoso et al. n.d.). Overall, why native forests are not properly managed is not explained by poor productivity, which could be the case, but by governance problems that will be visited later in this chapter.

The thousands (or probably millions) of native forests that have been degraded have their capacity to provide ecosystem services dramatically diminished (e.g. Vásquez-Grandón et al. 2018). According to international agreements (Bannister et al. 2018), Chile must restore 15% of its degraded ecosystems by the year 2020 (AICHI targets). Furthermore, Chile has committed to reforest at least 100,000 ha by the year 2030 using mainly native species (Paris Climate Agreement COP21, New York Declaration of Forests, Initiative 20 × 20). The new Chilean Forestry Policy (2015–2035; CONAF 2015) matches these international commitments, and goes even further, aiming at restoring 500,000 ha of degraded or fragmented ecosystems within priority areas. Considering the high productive capacity of native forests, plus the tremendous capacity of native species to regenerate through seeds and vegetative sources (most species have a strong sprouting capacity), meeting the management and restoration targets mentioned above is more a challenge for Chile in terms of governance, implementation, and long-term monitoring, rather than an ecological-silvicultural one. Native forests continue to be logged under unsustainable management (Lara et al. 2016), and, most importantly, trampling and browsing by cattle diminish regeneration density and growth, especially of dominant tree species in these systems (Zamorano-Elgueta et al. 2014). Again, management of native forests and restoration of degraded forests is not explained by ecological reasons, but by the lack of effective policies, regulations and monitoring that truly promote the recovery of these forests.

8.4 Industrial Exotic Plantations and the Urgent Need for Better Management

Industrial plantations of radiata pine and *Eucalyptus sp.* are managed in a manner that aims to maximize private returns, which means that they are harvested at early ages (with clearcutting), depending on site conditions, but approximately at 20–22 years in the former case and 12–15 years in the latter. In addition, they occupy continuous large areas, which especially in the coastal range have barely left any native forests except for the borders of stream channels (Fig. 8.1). Under these scenarios, these plantations need to be better managed at both the stand and the landscape scales if they are to contribute to more resilient landscapes. A recent study by McFadden and Dirzo (2018) addresses these challenges, and proposes six major strategic decisions that could greatly influence biodiversity outcomes. These include making changes in spatial and temporal planning, legacy retention, site preparation, regeneration, vegetation management and thinning and pruning. In summary, better decisions in this regard should be geared towards having larger areas of native forests within the plantation matrix, leaving intact riparian buffers as wide as possible, leaving legacy structures after all harvests, and promoting understory vegetation.

Plantations therefore should be managed in a different manner both at the stand and the landscape levels. In addition to the proposals given, for example, by McFadden and Dirzo (2018), changes could include extended rotations on some part of the land base (Curtis 1997), and restoring native forests not only in the 40,000 ha that the Forest Stewardship Council (FSC) determined that was the area of plantations established on replaced native forests from year 1993 on (AIFBN 2011) (onset of FSC in Chile), but also in many lands were replacements occurred before. All this would lead to a scenario with more friendly plantations within a matrix that would have not only better visual and ecological effects, but also a greater overall provision of ecosystem services. While these proposals seem reasonable and would not put a lot of pressure upon companies in terms of making radical changes in their business, they are not occurring. Of course, governance of the plantation-dominated landscapes has been left mostly to private companies, which will continue aiming to maximize their private profits.

8.5 A Better Governance of Forest Landscapes: Key to the Provision of Multiple Goods and Services

As mentioned above, central and south-central Chile require greater attention to their forest landscapes, especially in the sense of increasing their resilience to human and natural disturbances. Policies and regulations to move things in that direction need to consider native forests in their different states of conservation as well as plantations.

In regards to native forests, these are complex adaptive systems because they possess properties such as (e.g. Messier et al. 2013) heterogeneous structures and non-linear relationships, negative and positive feedback mechanisms, memory following major disturbances and changes, and individual components are constantly reacting to one another and outside influences, thus continually modifying the system and allowing it to adapt to altered conditions. At a landscape level, there is an enormous variability of forest ecosystems in Chile due to their distribution along a vast latitudinal range and the topographic variations in the west-east axis caused by the Andean and Coastal ranges. Thus, complexity occurs at the forest stand and the forest landscape levels, but at both scales it has been altered due to major transformations of the landscapes and mismanagement of the forests. To have more resilient landscapes and forests, therefore, the problem has to be tackled at a regional scale (i.e., central and south-central Chile). In addition, organizations, people, regulations and processes through which decisions are made relating to forests need to be in balance, but reaching that governance dream needs a broad platform to work on.

8.5.1 Building a Future Scenario

Folke et al. (2002) indicate that evidence has been accumulating in diverse regions all over the world suggesting that natural and social systems behave in nonlinear ways, exhibit marked thresholds in their dynamics, and that social-ecological systems act as strongly coupled, complex and evolving integrated systems. From there, two useful tools for resilience-building in social-ecological systems are structured scenarios and active adaptive management. These tools require and facilitate a social context with flexible institutions and multi-level governance systems that allow for learning and increasing adaptive capacity without foreclosing future development options. How far is Chile from possessing these tools?

Scenarios are needed to envision alternative futures and the pathways by which they might be reached. Thinking about these scenarios is necessary to implement actions consistent with increasing the possibilities to attain or approach them, including the avoidance of non-desired ones. Active adaptive management views policy as a set of experiments designed to reveal processes that build or sustain resilience. In Chile, there is a Forest Policy document (2015–2035; CONAF 2015), developed by a group of private and public institutions, that includes four main strategic objectives: (a) To establish public agencies consistent with the strategic importance of the forest sector for Chile, to implement a sustainable forest development; (b) To promote silviculture, industries and an integral use of forest resources to increase overall productivity and the provision of goods and services for economic development; (c) To develop the necessary conditions and instruments so that forest development diminishes technological and social gaps, improves life conditions of workers, and respects cultures and traditions of small land-owners and indigenous communities, and; (d) To conserve and increase the public forest heritage and to restore and protect forest resources. Four years after the release of this policy document, there has been no rearrangements or improvements in either of the two main public forest agencies (CONAF, the forest service, in charge of developing forest legislation and enforcing laws and regulations, and INFOR in charge of forest research). Chile continues having nominal areas of native forest management (Donoso et al. 2018); there is no progress in restoration of degraded forests (Bannister et al. 2018) and there has been no modification to the current main laws related to forest management. On the other side, 1.4 million hectares of Protected Areas have been added as the system of National Parks in Patagonia (CONAF 2019), but without adding new support for their administration (Petit et al. 2018). Actually, Petit et al. (2018) make the following question: is it sustainable to continue adding protected areas to the national system even though it is clear that the existing support is insufficient to meet the minimum requirements for full implementation?

Although some scenarios have been envisioned for the future of (especially) native forests, the business-as-usual scenario is the one that continues to dominate. Three topics illustrate this judgement. These are similar to those mentioned by Donoso and Otero (2005) when questioning whether Chile could be recognized as a forest country. First, on the silvicultural side, there is barely any native forest management, and exotic forest management is still oriented to maximize companies' benefits with short-rotation monocultures covering continuous tracts of the landscape. In both cases, when forest management complies with current regulations, there is almost no room for innovation and flexibility, but the trend is that most harvested native forests continue to be high-graded. Second, there has been no modernization of the public agencies in the forest sector. Third, the public system of protected areas continues to under-represent many ecological regions in Chile (Armesto et al. 1998; Petit et al. 2018) including those with more diversity and endemic species in the central portion of the country (a biodiversity Hotspot in central Chile; Myers et al. 2000). These gaps challenge managers and decision makers to look forward for new approaches discussed in the next section.

8.6 Promoting Adaptive Management: Flexible Institutions and Multi-level Governance

A better governance of Chilean forest ecosystems depends on the availability of updated information and on improving the structure, coordination and availability of resources of public agencies. As Weimer and Vining (2016) stress, public executives generally operate in environments characterized by great asymmetry of information, without knowing or understanding the interests of the stakeholders and other representatives as well. It is also crucial to let grass-root organizations have a stronger say on forest management, so that decision makers are aware of the changing scenarios that rural people are facing, and how those scenarios affect forest management or harvesting. These approaches should aim at increasing the possibilities of attaining resilient landscapes in the near future, and to assure equity and fairness.

Stronger institutions should be able to generate more and better information. Possible scenarios and risks for the short and medium term could be posed with a better knowledge of the changing ecological conditions and social-economic impacts on the territories. Those scenarios may consider extreme/rare events, and a range from the worst- to the best-case scenarios. Risk management has been a fundamental motivation for the development of social and governance structures over the last 10,000 years (McDaniels and Small 2003). Risk-informed decision-making has the potential to reduce risk while simultaneously reducing complying cost (Bier et al. 2004). Decisions should be informed by social values, considering value conflicts in all scenarios, since those are inherent on public decision-making, and they should be based on economic and non-economic indicators. There is a great variety of indicators not tied to macroeconomic data that provide measures of various dimensions of social welfare (Weimer and Vining 2016).

When scenarios and objectives are defined, problems and possible solutions emerge in an organized manner. Then, it is possible to design a set of feasible actions to address them, from *status quo* to the most complete group of actions, depending on the impacts, resources, efforts, time constraints and others. Structured decision-making is a useful organized approach for making choices in complex decision situations, such as managing renewable natural resources, when complete and accurate information is available. It allows to undertake efficient actions given any possible scenario and resources availability, since possibilities and their outputs and outcomes were already anticipated.

A sustainable forest sector in Chile depends on innovating in management, institutions and regulations. This would lead to more resilient and productive forests and landscapes. These arrangements must consider the beliefs, expectations and needs of stakeholders (Moorman et al. 2013), as well as respond to international commitments to mitigate and adapt to climate change.

8.7 Concluding Remarks: Building Resilient Landscapes

Today uncertainty is a major consideration for the management of natural resources. Natural or anthropogenic disturbances are constant drivers of change in land-use, diversity and productivity of forest ecosystems. To cope with these current and future uncertainties, countries need to have strong institutions with clear objectives (structured scenarios), and allow for flexibility in the management of resources as a means to cope with adaptation when necessary. In Chile this means that harvesting in native forests has to be hand in hand with silvicultural approaches aimed to maintain or promote mixed-species forests that will have greater diversity and productivity, thus providing a larger suite of ecosystem goods and services (Franklin et al. 2018). It also means that especially large forest companies must end with the model of only maximizing private profits (at the expense of native forests, local communities, and the loss of ecosystem services), and move towards responsible and socially-concerned ways of doing things (which also needs to be enforced), thus contributing to more diverse and resilient forest stands and landscapes, greater respect for neighboring local communities, and better collaboration with small- and mid-sized forest-based companies. As stated by McDaniels and Small (2004): "Throughout the last century, economists have characterized the entire rationale for government as based in the support for collective efforts that cannot be accomplished through private markets". Collective efforts are needed, where different views are integrated, and the decision-making is not (only) market or economic oriented.

Building resilient landscapes in south-central Chile greatly depends on better ways of conserving, managing and restoring native forests (14 million ha), but also on better ways of managing plantations (three million ha). Chile has a huge opportunity to become a model for sustainable forest management, i.e. a country where both native forests are managed to increase their provision of goods and services thanks to strong institutions, well-prepared professionals, and concerned private stewards; and plantations are managed in ways that continue to sustain profits for their owners, but where these are limited by laws and regulations aimed to promote better ecological, economic, and social environments.

References

Armesto JJ, Rozzi R, Smith-Ramírez C, Arroyo MTK (1998) Conservation targets in south American temperate forests. Science 282(5392):1271–1272

Agrupación de Ingenieros Forestales por el Bosque Nativo (2011). Hacia un Nuevo Modelo Forestal

Bannister JR, Vidal OJ, Teneb E, Sandoval V (2012) Latitudinal patterns and regionalization of plant diversity along a 4270-km gradient in continental Chile. Austral Ecol 37:500–509

Bannister JR, Vargas-Gaete R, Ovalle JF et al (2018) Major bottlenecks for the restoration of natural forests in Chile. Restor Ecol. https://doi.org/10.1111/rec.12880

Bier M, Scott F, Jacobs H, James L, Small M (2004) Risk of extreme and rare events: lessons from a selection of approaches. In: McDaniels T, Small MJ (eds) Risk analysis and society: an interdisciplinary characterization of the field. Cambridge University Press, Cambridge

Camus P (2004) Los bosques y la minería del norte chico, siglo XIX. Un mito en la representación del paisaje chileno. Historia 37(2):289–310

CONAF (2015) Política Forestal 2015–2035. http://www.conaf.cl/wp-content/files_mf/1462549405politicaforestal201520351.pdf

CONAF (2019) Press Note: Red de Parques Patagónicos protegerá ecosistemas y especies amenazadas únicas en el mundo y alcanza récord de aumento de superficie del SNASPE desde 1969. http://www.conaf.cl/red-de-parques-patagonicos-protegera-ecosistemas-y-especies-amenazadas-unicas-en-el-mundo-y-alcanza-record-de-aumento-de-superficie-del-snaspe-desde-1969/

Cubbage F, MacDonagh P, Sawinski Júnior J et al (2007) Timber investment returns for selected plantations and native forests in South America and the southern United States. New For 33(3):237–255

Curtis RO (1997) The role of extended rotations. In Khom KA, Franklin JF (Eds.) Creating a forestry for the 21st century, Island Press, Washington, D.C., USA, pg. 165–170

Donoso P (2012) Ingeniería Forestal en crisis... o no? REVISTA Bosque Nativo 50:14–15

Donoso PJ (2013) Necesidades, opciones y futuro del manejo multietáneo en Chile. En : Donoso PJ, Promis A (Editores), Silvicultura en Bosques Nativos. Avances en la investigación en Chile, Argentina y Nueva Zelandia. Estudios en Silvicultura de Bosques Nativos, Vol. 1. Ed. Marisa Cuneo, Valdivia, Chile. 253 p. https://sites.google.com/site/alvaropromis/Home/libro-silvicultura-bosques-nativos

Donoso P, Otero L (2005) Hacia una definición del país forestal: ¿Dónde se sitúa Chile? Bosque 26(3):5–18

Donoso P, Ponce D, Salas C (2018) Opciones de manejo para bosques secundarios de acuerdo a objetivos de largo plazo y su aplicación en bosques templados del centro-sur de Chile. In: Donoso PJ, Promis A, Soto DP (eds) Silvicultura en Bosques Nativos. Experiencias en silvicultura y restauración en Chile, Argentina y el oeste de los Estados Unidos. Editorial College of Forestry, Oregon State University, EE.UU

Donoso PJ, Ojeda PF, Schnabel F, Nyland RD (n.d.) Initial responses in growth, yield and regeneration following selection cuttings with varying residual densities in evergreen hardwood-dominated temperate rainforests in Chile. Submitted to Forestry

Echeverría C, Coomes D, Salas J et al (2006) Rapid deforestation and fragmentation of Chilean temperate forests. Biol Conserv 130(4):481–494

Folke C, Carpenter S, Elmqvist T et al (2002) Resilience and sustainable development: building adaptive capacity in a world of transformations. Ambio 31(5):437–440

Franklin JF, Johnson KN, Johnson DL (2018) Ecological forest management. Waveland Press, Inc., Long Grove

Lara A, Veblen TT (1992) Forest plantations in Chile: a successful model? In: Mather A (ed) Afforestation: policies, planning and progress. Belhaven Press, London, pp 118–139

Lara A, Zamorano C, Miranda A et al (2016) Bosque Nativo, Capítulo 3 Informe País. Estado del Medioambiente en Chile 1999-2015:117–220

Loguercio GA, Donoso PJ, Müller-Using S et al (2018) Silviculture of temperate mixed forests from South America. In: Bravo-Oviedo A, Pretzsch H, del Río M (eds) Dynamics, silviculture and management of mixed forests. Springer International Publishing AG, Cham, pp 271–317

Manuschevich D, Beier CM (2016) Simulating land use changes under alternative policy scenarios for conservation of native forests in south-Central Chile. Land Use Policy 51:350–362

McDaniels T, Small M (2004) Risk analysis and society: an interdisciplinary characterization of the field. Cambridge University Press, Cambridge. https://doi.org/10.1017/CBO9780511814662

McFadden TN, Dirzo R (2018) Opening the silvicultural toolbox: a new framework for conserving biodiversity in Chilean timber plantations. For Ecol Manag 425(1):75–84

Messier C, Puettmann KJ, Coates KD (2013) Managing forests as complex adaptive systems. The Earthscan Forest Library, New York, USA. 353 p

Morin X, Fahse L, Scherer-Lorenzen M, Bugmann H (2011) Tree species richness promotes productivity in temperate forests through strong complementarity between species. Ecol Lett 14:1211–1219
Moorman M, Nelson S, Moore S, Donoso P (2013) Stakeholder perspectives on adaptive co management as a Chilean conservation management strategy. Soc Nat Resour 26:1022–1036
Myers N, Mittermeier RA, Mittermeier CG et al (2000) Biodiversity hotspots for conservation priorities. Nature 403:853–858
Niklitschek M (2007) Trade liberalization and land use changes: explaining the expansion of afforested land in Chile. For Sc 53(3):385–394
Petit IJ, Campoy AN, Hevia MJ et al (2018) Protected areas in Chile: are we managing them? Rev Chil Hist Nat 91:1. https://doi.org/10.1186/s40693-018-0071-z
Ramírez F (2003) La guerra contra los "montes" y la extracción de los "palos": una aproximación histórico-ecológica a los procesos de degradación de los bosques nativos del sur de Chile. Ponencia presentada al Simposio de Historia Ambiental Americana, realizado en Santiago de Chile, 14–19 de julio de 2003
Salas C, Donoso PJ, Vargas R, Arriagada CA, Pedraza R, Soto DP (2016) The forest sector in chile: an overview and current challenges. J For 114(5):562–571. https://doi.org/10.5849/jof.14-062
Salas C, Fuentes-Ramírez A, Donoso PJ et al (2018) Crecimiento de bosques secundarios y adultos de Nothofagus en el centro-sur de Chile. In: Donoso PJ, Promis A, Soto DP (eds) Silvicultura en Bosques Nativos. Experiencias en silvicultura y restauración en Chile, Argentina y el oeste de los Estados Unidos. Editorial College of Forestry, Oregon State University, EE.UU
Vásquez-Grandón A, Donoso PJ, Gerding V (2018) Degradación de los bosques: Concepto, proceso y estado – Un ejemplo de aplicación en bosques adultos nativos de Chile. In: Donoso PJ, Promis A, Soto DP (eds) Silvicultura en Bosques Nativos. Silvicultura en bosques nativos. Experiencias en silvicultura y restauración en Chile, Argentina y el oeste de Estados Unidos, Editorial College of Forestry, Oregon State University, EE.UU, pp 175–196
Veblen TT, Alaback PB (1995) A comparative review of Forest dynamics and disturbance in the temperate rainforests of north and South America. In Lawford R, Alaback PB, Fuentes E (Eds.) High-latitude rainforests and associated ecosystems of the west coast of the Americas. Climate, hydrology, ecology, and conservation. Springer-Verlag, New York, U.S.A., pg. 173–213
Veblen TT, Schlegel FM, Oltremari JV (1983) Temperate broad-leaved evergreen forests of South America. In: Ovington JD (Ed.) Temperate broad-leaved forests, Amsterdam, The Netherlands, Elsevier, pg. 5–31
Vila M, Carrillo-Gavila A, Vayreda J et al (2013) Disentangling biodiversity and climatic determinants of wood production. PLoS One 8(2):e53530
Weimer D, Vining AR (2016) Policy analysis: concepts and practice, 5th ed. ISBN 9780205781300 (pbk) ISBN-13: 978–0205781300 ISBN-10: 0205781306. New York, U.S.A., Routledge., 473p
Zamorano-Elgueta C, Cayuela L, Rey-Benaya JM et al (2014) The differential influences of human-induced disturbances on tree regeneration community: a landscape approach. Ecosphere 5/7:1–17

Chapter 9
On Ecosystem Dynamics for the Conservation of Wetlands and Forest

Milan Stehlík, Jozef Kiseľák, and Jiří Dušek

9.1 Introduction

In general, living systems show high dynamics of its functions, processes, and overall behaviors. Ecosystems we can see as a dynamics complex systems changing within the course of time. Ecosystem dynamic is important if we want to sustainably manage ecosystem abilities and their functions. Forest (woodlands) ecosystems are large areas dominated by the phanerophytes plant life form according to (Raunkiær 1905). Phanerophytes are woody plants that grow taller than 0.25–0.5 m or whose shoots do not die back periodically that limit its height (Ellenberg and Mueller-Dombois 1967). Ecosystems created by phanerophytes (forests) are very dynamical ecosystems where abilities and functions spread not only in time but also in its diverse space compartments creating three-dimensional systems by individual trees (Bohn and Huth 2017). Just the structure plays major roles in dynamic of ecosystem and diversity of the forest ecosystems (Spies 1998). Forest ecosystems are usually

M. Stehlík (✉)
Department of Applied Statistics and Linz Institute of Technology, Johannes Kepler University in Linz, Linz, Austria

Institute of Statistics, University of Valparaíso, Valparaíso, Chile

Department of Statistics and Actuarial Science, The University of Iowa, Iowa City, Iowa, USA
e-mail: milan.stehlik@jku.at

J. Kiseľák
Institute of Mathematics, P.J. Šafárik University, Košice, Slovakia

Department of Applied Statistics and Linz Institute of Technology, Johannes Kepler University Linz, Linz, Austria
e-mail: jozef.kiselak@upjs.sk

J. Dušek
Global Change Research Institute CAS, Brno, Czech Republic
e-mail: dusek.j@czechglobe.cz

F. Fuders and P. J. Donoso (eds.), *Ecological Economic and Socio Ecological Strategies for Forest Conservation*, https://doi.org/10.1007/978-3-030-35379-7_9

at the end of the process of sequential vegetation changes (succession) which is associated with response to its conditions and environment. The end of the succession process is called climax stage, and this stage is an equilibrium stage of old-growth forest. The final climax old-growth stage of forest is reached after completion of previous succession stages such as stand of initiation, stem exclusion, and understorey reinitiation within a certain time (Clements 1905; Tansley 1935; Kuuluvainen 2016). Climax stage is in equilibrium with conditions of environment by its definition, but it does not mean that the forest ecosystem in climax stage is without any dynamic. Forest ecosystems must constantly balance individual biogeochemical processes in relation with actual environmental conditions. This balance results in a maximum of biomass and symbiotic function between individual and groups of organisms (Odum 1969) with minimum loss of energy during unfavorable growth conditions (e.g., periods of dormancy in winter period, periods of drought). We try to show statistical backgrounds and possibilities how to use an approach of the recurrence plots in the description of dynamics in selected processes with using available data of forest and wetland ecosystem. The description of the dynamics is important mainly for ecosystem conservation processes which must be based on a sustainable management that focuses on the development of abilities and functions of forest and wetland ecosystems.

9.1.1 The Origin of the Data and Their Measurement

The data were obtained within the monitoring of carbon cycle in the sedge-grass marsh which is a part of a large wetland complex called the "wet meadows." The "wet meadows" situated near the town of Třeboň, South Bohemia, Czech Republic, and the location of measuring instruments is 49° 01′ 29″ N, 14° 46′ 13″ E. The monitored site is a flat area about 1 ha in size and 426 m above see level situated in the inundation area of a large human-made lake (Rožmberk fishpond, 5 km^2 water surface). More detail of the monitored site is summarized in Dušek et al. (2013, 2017).

Releasing of gas (emissions) was measured by the developed automatic non-steady-state flow-through chamber system (Czech utility model: UV 3237) (Dušek et al. 2012). Changing concentrations of gases inside the closed chamber were measured by a fast CH_4 analyzer (DLT-100, Los Gatos Research Inc., USA) in 1 Hz frequency. Final calculation of gas fluxes is based on the linear increase of gas concentration inside the chamber during the period of it closing, taking into account the chamber volume (780 L) and surface area covered (0.785 m^2). Final flux includes corrections to the current air temperature and ambient air pressure by the physics of ideal gas law and also for water vapor. For the present analyses was selected raw data measured in wetter part of the sedge-grass marsh during 1 week at the end of May and beginning of June (2013). In this period, emissions of both gases were quite stable without fluctuation.

9.2 Theoretical Background

9.2.1 DS of t-Score Functions

In statistics, score function $S(x;\theta) := \frac{\partial}{\partial\theta} \ln f_X(x;\theta)$ indicates how sensitive a likelihood function is to its parameter θ. Scoring algorithm, also known as Fisher's scoring, is a form of Newton's method used in statistics to solve maximum likelihood equations numerically. Notice that this is nothing else than a dynamical system since $\theta_{m+1} = \theta_m + \mathcal{J}^{-1}(\theta_m)S(\theta_m)$, where J is the observed information matrix.[1] Under certain regularity conditions, it can be shown that $\theta_m \to \theta^*$

The transformation-based score (Fabián 2001; Stehlík et al. 2010) or shortly the t-score for the density f (with parameter θ) and suitable mapping η is defined as

$$T_\eta(x;\theta) = -\frac{1}{f(x;\theta)}\frac{d}{dx}\left(\frac{1}{\eta'(x)}f(x;\theta)\right).$$

It is a relative change of a "basic component of the density," i.e., density divided by the Jacobian of η. Moreover it can be understood as a suitable function for using the generalized moment method for estimation of parameters of heavy-tailed distributions. If $(X_1,\ldots,X_n)$ be iid sample from F with probability density function f, then as a measure of central tendency can be suggested the zero of its t-score, i.e., $x^* : T_\eta(x;\theta) = 0$. Then $S(x;\theta) = \eta'(x^*)T(x;\theta)$ is called score function for t-mean. One can use the score moment estimates

$$\frac{1}{N}\sum_{i=1}^{N} S^k(x_i;\theta) = \mathbb{E}_\theta[S^k].$$

Since the scalar score is centered around the t-mean, it yields the moment estimation equations for θ in the form $\hat{\theta} : \frac{1}{n}\sum_{i=1}^{n} T_\eta(x_i;\theta) = 0$ (it is strongly consistent and asymptotically normal).

For generalized t-Hill estimator, i.e., Pareto distribution with probability density function

$$f_X(x) = \begin{cases} \frac{\theta x_{\mathrm{m}}^{\theta}}{x^{\theta+1}} & x \geq x_{\mathrm{m}}, \\ 0 & x < x_{\mathrm{m}}, \end{cases}$$

where $\theta > 0$ is a shape parameter (the tail index) and x_m is the (necessarily positive) minimum possible value of X, and $\tilde{\eta}(x) = \ln(x - x_m),\ x > x_m$, we have the t-score

[1] $\mathcal{J}(\theta_0) = -\sum_{i=1}^{n} \nabla\nabla^\top\big|_{\theta=\theta_0} \log f(Y_i;\theta)$

$$T_\eta(x;\theta) = \theta\left(1 - \frac{x_m\,(\theta+1)}{\theta x}\right), x \geq x_m.$$

Notice that in case of Pareto, one has score function $S(x;\theta) = \frac{1}{\theta} + \ln(x_m) - \ln(x)$. Thus we can define the score

$$S(x;\theta,\beta) = \begin{cases} \theta\left(1 - x_m \frac{\theta+\beta-1}{\theta x^{\beta-1}}\right), & \beta \neq 1, \\ \frac{1}{\theta}\left(1 + \theta\,\ln(x_m/x)\right), & \beta = 1, \end{cases} \tag{9.1}$$

where $\beta > 0$ is tuning parameter. For $\beta = 2$, we receive t-Hill.

Here an important inverse problem arises in general. For a given score $\tilde{S}$ does there exist one or several sufficiently smooth functions η such that equation

$$T_\eta = S$$

holds? In Stehlík et al. (2017) is shown a particular answer, how a DS given by a t-score function for some class of monotonic data transformations generates consistent extreme value estimators.

Here we introduce t-scoring algorithm for $V(\theta) = \sum_{i=1}^{N} T_\eta(x_i;\theta)$. Using a Taylor expansion of the function, $V(\theta)$, about θ_0 we obtain $V(\theta) \approx V(\theta_0) + \frac{\partial V}{\partial\theta}(\theta_0)(\theta - \theta_0)$. Using that $V(\theta^*) = 0$ and rearranging gives us

$$\theta^* \approx \theta_0 - \left[\frac{\partial V}{\partial\theta}(\theta_0)\right]^{-1} V(\theta_0).$$

We therefore use iteration

$$\theta_{m+1} = \theta_m - \left[\frac{\partial V}{\partial\theta}(\theta_m)\right]^{-1} V(\theta_m) := h(\theta_m) \tag{9.2}$$

For simplicity, we consider 1D case, i.e., we assume that we have two parameters, but only one is known.

Example 1 For Pareto distribution with fixed threshold x_m and $T_\eta := S$ given by (9.1) with $b = 1$, we have $\theta_{m+1} = h(\theta_m)$, where $h(z) = z(2 - az)$, where

$$a = \frac{1}{N}\sum_{i=1}^{N} \ln(x_i/x_m).$$

Here we demand that $z \leq 1/a$ (from inverse $1/a + \sqrt{1-az}/a$ we have rank of map f), and one can compute that Lyapunov exponent is negative for $a > 0$. It can be shown that from map $z(2-az)$, $a > 0$ chaos does not develop and that it converges to $1/a$, i.e., to MLE for θ.

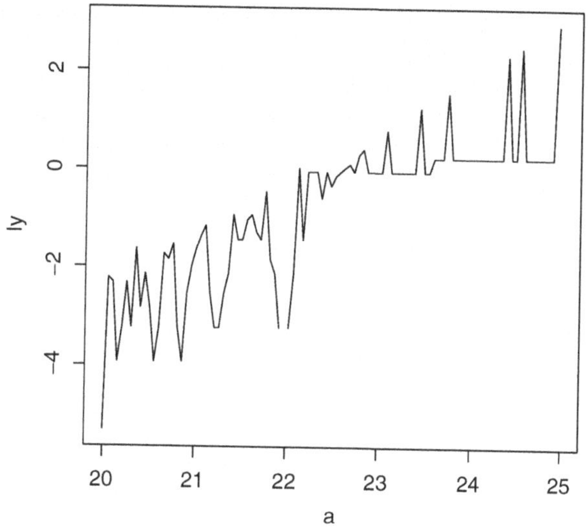

Fig. 9.1 Lyapunov exponents of $a \in [20, 25]$

Example 2 There is a direct connection between θ and x_m through estimation $1/\hat{\theta} = \frac{1}{N}\sum_{i=1}^{N} \ln(x_i/\hat{x}_m)$. Nevertheless, one has to be careful if one parameter is fixed. In the sense of convergence, it is not a bijection. For Pareto distribution with fixed $\theta > 0$ and $T_\eta := S$ given by (9.1) with $b = 1$, we have $x_{m_{k+1}} = h(x_{m_k})$, where $h(z) = z(1 + a - \ln(z))$ and

$$a = \frac{1}{N}\sum_{i=1}^{N} \ln(x_i) - \frac{1}{\theta}.$$

We have to force $z \leq e^{1+a}$. Notice that for such a map chaos might occur when data generate specific value of a. See Fig. 9.1 for $a \in [20, 25]$ where Lyapunov exponents are plotted.

Previous example shows that the mapping η might act as some kind of stabilizer if the algorithm with standard score exhibits chaotic behavior.

9.2.2 *The Recurrence Plots*

Recurrence is a fundamental property of DS, which can be exploited to characterize the system's behavior in phase space. A *recurrence* is a time the trajectory returns to a location it has visited before, i.e., when the distance between two points is below a certain threshold. If only a time series is available, the phase space can be reconstructed by using a time delay embedding (see Takens' theorem): $\mathbf{x}(i) = (X(i), X(i+\tau), \ldots, X(i+\tau(m-1)))$, where $X(i)$ is the time series, m the embedding dimension, and τ the time delay. In Eckmann et al. (1987), *recurrence plots* are introduced as a new diagnostic tool for measuring the time constancy of DS. The recurrence plot displays pairs of times at which the trajectory is at the same place, i.e., the set of (i, j) with $\mathbf{x}(i) = \mathbf{x}(j)$. Afterward, one has to count as a recurrence any time the trajectory gets sufficiently close (within ε) to a point it has been previously. It can be recorded by the following function: $0 < \varepsilon \ll 1$,

$$R(i, j) = \text{Heaviside}(\varepsilon - \|\mathbf{x}(i) - \mathbf{y}(j)\|), \tag{9.3}$$

$\mathbf{x}(i), \mathbf{y}(i) \in \mathbb{R}^m, \quad i = 1, \ldots, N_{\mathbf{x}}, \; j = 1, \ldots, N_{\mathbf{y}}$, and the recurrence plot puts point at coordinates (i, j) if $R(i, j) = 1$.

9.2.3 The Recurrence Quantification Analysis

The recurrence quantification analysis (RQA) (Zbilut and Webber 1992; Marwan et al. 2002) allows us to describe the recurrence plots in a quantitative way. It is a method of nonlinear data analysis which quantifies the number and duration of recurrences of a DS presented by its state space trajectory. It can also be derived from recurrence plots. The main advantage of recurrence plots is that they provide useful information even for short and nonstationary data, where other methods fail.

For computations, we have used procedure[2] `crqa()` from Coco et al. (2018) in software (R Core Team 2018). In order to do RQA on continuous data, we need to figure out what could the key parameters be. We cautiously use `optimizeParam` to find[3] optimal embedding dimension $m = 4$, optimal lag (delay) $\tau = 16$, and radius $\varepsilon = 0.00256$. Procedure `crqa()` returns a list with different measures extracted from the recurrence plot; see e.g., (Marwan et al. 2007). The density of recurrence points in a recurrence plot is the *recurrence rate*[4] $\mathbf{RR} = \frac{1}{N^2}\sum_{i,j=1}^{N} R(i, j)$. The percentage of recurrence points which form diagonal lines is called *determinism*,[5] $\mathbf{DET} = \frac{\sum_{l=l_{\min}}^{N} lP(l)}{\sum_{l=1}^{N} lP(l)}$, where $P(l)$ is the frequency distribution of the lengths l of the diagonal lines. The proportion of recurrent

[2]Cross-recurrence measures of two time series, time delayed, and embedded in higher dimensional space.

[3]If the embedding parameters are estimated from both time series, but are not equal, the higher embedding should be chosen. It has been observed by Zbilut (2005) that a sufficiently large embedding would be sufficient to contain all relevant dynamics.

[4]The recurrence rate corresponds with the probability that a specific state will recur (the percentage of recurrent points falling within the specified radius).

[5]It is a proportion of recurrent points forming diagonal line structures and is closely related with the predictability of the DS, because white noise has a recurrence plot with almost only single dots and very few diagonal lines, whereas a deterministic process has a recurrence plot with very few single dots but many long diagonal lines.

points which form vertical lines is called *laminarity*, $\mathbf{LAM} = \frac{\sum_{v=v_{\min}}^{N} vP(v)}{\sum_{v=1}^{N} vP(v)}$, where $P(v)$ is the frequency distribution of the lengths v of the vertical lines, which have at least a length of $v_{\min}$. $\mathbf{L} = \frac{\sum_{l=l_{\min}}^{N} l\,P(l)}{\sum_{l=l_{\min}}^{N} P(l)}$ defines averaged diagonal line length, and it is related with the predictability time of the DS and the *trapping time*, measuring the average length of the vertical lines, $\mathbf{TT} = \frac{\sum_{v=v_{\min}}^{N} vP(v)}{\sum_{v=v_{\min}}^{N} P(v)}$. The length of the longest diagonal line segment in the plot, excluding the main diagonal $\max \mathbf{L} = \max(\{l_i;\ i = 1, \ldots, N\})$ and the total number of lines in the recurrent plot are also the measures of the RQA.

9.3 Results

Here we refer to Marwan et al. (2007). In the case of original methane data and transformed (by t-score) CH_4, see Figs. 9.2 and 9.3, respectively. Estimated parameters of Pareto distribution are $\hat{x}_m = 0.002477, \hat{\theta} = 0.51742$ (using score and also mle). The recurrent points are marked with blue color, whereas the nonrecurrent points are left blank. Single, isolated recurrence points reflect random, stochastic behavior (strong fluctuation in the process). In Figs. 9.4 and 9.5, the blue points are the temporal "coordinates" where the trajectories get sufficiently close according to the radius parameter. Evidently 9.4 shows considerably more "stochasticity." High values of **DET** might be an indication of determinism in the studied system; see

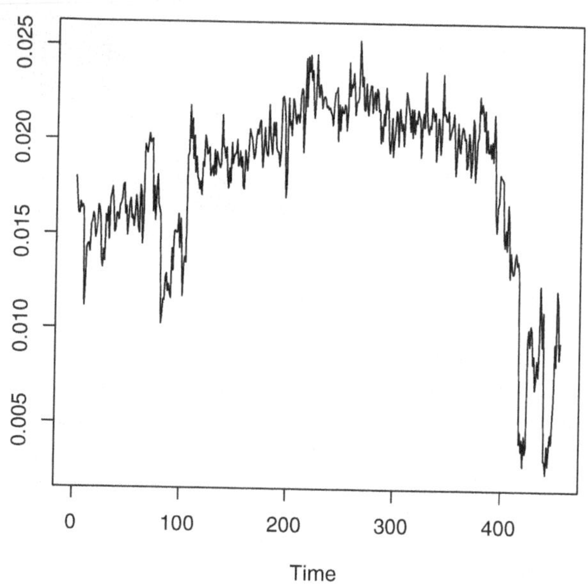

Fig. 9.2 Time series of CH_4

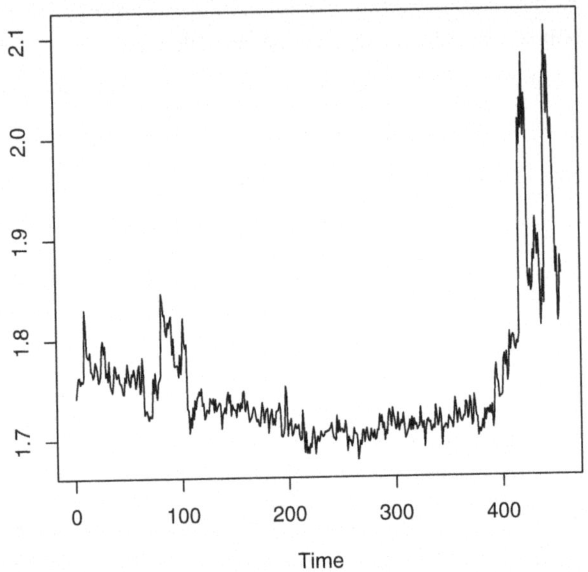

Fig. 9.3 t-score values of time series of CH_4

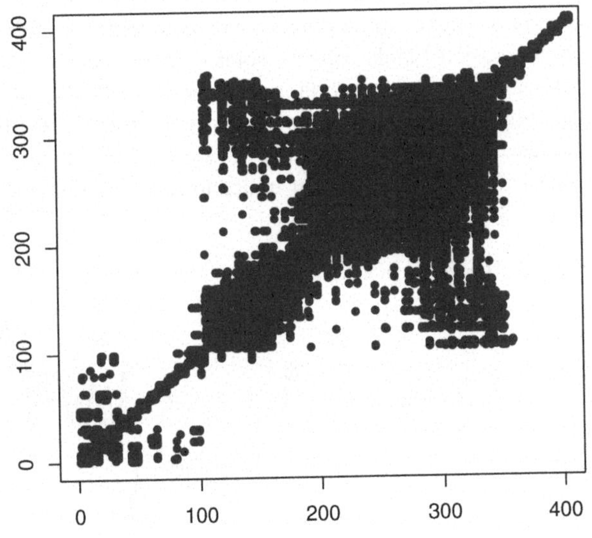

Fig. 9.4 Recurrence plot of original data

Table 9.1. On the other hand, values **DET** and **L** could indicate original series to be more stable but only a little more. Moreover, fading to the upper left and lower right corners implies non-stationarity of the data. If diagonal lines occur beside single isolated points, there could be chaotic part of the process. Notice however that it is just a necessary condition, not a sufficient one, and the values are not very high, which indicates that chaotic part and stochastic part might be included. **RR** was higher in original series; this means there is more sharing of patterns.

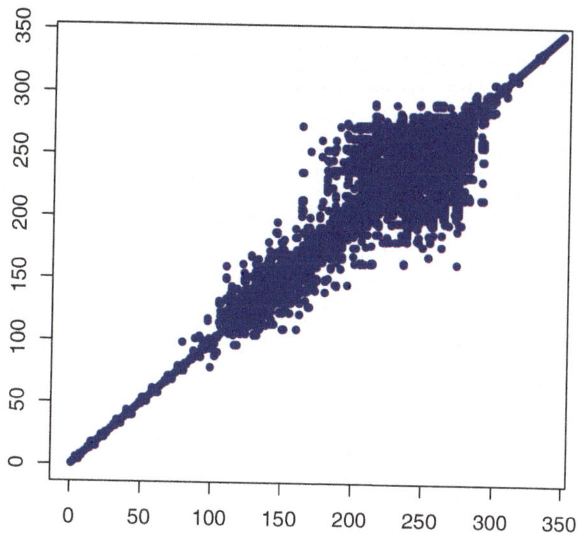

Fig. 9.5 Recurrence plot of t-score

Table 9.1 Recurrence-based measures for original data and t-score

CH_4	RR	DET	NRLINE	maxL	L	LAM	TT
Original	7.36%	55.9%	2371	408	2.89	72.4%	3.52
t-score	2.58%	32%	259	348	3.86	53.8%	2.72

Moreover, chaos-chaos transitions, as band merging points, inner crises or regions of intermittency[6], represent states with short laminar behavior and cause vertically and horizontally spread areas in the recurrence plot. Vertical (and horizontal) lines occur much more frequently at supertrack crossing points (chaos-chaos transitions) than in other chaotic regimes. The measures characterizing this information are laminarity and trapping time (**LAM** and **TT**). This practical example also indicates that in some cases, t-score could be suitably chosen in the sense of stabilization of information. However, both data sets should be used to obtain full information.

9.4 Discussion

Forest conservation is an important topic, which also closely relates to the wetland conservation. In Chile, there is no specific statutory rules or regulations on wetlands, and current legal standards do not protect equally the different types of inland wetlands, being swamp forests, peatlands, and brackish Andean lakes less protected;

[6]Irregular alternation of phases of apparently periodic and chaotic dynamics (Pomeau-Manneville dynamics), or different forms of chaotic dynamics (crisis-induced intermittency).

see Möller and Muñoz-Pedreros (2014). As we can see in Cai et al. (2018), wetland restoration is integral part of forest conservation, and one of the inputs is measure of chaos. This measure of chaos is developed in our paper. According to Cai et al. (2018), the purpose of the optimization models is to find an optimal restoration measures that can minimize the total investment in wetland restoration projects and obtain additional ecological environment and socioeconomic benefits. The optimization model can also decrease the influence of interval uncertainty in the system by expressing the executed solution as interval numbers with an upper bound and a lower bound. For practical computing of lower/upper bounds, our chaos measurement model in wetland is useful. In White and Fennessy (2005), a GIS-based model is developed to predict the suitability for wetland restoration for all locations in the Cuyahoga River Watershed (2107 km^2), in Northeast Ohio (USA). The multicriterial evaluation method is developed, and our model of chaos of gass (methane) emissions can be one of the important inputs to the system. Evaluating ecological vulnerability (EV) is significant for protecting and promoting ecosystem stability. In He et al. (2018), authors present a prototype framework that can assess EV with integrating spatial analysis of GIS method and multi-criteria decision analysis. They achieve EV map on which decision makers can visually see the results in different regions. Among others they use simple aggregation of the four group indices into general EV value by

$$EV = G_1W_1 + G_2W_2 + G_3W_3 + G_4W_4 \tag{9.4}$$

where $G_i = \sum_{j=1}^{n_i} E_j w_j$ are global values for ecological situation, ecological infrastructure, environmental situation, and human health, respectively, W_i are the corresponding weights, and w_j is the local weight of each attribute E_j (a.o. average annual concentration of toxic gases, nitrogen dioxide, or sulfur dioxide). Our model can be helpful with chaotic time series prediction for CH_4 (e.g., the mean function), which can be one of the attributes in EV. See also Stehlík et al. (2016)

Acknowledgements Milan Stehlík acknowledges the support of project Fondecyt Regular No. 1151441 and LIT -2016-1-SEE-023 mODEC and WTZ Project HU 11/2016. This work was supported by the Slovak Research and Development Agency under the contract no. APVV-17-0568. Data were obtained within the monitoring of carbon cycle in the sedge-grass marsh supported by the projects of the Ministry of Education, Youth and Sports of Czech Republic within the National Sustainability Program I (NPU I) LO1415 and by the Project CzeCOS/ICOS No. LM2015061. We thank the editor and reviewers, whose insightful comments helped us to sharpen the paper considerably.

References

Bohn FJ, Huth A (2017) The importance of forest structure to biodiversity–productivity relationships. In: Royal Society Open Science 4.1. https://doi.org/10.1098/rsos.160521. eprint: http://rsos.royalsocietypublishing.org/content/4/1/160521.full.pdf

Cai B, Zhang Y, Wang X et al (2018) An optimization model for a wetland restoration project under uncertainty. Int J Environ Res Public Health 15(12). https://doi.org/10.3390/ijerph15122795. ISSN: 1660-4601

Clements F (1905) Research methods in ecology. University Publishing Company, Lincoln
Coco MI, RD with contributions of James D Dixon, Nash J (2018) CRQA: Cross-recurrence quantification analysis for categorical and continuous time-series. R package version 1.0.7
Dušek J, Stellner S, Komárek A (2013) Long-term air temperature changes in a Central European sedge-grass marsh. Ecohydrology 6(2):182–190. https://doi.org/10.1002/eco.1256. ISSN: 19360584.
Dušek, J, Hudecová Š, Stellner S (2017) Extreme precipitation and longterm precipitation changes in a Central European sedge-grass marsh in the context of flood occurrence. Hydrol Sci J 62(11):1796–1808. https://doi.org/10.1080/02626667.2017.1353217. ISSN: 0262-6667, 2150-3435.
Dušek J, Stellner S, Pavelka M (2012) Utility model no. 024073. Patent
Eckmann JP, Kamphorst SO, Ruelle D (1987) Recurrence plots of dynamical systems. EPL (Europhys Lett) 4(9):973
Ellenberg H, Mueller-Dombois D (1967) A key to Raunkiaer plant life forms with revised subdivisions. Ber Goebot Inst ETH Stiftg Rubel Zurich 37:56–73
Fabián Z (2001) Induced cores and their use in robust parametric estimation. Comput Stat Theory Methods 30(3):537–555. eprint: https://doi.org/10.1081/STA-100002096
He L, Shen J, Zhang Y (2018) Ecological vulnerability assessment for ecological conservation and environmental management. J Environ Manag 206:1115–1125. https://doi.org/10.1016/j.jenvman.2017.11.059. ISSN: 0301-4797
Kuuluvainen T (2016) Conceptual models of forest dynamics in environmental education and management: keep it as simple as possible, but no simpler. For Ecosyst 3(1):18. https://doi.org/10.1186/s40663-016-0075-6. ISSN: 2197-5620
Marwan N, Romano MC, Thiel M et al (2007) Recurrence plots for the analysis of complex systems. Phys Rep 438(5):237–329. https://doi.org/10.1016/j.physrep.2006.11.001. ISSN: 0370-1573
Marwan N, Wessel N, Meyerfeldt U et al (2002) Recurrence-plot-based measures of complexity and their application to heart-rate-variability data. Phys Rev E 66(2):026702. https://doi.org/10.1103/PhysRevE.66.026702
Möller P, Muñoz-Pedreros A (2014) Legal protection assessment of different inland wetlands in Chile. Rev Chil Hist Nat 87(1):23. https://doi.org/10.1186/s40693-014-0023-1. ISSN: 0717-6317
Odum EP (1969) The strategy of ecosystem development. Science 164(3877):262–270. https://doi.org/10.1126/science.164.3877.262. ISSN: 0036-8075. eprint: http://science.sciencemag.org/content/164/3877/262.full.pdf
R Core Team (2018) R: a language and environment for statistical computing. R Foundation for Statistical Computing, Vienna
Raunkiær C (1905) Types biologiques pour la géeographie botanique. Kongelige Danske Videnskabernes Selskabs Forhandlinger 5:347–438
Spies TA (1998) Forest structure: a key to the ecosystem. Northwest Sci 72(2):34–39
Stehlík M, Aguirre P, Girard S et al (2017) On ecosystems dynamics. Ecol Complex 29:10–29. https://doi.org/10.1016/j.ecocom.2016.11.002. ISSN: 1476-945X
Stehlík M, Potocký R, Waldl H et al (2010) On the favorable estimation for fitting heavy tailed data. Comput Stat 25(3):485–503. https://doi.org/10.1007/s00180-010-0189-1. ISSN: 1613-9658
Stehlík M, Dušek J, Kiseľák J (2016) Missing chaos in global climate change data interpreting? Ecol Complex 25:53–59. https://doi.org/10.1016/j.ecocom.2015.12.003
Tansley AG (1935) The use and abuse of vegetational concepts and terms. Ecology 16(3):284–307. https://doi.org/10.2307/1930070. eprint: https://esajournals.onlinelibrary.wiley.com/doi/pdf/10.2307/1930070
White D, Fennessy S (2005) Modeling the suitability of wetland restoration potential at the watershed scale. Ecol Eng 24(4):359–377. Wetland creation. https://doi.org/10.1016/j.ecoleng.2005.01.012. ISSN: 0925-8574

Zbilut JP (2005) Use of recurrence quantification analysis in economic time series. In: Salzano M, Kirman A (eds) Economics: complex windows. Springer Milan, Milano, pp. 91–104. https://doi.org/10.1007/88-470-0344-X_5. ISBN: 978-88-470-0344-6

Zbilut JP, Webber CL (1992) Embeddings and delays as derived from quantification of recurrence plots. Phys Lett A 171(3):199–203. https://doi.org/10.1016/0375-9601(92)90426-M. ISSN: 0375-9601

Part III
Brazil

Chapter 10
Transdisciplinary Case Study Approaches to the Ecological Restoration of Rainforest Ecosystems

Abdon Schmitt Filho and Joshua Farley

10.1 Introduction

Forests provide critical ecosystem services including the regulation and purification of water and air, wildlife habitat, erosion control, carbon storage, climate regulation, and the renewable provision of valuable raw materials, which are vital to humans and innumerable other species (Alarcon et al. 2015; Myers 1997). About half of tropical forests have been lost, most since the middle of last century, with 2016 and 2017 the worst years in history for tree cover loss (Weisse and Goldman 2018). Tree cover in temperate areas has been more stable over this period, but climate change poses a major new threat to temperate forests, with significant increase in loss to insects, drought and forest fires, as well as shifts in species composition (Millar and Stephenson 2015). Deforestation currently accounts for an estimated 6–17% of global CO_2 emissions (Baccini et al. 2012), and we run the risk of positive feedback loops in which deforestation drives climate change which then exacerbates forest loss (Reyer et al. 2015). Ecological restoration can help counteract these losses and capture carbon in forest soils and tree biomass (Nave et al. 2018). Agriculture has long been the greatest global threat to forests and other ecosystems, but is also essential for human welfare, and is heavily dependent on the ecosystem services forests provide (De Schutter and Vanloqueren 2011; Godfray 2011; Godfray et al. 2010; Rockstrom et al. 2009). A growing and increasingly affluent human population is increasing the demand for food production, thus posing new threats to global forests. One of the most serious challenges humanity currently

A. Schmitt Filho (✉)
Silvopastoral System and Ecological Restoration Laboratory, Department of Animal Science and Rural Development, Federal University of Santa Catarina, Florianópolis, Santa Catarina, Brazil

J. Farley
Community Development and Applied Economics, University of Vermont, Burlington, VT, USA

F. Fuders and P. J. Donoso (eds.), *Ecological Economic and Socio Ecological Strategies for Forest Conservation*, https://doi.org/10.1007/978-3-030-35379-7_10

faces is how to rehabilitate rural landscapes and protect and restore forests and other threatened ecosystems while ensuring the continued provision of adequate food supplies (Farley et al. 2015; Schmitt Filho et al. 2013).

This global challenge plays out at smaller scale in Brazil's Atlantic Forest biome (Fig. 10.1). This biome, extending from the country's tropical Northeast to the temperate South, is among the most biodiverse and endangered biodiversity hotspots on the planet. Estimates of remaining forest cover range from 7% to 28% (Rezende et al. 2018; Ribeiro et al. 2009; SOS Mata Atlantica 2009), depending primarily on the size of vegetation fragments counted as forest. Much of the remaining forest is secondary growth and suffers from significant edge effects, species homogenization, and biodiversity loss (Arroyo-Rodríguez et al. 2017; Joly et al. 2014). The emptiness of forests remnants due lack of keystone animal species is also an eminent threat (Redford 1992; Wilkie et al. 2011; De Coster et al. 2015). Deforestation rates remain high (INPE 2017) while climate change and other ecological disturbances pose new threats with highly uncertain outcomes (Scarano and Ceotto 2015). Perhaps of greatest concern, ecologists have identified an ecological threshold at approximately 30% forest cover, below which there may be a catastrophic loss of vertebrates,

Fig. 10.1 Despite of its biodiversity and high levels of endemism, the Atlantic Forest has experienced continuous habitat loss since colonization (Mittermeier et al. 2011). Only 7–28% of the Forest remains today, depending on fragment size (Rezende et al. 2018; Ribeiro et al. 2009; SOS Mata Atlantica 2009). Few tropical biodiversity hotspots are "hotter" in terms of existing threats and conservation value (Mittermeier et al. 2011; Scarano et al. 2015). (Photo: LASSre 2011)

plant families and presumably other species (Banks-Leite et al. 2014; Lima and Mariano-Neto 2014) threatening the continued viability of the ecosystem and the vital ecosystem services it generates. Fortunately, there can be a long lag time between habitat loss and expected extinctions, known as an extinction debt, offering a window of opportunity to restore forest cover and the populations of threatened species (Kuussaari et al. 2009; Metzger et al. 2009). While apparently few species have gone extinct to date (Joly et al. 2014), the biome accounts for 60% of all endangered species in Brazil (Rezende et al. 2018).

It is difficult to overstate the importance of conserving Brazil's Atlantic Forest. It is the second largest forest ecosystem in South America and likely the most biodiverse. Some two-thirds of Brazil's population lives within its original boundaries and is highly dependent on the life sustaining ecosystem services it provides. The region provides 70% of Brazil's GDP, over half the country's horticultural land (Scarano and Ceotto 2015), 62% of its hydroelectric power, and drinking water for 75% of the population. The forest regulates rainfall patterns, helps control flooding and landslides, and likely has important impacts on the productivity of coastal marine systems (Joly et al. 2014). By the nineteenth century, conversion of forests to agriculture around the city of Rio de Janeiro was already threatening the city's water supply, leading the Emperor D. Pedro II to proclaim much of surrounding watershed as Brazil's first conservation area – now Tijuca National Park – and fund its reforestation (Freitas et al. 2006). Water provision and regulation remains one of the most valued local ecosystem services (Alarcon et al. 2016), while other critically important services extend to regional, national and global scales (Farley et al. 2010). Irreversible degradation or loss of the ecosystem would have immeasurably large costs to this and future generations (Fig. 10.2).

The main driver of forest loss in the system has been and continues to be agriculture, which is the predominant land use in degraded and fragmented areas throughout the biome (Dean 1997). While agriculture accounts for less than 5% of Brazil's GDP, this number is highly deceiving. Food is arguably the most essential and least substitutable of all commodities and exhibits extremely inelastic demand. This means that a small decrease in the supply of food will lead to an enormous increase in price. For example, during the food crises of 2007–2008 and 2011–2012, a combination of droughts, increased biofuel production and speculation caused global grain prices to double (Farley et al. 2015; Lagi et al. 2011, 2012). Brazil is a globally important food producer, and sacrificing agricultural land for ecological restoration could have global repercussions (Scarano and Ceotto 2015); if done on a global scale, it could easily result in catastrophic price increases. Furthermore, small family farmers account for 84% of the farms in the biome occupying 24% of the farmland; small farmers throughout Brazil provide 72% of the country's food base (Ribeiro et al. 2011). In Santa Catarina – the focus of this chapter – small family farmers account for 87% of the farms, covering 44% of the farmland, but are responsible for 70% of all agricultural production and 82% of agricultural employment (Mattei 2019). If these farmers reforest enough of their land to avoid ecological collapse, many will no longer have enough agricultural land to sustain their families.

Fig. 10.2 Today the Atlantic Forest is a collection of fragile and biologically impoverished remnants distributed in 245,173 fragments, 83.4% smaller than 50 ha (Mittermeier et al. 2011; Scarano and Ceotto 2015). (Photo: LASSre 2016)

In summary, it appears that Brazil's Atlantic Forest has already crossed a critical ecological threshold. Failure to restore forest cover to above 30% risks a catastrophic loss of biodiversity and ecosystem services. Since climate change also threatens the Atlantic Forest, it is quite possible that more than 30% forest cover must be restored to ensure resilience. At the same time, restoration of this much forest cover threatens to push small family farmers across an economic threshold of poverty. The Atlantic Forest is a microcosm of a global problem, so viable efforts to address this challenge must be scalable. This chapter will use the lens of ecological economics to examine a case study of a participatory action research project that uses agroecology to promote rehabilitation and restoration of the Atlantic Forest biome and the ecosystem services it generates while enhancing the livelihood of small family farmers and ensuring the continued provision of food.

Agroecology is a broad field defined differently by different practitioners. In the context of this case study, agroecology is a science and practice that strives to adopt ecological principles to agricultural production with the goals of increasing productivity and enhancing socio-economic system stability and sustainability (Altieri 1995). It extends the domain of scientific knowledge to include local knowledge and indigenous practices (Gliessman 2015). Since reliance on non-renewable resources is inherently unsustainable, it seeks to restore natural ecological functions to replace purchased inputs, many of which are expensive toxic chemicals, but does not necessarily reject all use of fertilizers and pesticides. It recognizes that humans are part of the ecosystem, and every ecosystem is different (Moore et al. 2014; Pahl-Wostl 2009). What makes an agroecosystem successful is place dependent (Virapongse et al. 2016). Agroecology is also a movement designed to build a sustainable and equitable agricultural system that empowers farmers by putting them in control of the production process (Méndez et al. 2017). In industrial or conventional agriculture, hybrid seeds, chemical inputs, machinery and farming practices have been designed by corporations and universities, protected by intellectual property rights, and "safeguarded" by lobbying. Farmers must purchase the inputs and have little say in the practices. In distinct contrast, agroecology favors participatory action research in which farmers, scientists and extension agents work together to determine what technologies and practices will best achieve both farmer and community goals. Practices spread from farmer to farmer with no intellectual property rights. Agroecology is inherently transdisciplinary, integrating agronomy, ecology, economics, sociology, anthropology and ethics (Altieri 2002; Gliessman 2015; Méndez et al. 2017). With its recognition that the agricultural economy is sustained and contained by the global ecosystem and its goals of sustainability, just distribution and efficient production, it has much in common with ecological economics.

The project is taking place in Santa Catarina's Encostas da Serra Geral, a mountainous region marking the transition from the coastal regions to the interior plateau (Geremias 2011; Schmitt Filho et al. 2010; Schröter et al. 2015). The epicenter of the project is Santa Rosa de Lima, the State's agroecology capital (Moreno-Peñaranda and Kallis 2010).

The remainder of this chapter is organized as follows. We first provide a site description, including an agricultural history of the community. We then explain how our project uses agroecology to promote restoration of the landscape, including the forest, and the ecosystem services it provides. This section also discusses the legal mandate of Brazil's New Forest Code (NFC). We then turn to ecological economics to understand the challenges of managing land to provide both public and private benefits in a market system before discussing our efforts to apply participatory action research to the policy problem. We provide a general outline of the Payments for Ecosystem Services (PES) scheme emerging from these efforts. We hope our research can contribute to global efforts to synergically integrate agroecology, agroecosystem rehabilitation and ecological restoration.

10.2 Site Description and History

Santa Rosa de Lima is a small municipality (202,004 km^2, ~2130 inhabitants) in the Encostas da Serra Geral (coastal mountain slopes), Southern Brazil (Fig. 10.3). The mountainous terrain is poorly suited to large-scale mechanized agriculture or annual crops, and most land is currently in family farm parcels averaging 18 ha

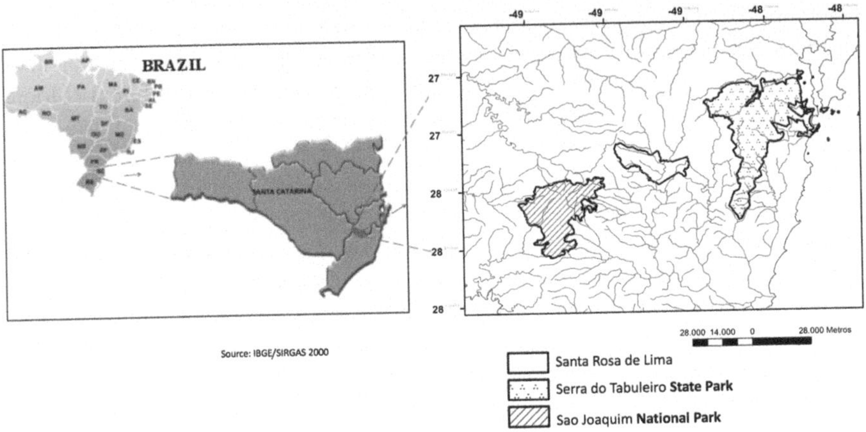

Fig. 10.3 The study site: Santa Rosa de Lima, Santa Catarina, Brazil. (LASSre 2018)

(IBGE 2017). The region was originally farmed by the Tupi-Guarani, semi-nomadic horticulturalists whose slash and burn practices created a resilient mosaic of cleared areas, second growth and natural forest, but were largely exterminated by disease, European colonists, State, and Church (Barreto and Drummond 2017; Dean 1997; Denevan 1992). European settlers arrived in Santa Rosa de Lima at the start of the twentieth century, where they practiced highly diversified, primarily subsistence slash-and-burn agriculture on small family farms, with excess production and pig lard sold for a limited cash income (Moreno-Peñaranda and Kallis 2010). Farmers used few purchased inputs, relying instead on nutrient cycling, biological pest control, and other ecosystem services.

In the 1950s and 1960s, extensive forest was cleared for timber and charcoal and farmers began planting exotic plantations of eucalyptus and pine. Farmers also began growing tobacco, helped by extension packages of training and credit provided by tobacco companies who also offered a guaranteed market. By the late 1980s, most farmers were combining tobacco production with traditional subsistence practices (Moreno-Peñaranda and Kallis 2010). Unfortunately, tobacco quickly depleted soil nutrients and accelerated erosion. Purchased pesticides and fertilizers made from non-renewable resources were required to replace the resulting loss of ecosystem services but had negative impacts on human health and the environment. Coupled with falling tobacco prices in the 1980s, farmers began looking for better alternatives. Many turned to dairy and some to Eucalyptus monocrop plantations (Schröter et al. 2015).

In the late 1990s, local family farmers began working with UFSC professors on agroecological production, including organic vegetable and pasture-based dairy, as an alternative to tobacco and conventional agriculture. Together they created the "Associação dos Agricultores Agroecológicos da Encosta da Serra Geral - AGRECO" (http://www.agreco.com.br/) – an agroecological cooperative (Geremias 2011; Schmitt Filho et al. 2010). AGRECO managed organic certification for participating farmers and facilitated marketing and development of local, value-added enterprises. It also attracted an influx of agro-tourists further stimulated by the organization of an agrotourism cooperative – "Acolhida na Colonia" (welcomed by the community http://acolhida.com.br/contato/).

In 1998, the former Voisin Grazing Group – GPVoisin/UFSC (since renamed and hereafter referred to as the Silvopastoral System and Ecological Restoration Lab (LASSre/UFSC)) began working with community farmers, professors and university students to implement and disseminate agroecological pasture based dairy through Management Intensive Grazing – MIG (Geremias 2011; Murphy 2010; Schmitt Filho et al. 2010; Schröter et al. 2015). In MIG, farmers divide their pastures into numerous paddocks, which are grazed intensively for short periods then allowed to rest until pasture returns to optimal grazing height. This practice maximizes pasture production, avoids the selective grazing of early regrowth, disrupts the life cycle of some parasites, and restores soil biocenosis (Battisti et al. 2018; Dorsey et al. 1998; Savory 2016; Sovell et al. 2000; Teague et al. 2011) disrupted by tobacco farming. Farmers were initially skeptical of MIG and reluctant to adopt it. LASSre/UFSC organized numerous field days and workshops to help farmers understand the practice and tailor it to farmers' needs. One farmer with highly degraded land and little to lose agreed to try it. His success helped persuade other farmers to adopt MIG as well (Schröter et al. 2015). Subsequent research found that MIG farmers in the region on average sustain higher stocking rates on better pasture, produce more milk per cow, and increase their incomes while enhancing ecosystem services (Alvez et al. 2014; Back et al. 2009; Farley et al. 2012; Maurer et al. 2009; Schmitt Filho et al. 2013; Surdi et al. 2011). MIG proved so successful that it has received state-wide government support since 2006, and it is quickly becoming the standard approach to grazing in the state (Jochims and Leopoldino da Silva 2016; Schmitt Filho et al. 2008).

10.3 Agroecology and Reforestation

Though agroecology was achieving notable economic and environmental success in the region, it was having limited impact on forest cover (Amazonas et al. 2016). Not only did failure to restore forests risk the continued provision of ecosystem services on which agriculture depends, it was also illegal. Adopted in 1965, Brazil's forest code mandated permanent forest cover in critical ecological areas such as 30 m borders in riparian zones (and more for larger rivers), hill tops and steep slopes, known collectively as the Areas of Permanent Protection (APP). It also required forest cover over an additional 20% of rural Atlantic Forest properties, known as the legal reserve (RL). If fully enforced, the law would have resulted in ~30% forest cover over the entire Atlantic Forest (Metzger 2010). Adopted in 2006, the Atlantic Forest law offered additional protection for the biome, including a prohibition on deforestation. Neither law was well enforced. Native forest cover in Santa Rosa fell from 72% in 2002 to 51% in 2010 (Amazonas et al. 2016), forest cover in the state as a whole is only 23%, and deforestation in the state and the biome as a whole has recently accelerated (INPE 2017).

Nonetheless, in 2009, the governor of Santa Catarina announced that the forest code was too stringent, and if fully enforced, would drive the state's small family farmers into poverty. Stating that Santa Catarina must choose between larger farms or larger slums, he unilaterally declared that Santa Catarina would adopt a more lenient forest code and no longer respect the national code (Globo 2009). This triggered a major national debate over the code, resulting in the adoption of a new forest code (NFC) in 2012. This more lenient forest code offered an amnesty for deforestation that had taken place before July, 2008, reduced riparian buffer widths from high water levels to mean water levels, further reduced buffer widths for small family farmers, and also allowed small family farmers to count agroforestry systems as part of their APPs, among other changes. Numerous publications describe the law in detail (Alarcon et al. 2015; Presidência da República do Brasil 2012; Soares-Filho et al. 2014; Sparovek et al. 2012). While overall the new forest code reduces reforestation requirements, it has also led to significant steps to enforce the code. For examples, farmers were required to submit an environmental registry of their properties that shows the current level of compliance with the code and the steps they will take to achieve full compliance. LASSre has worked with numerous farmers in the region to complete the environmental registry and found that the vast majority of farms will require extensive restoration. Even with the less stringent reforestation requirements, the upfront costs of ecological restoration and the loss of productive farmland presents a serious challenge for local farmers and policy makers.

LASSre is therefore using participatory action research (Méndez et al. 2017) to work with farmers, local communities, students, state extension agents, non-governmental organizations, and local and state governments to develop agroecological practices that could synergically improve forest cover and ecosystem services, rehabilitate pastureland, comply with the forest code, avoid ecological thresholds and improve farmer livelihoods. The project integrates agronomy, forestry, ecology, economics, policy analysis, community development, and other tools and disciplines to achieve its goals. We first discuss the agroecological practices being adopted and the goals we hope to achieve, followed by a discussion of the ecological-economic theory relevant to the provision of ecosystem services, and concluding with our efforts to build effective policies to advance this initiative.

Our participatory research has led to the development of two specific agroecological systems designed to achieve our goals: Multi-Function Riparian Forests (MultRF) and High Biodiversity Silvopastoral Systems (SPSnuclei) based on the reforestation strategy of applied nucleation (Corbin and Holl 2012). Both systems were designed to make ecological restoration and pastureland rehabilitation economically viable through the provision of non-timber forest products from native species in the context of smallholder agriculture and food sovereignty.

10.4 Multi-function Riparian Forests: MultRF

In December 2015, we initiated a Mult-RF pilot project in Rio dos ´indios, in the municipality of Santa Rosa de Lima, Santa Catarina, Latitude S28 ° 02.772 and Longitude O49o11'10.6,. The Rio dos ´indios has two identified sources, one at Lat. 27 ° 58.815 S and Long. 049 ° 16.023, W and the other at Lat. 27 ° 58,875 S and Long. W049 ° 16.019, W, with the mouth at Lat. 28 ° 03.769'S, ending at the Braço do Norte River. The river has an extension of approximately 17 km (Fig. 10.4).

The design and implementation process were participatory, starting with contacting local leaders of the Associação Mata Verde – Microbacias Rio dos Indios. Subsequently, meetings were held with the farmers to discuss and design the Multifunctional Riparian Forest System (MultRF). Seventy-seven families enrolled in the project, however it was possible to restore only those located in the first 3000 m from mouth of the Rio dos Indios (Amazonas et al. 2017; Carvalho Filho et al. 2016).

The MultRF is implemented in riparian corridors with a minimum of 8 m wide on either side of waterways to comply with Brazil's New Forestry Code. Riparian zones are fenced off and are allowed to fill in through natural succession. The methodology is based on functional diversity (Laureto et al. 2015; Petchey and Gaston 2006) to maximize ecological restoration through successional processes.

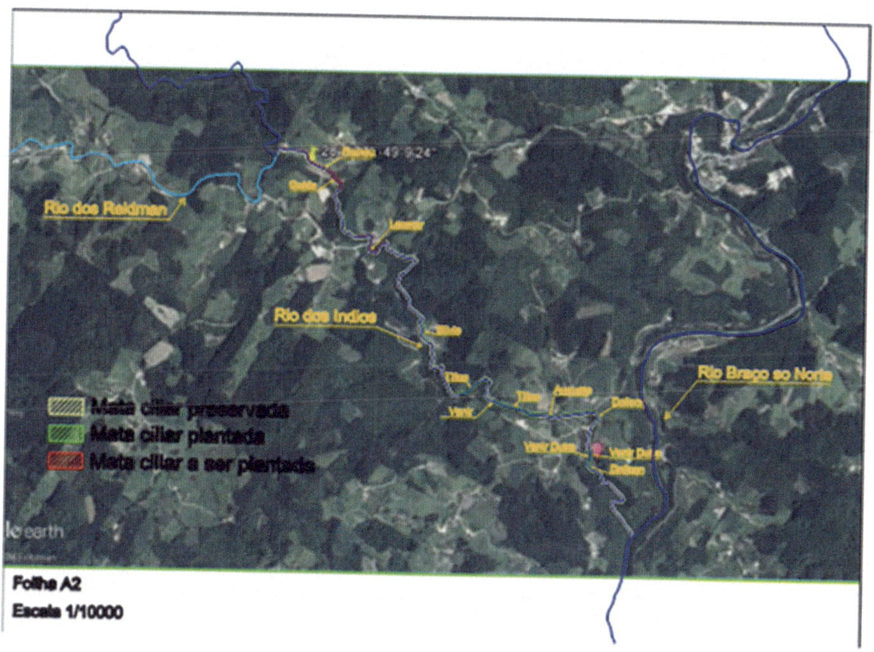

Fig. 10.4 Pilot Project area of Mult Function Riparian Forest (MultRF) – Farmers that implemented MultFR in the watershed of Tubarão River and Complexo Lagunar. (LASSre 2014)

Species selected was based on functional diversity and economic benefits (Secretaria do Meio Ambiente do Estado-SP 2011).

Five functional groups (FG) where carefully selected. Functional Goup 1 (FG1), the Provisioning/Primary Production Group, was composed of banana plants (*Musa sp.)* and rose pepper (*Schinus terebinthifolius*). Banana was the only non-native species, included to provide short-term income. Rose pepper is a valuable spice, produces good honey, and compounds in the tree show promise as pesticides and as antibiotics to which bacteria may be unable to develop resistance (Orwa et al. n.d.; Uliana et al. 2016). These two species are responsible for the initial revenue from non-timber forest products (NTFP) and are responsible for the farmers' initial engagement in the participatory ecological restoration process (Carvalho Filho et al. 2016).

Functional Group 2 (FG2), the Landscape Restructuring/Succession Facilitator group, is composed of fast-growing pioneer species that rapidly transform the landscape and create the necessary conditions for the emergence of an understory. The main species used is Bracatinga (*Mimosa scabrella* – Fabaceae), supplemented by five additional species (*Inga senssilis* – Fabaceae; *Anadenanthera macrocarpa* – Fabaceae; *Piptadenia gonoacantha* – Mimosoideae; *Citharexylum myrianthum* – Verbenaceae; *Nectandra lanceolata* - Lauraceae) – all species known by the locals for their fast growth, dense shade and economic value (Carvalho Filho et al. 2016). *Bracatinga* for example is among the fastest growing trees in the Atlantic Forest. It fixes nitrogen, produces fodder for animals, provides abundant flowers for honey bees when most trees are not in flower, provides exceptional shade cover in 3 years, and can be harvested for timber or charcoal after other species have become established (Orwa et al. n.d.). It could potentially replace exotic Eucalyptus or Pinus species as a source of timber and wood-pulp.

Functional Group 3 (FG3), the Key stone Species/Provisioning group, is based on the Juçara Palm (*Euterpe edulis*). *E. edulis* is a keystone species almost exterminated for heart of palm (Fig. 10.5). It produces a valuable berry nearly identical to that of its close cousin, açai – *Euterpe oleracea* (Carvalho Filho et al. 2016). Demand and processing capacity in the region greatly exceed supply of the fruit, international markets are growing, and it earns a substantial premium over açai berries (Fadden 2005; Homma et al. 2006; Silva Filho 2005; Trevisan et al. 2015).

Functional Group 4 (FG4), the Enrichment group, consists of two nuclei of six late secondary and climax trees per 100 m to enrich the forest with high biodiversity at an acceptable cost. Functional Group 5 (FG5), the Pollinators group, consists of two hives of native *Meliponas* bees per 100 m of riparian forest (Buratto 2019; Buratto et al. 2018a; Carvalho Filho et al. 2016).

The spatial distribution of species across each site is designed to offer a rapid and cost-effective successional framework for restoration of riparian ecosystems. The design of our working module was deeply participatory and based on the forest cover criteria established by the NFC specifically for family farms in the region (Alarcon et al. 2015; Sparovek et al. 2012). The working module dimension is 8 m width by 100 m length for each side of a waterway. The implementation process for each 8 × 100m module starts with ten rose peppers planted as living fence posts 8 m

Fig. 10.5 The Juçara Palm (*Euterpe edulis*) that produce the Southern açai is a keystone species almost exterminated for heart of palm. It produces a valuable berry nearly identical to that of its close cousin, açai – *Euterpe oleracea* (Carvalho Filho et al. 2016; Buratto et al. 2018c; Buratto 2019). Demand and processing capacity in the region greatly exceed supply of the fruit, international markets are growing, and it earns a substantial premium over açai berries (Fadden 2005; Homma et al. 2006; Silva Filho 2005; Trevisan et al. 2015). LASSre 2014

from the river, followed by 20 bananas (FG1) and 20 fast growing pioneer species (FG2), all planted inside the fenced area. In the second and third years, two juçara palms are planted next to each banana plant (FG3) and two high diversity nuclei (FG4) are placed equidistantly in the module jointly with two native bee hives (FG5) (Carvalho Filho et al. 2016).

The economic analysis of a MCmult module for a period of 10 years estimated an Internal Rate of Return (IRR) of 27%, four times greater than the current Selic rate of 6.5% (Banco Central do Brasil 2018), and a payback of 6 years. The MultRF system promises better financial returns than savings accounts and most other investment options available to farmers, and simultaneously complies with the NFC requirements for environmental regularization of the property. Banana and rose peppers strategically provide a revenue stream in the early years to motivate farmer involvement before the establishment of dense canopy cover and the growth of Juçara Palm for açai berry (Buratto 2019; Buratto et al. 2018a, b, c; Carvalho Filho et al. 2016; Kohn 2017).

In addition to its economic benefits and compliance with the law, the MultRF provides multiple ecosystem services, such as water regulation and purification, and control of "borrachudos" (*Simulium spp.*), biting black flies that torment farmers and cattle. If the economic benefits prove adequate, we anticipate farmers will consider expanding the MultRF to 30 m riparian buffers, as required by the pre-2012 forestry code. This would increase the provision of many critical ecosystem services and help to achieve the 30% forest cover required to avoid collapse of the Atlantic Forest Biome (Banks-Leite et al. 2014; Metzger 2010; Silva et al. 2011).

Brazil urgently needs policies that enable the ecological restoration of heavily anthropized rural landscapes, especially in the Atlantic Forest. Strategies that synergistically provide ecological restoration and agriculture products are necessary conditions for the enthusiastic immersion of the stakeholders (Schmitt Filho et al. 2017). The involvement of multiple stakeholders facilitated the emergence of a practical approach with high value non-timber forest products. Economic returns and the reduction of environmental liabilities can make MultRF a public policy option to ensure Brazil is able to meet large scale restoration targets required to comply with the Intended Nationally Determined Contributions (INDCs) under the United Nations Framework Convention on Climate Change (UNFCCC) for reductions in greenhouse gas emissions (Brancalion et al. 2012; Pinto et al. 2014).

10.5 High Biodiversity Silvopastoral Systems: SPSnuclei

Applied nucleation for reforestation involves small clumps of trees dotted across the target landscape. Areas between the clumps are allowed to fill in on their own. The process is designed to mimic natural succession, and to be more effective and less expensive than other restoration techniques (Corbin and Holl 2012). Based on this approach, the SPSnuclei system utilizes 40 equidistantly placed nuclei per hectare, corresponding to 10% of the pasture area (Fig. 10.6). The system provides shade, revenue from non-timber forest products (NTFP), biodiversity (50 native tree species/ha), connectivity, and an improved landscape matrix, in addition to numerous ecosystem services (Schmitt Filho et al. 2017).

The 5 × 5 m nuclei of native agroforest are integrated into treeless pastureland. Grazing by cattle prevents natural succession within the pastures, but the fenced off nuclei significantly increase tree cover and its accompanying benefits. This adopted nuclei density was the most suitable for the project landscape and the stakeholders' goals of achieving 30% shade cover by the third year, additional income from a multifunction agroecosystem and the potential for repaying initial investments by the sixth year. As in the MultRF, tree selection was based on functional diversity (Laureto et al. 2015; Petchey and Gaston 2006) and economic benefits (Secretaria do Meio Ambiente do Estado-SP 2011). The chosen species and planting sequence are similar enough to the MultRF (Schmitt Filho et al. 2017, 2018b).

In the first year, each nucleus is planted with four rose peppers as a living post in each of four corners (5 × 5 m), six fast growing pioneer trees from six species to help capture the site, and four banana plants on the edges between two rose peppers. Once shade cover has been established by year 3, the nuclei are enriched with two juçara palms (*E. edulis*) placed next to each banana plant and two climax species

Fig. 10.6 Five year-old high biodiversity silvopastoral system (SPSnuclei) at Rosangela's farm in Santa Rosa de Lima. SPSnuclei shows a discounted payback of 5 years and an internal rate of return of 57% (Schmitt Filho et al. 2018a). (Photo: LASSre 2017)

in the center of the nucleus under the full shade in a process known as successional agroforestry (Young 2017). The Climax species are different for each nucleus, with a minimum diversity of 50 tree species per hectare (Schmitt Filho et al. 2017, 2018b).

The SPSnuclei cover 10% of the pastureland, but within 2 years they provide shade cover over approximately 30% of the pasture area. Shade alleviates heat stress and can therefore increase milk production by as much as 20% during the hot months of late spring, summer and early fall (Craesmeyer et al. 2017; Deniz et al. 2018; Kretzer et al. 2018). Shade cover is one of the farmers' primary motivations for adopting SPSnuclei (Joseph et al. 2016). Rose peppers and bananas begin producing fruit in 2 years, and *E. edulis* within seven. Estimates of the internal rate of return (IRR) for *E. edulis* agroforestry systems with 625–1000 trees per hectare range from 21% (Andrade 2016) to 67.5% (Silva Filho 2005). One hectare of SPSnuclei when fully established includes 320 *E. edulis*, 160 Rose pepper, and 160 bananas per nuclei. Ten native beehives per hectare to provide honey and pollination are also part of system. Our economic analysis estimated that over a 10-year period, a hectare of SPSnuclei generates a Net Present Value (NPV) of $15,700 from banana, rose pepper, honey and açai (after the eighth year). The estimated IRR was 57%, significantly higher than the minimum rate of attractiveness (MRA) of

7%, as measured by the Selic. Implementation costs could be recovered in 3–4 years (Schmitt Filho et al. 2018a).

Ongoing research confirms that the agroforest nuclei sequester carbon in soils (Battisti et al. 2018) and biomass (Silva et al. 2018), improve nutrient cycling (Battisti et al. 2018), provide habitat for pollinators and other species (Simioni et al. 2019), improve dairy cows ambience and welfare (Deniz et al. 2018), and regulate the microclimate (Kretzer et al. 2018). We anticipate numerous other ecological benefits as well.

Our analysis suggests that under suitable conditions, the system should readily spread farmer to farmer with adaptations for specific environmental, economic and social conditions. Continual innovations can improve both economic and ecological benefits. The locus of control and power will remain with the farmers. One serious challenge however is that interest rates on debt in Brazil are among the highest in the world, often surpassing 300% annually on credit cards (Trevisani 2018). The opportunity cost of any investment is therefore exceptionally high for indebted farmers. Furthermore, the benefits of these systems extend well beyond the farmers' fields, and there are currently no well-established mechanisms for compensating farmers for these benefits. If adoption is left entirely up to the farmers and their limited resources, the systems is unlikely to spread at the scale and speed necessary to have a meaningful impact on the Atlantic Forest biome. We now turn to ecological economics for an improved theoretical understanding of the challenges we confront and possible solutions.

10.6 The Ecological Economics of Ecosystem Services and Agroecology

Society must develop and implement food systems that produce enough to feed the world while synergically restoring and protecting forests and other ecosystems. Our current economic system does little to advance this goal.

Early modern economists, specifically the physiocrats and classical economists, recognized that land and other natural resources inputs were essential to all economic production. The physiocrats in particular viewed agriculture as the foundation of the economic system (Mazzucato 2018). As modern neoclassical economics developed however, there was greater and greater emphasis on the role of capital, technology and labor, until land and natural resources were finally removed from the production function all together (Farley 2012b). If nature contributed nothing to production, there was no need for economists to account for it. Worries over natural resources constraints returned in the 1950s, but a series of economic studies in the 1950s and 1960s convinced most economists and policy makers that resource scarcity would drive price increases which in turn would reduce demand and stimulate the innovation of substitutes. Prices depended on relative scarcity, which was therefore important, but technology ensured that absolute scarcity was not a problem (Barnett and Morse 1963; The President's Materials Policy Commission 1952). Economists largely ignored concerns about the world's ability to feed growing populations and viewed the green revolution as proof they were correct. They were similarly dismissive of Meadow's et al.'s (1972) Limits to Growth (Beckerman 1972; Solow 1972). Pollution and environmental amenities, which later came to be known as ecosystem services, propelled by Rachel Carson's Silent Spring (1962) and similar works, were seen as a greater challenge than resource scarcity since they had no market prices to signal their growing scarcity. Again, however, a mainstream consensus emerged that if ecological costs were internalized in prices, markets and technological advance would solve these problems as well (Simpson et al. 2005). Economists remain relatively complacent about both food production and environmental degradation.

Ecological Economics in contrast views the economy as embedded in a finite global ecosystem. All economic production uses energy, labor and capital to physically convert finite stocks of raw materials from nature into economic products. Most energy comes from finite stocks of fossil fuels. We choose how fast to extract and burn fossil fuels, mine metals and other minerals, clear our forests, and fish our oceans. The first law of thermodynamics guarantees that matter-energy cannot be created or destroyed, while the second guarantees that entropy, or disorder, increases. This means that all economic products wear out, break down, and return to nature as waste, and the combustion of fossil fuels unavoidably generates waste emissions. If waste flows exceed the capacity of ecosystems to break down and absorb them, they accumulate as waste stocks. If renewable raw material stocks are extracted faster than they can reproduce, their stocks are depleted. Stocks are the summation of flows. These resources are therefore known in ecological economics as stock-flows (Daly and Farley 2011; Georgescu-Roegen 1971).

All stock-flow resources are rival, meaning that use by one person leaves less for others. When a rival resource is scarce, meaning that there is not enough available for all desired uses, economic and ecological competition for the resource is unavoidable. Most stock flow resources can be made excludable, which means it is possible to assign and enforce property rights that allow one economic actor to use the resources and to prevent others from doing so. Property rights are required for markets to exist. Markets use the price mechanism to ration access to resources, and when resources are rival, rationing prevents over use (Daly and Farley 2011; Georgescu-Roegen 1971).

The raw materials converted into economic products alternatively serve as the structural building blocks of ecosystems, which consist of a particular configuration of stock-flow resources. Healthy ecosystems are funds that generate a flux of ecosystem services. While forests are physically converted into timber, fuel and other economic products, ecosystems are not physically converted into the ecosystem services they provide. A forest can filter water, regulate water flows, sequester carbon and other pollutants, provide a space for recreation, serve as habitat for innumerable species and reproduce themselves without being physically transformed into something else in the process. Ecosystem funds generate ecosystem services at a rate over time that depends on the size and health of the ecosystem in question (Farley 2012a; Farley and Costanza 2010). We cannot save up services for later use because the flux of services cannot be stockpiled. Labor and capital have similar properties: they produce economic products at a rate over time that cannot be stockpiled. If a cook can make one pizza every 15 min and an oven can cook one in 15 min, we can't stockpile this capacity by leaving both unused for an hour than producing five pizzas in the following 15 min. They are not physically transformed into the pizzas they make, but merely worn out over time by the production process. Labor must get food and rest, and capital must be maintained. Ecosystems in contrast are continually restored by solar energy, which itself is available at a fixed rate over time, unlike the fossil fuels we can extract as fast as we like. Labor, capital, knowledge and ecosystems are all funds that generate a flux of services, and hence are known as fund-service resources (Daly and Farley 2011; Georgescu-Roegen 1971).

Ecosystem services are conventionally defined as the benefits nature provides to people (Fisher et al. 2008; Millennium Ecosystem Assessment 2005), but this definition lumps together the stock-flow and fund-service elements of natural capital, obscuring important distinctions. The stock-flow elements are called provisioning services. Analytically, it is much more useful to treat provisioning services as the capacity of an ecosystem to reproduce itself. Trees and fish can reproduce at a rate over time and are not physically transformed into their offspring. In this chapter, we classify all ecosystem services as fund-service (Farley and Costanza 2010). When the market economy first emerged, human made products were scarce and ecosystem goods and services relatively abundant and could reasonably be ignored. Now human made products are abundant and natural capital is scarce, and we must invest in the latter (Daly 2014).

Because fund-services are not physically converted into what they produce, many are non-rival, meaning that use by one person does not leave any less for others. For example, enjoying a beautiful view does not leave less available for someone else, or if a forest regulates water flows and prevents flooding, the benefits one person gains does not leave less for others. All non-rival resources are fund-services. Using prices to ration access to non-rival resources is inefficient, since it reduces use and hence benefits without reducing societal costs. For example, if someone develops a set of agroecological practices that increase yields without the application of fertilizers or pesticides, or a vaccine that cures a contagious disease, it might be possible to patent them and restrict use to those who pay royalties. However, society as a whole suffers when use is restricted. Many ecosystem services are also non-excludable. It is basically impossible to establish property rights to a stable climate, flood regulation, clean air, wild pollinators, sunshine, rainfall and so on. Markets are impossible for non-excludable resources, and inefficient for non-rival ones (Daly and Farley 2011; Farley and Costanza 2010; Farnsworth et al. 1983).

The concepts of rivalry, excludability, stocks and funds help explain the economic and political challenges to creating agroecosystems that help protect and restore forests. Many of the ecosystem services provided by intact forests and well-designed agroecosystems are both non-rival and non-excludable. The owner of the forest is therefore rarely compensated for the provision of these services. In contrast, the timber harvested from the forests is a rival and excludable market good. Even if the ecosystem services provided by the forest are far more valuable to society than the timber it could produce, markets will incentivize logging over conservation. A forest owner may choose to log sustainably, relying on forest's provisioning service. However, it is quite possible that profits from deforestation can be invested in other economic activities with rates of return exceeding the regrowth rate of the forest, in which case unsustainable harvests may maximize market benefits for the owner. Forests can also be cleared for agriculture. Agricultural land is a fund service, but the products it generates are marketable stock-flow resources, which creates an additional incentive for forest conversion.

The MultRF and SPSnuclei are designed to help restore critical ecosystem services that are often non-rival, non-excludable or both. Forest restoration sequesters carbon, with global benefits of increased climate stability. It increases habitat for numerous forest species, including migratory songbirds that help control insect pests in other regions of Brazil and other countries and pollinators that benefit adjacent farmers. It purifies water, reduces sedimentation, and regulates water flows, providing downstream benefits by lowering the costs of providing drinking water and hydroelectricity, and decreasing floods and droughts (Daily 1997; Millennium Ecosystem Assessment 2005; TEEB 2008). While the farmers ideally benefit from improved nutrient cycling, micro-climate regulation and increased provisioning services, these are only a small fraction of the ecosystem services provided, which presents at least two serious problems. First, investing in these agroecosystems has significant upfront costs. Farmers cannot be certain that all the trees they plant will survive, or that if they do survive, the additional benefits will be worth the costs. This is especially true because it takes several years for all the benefits from agroforestry and silvopastoral systems to materialize, while the costs are immediate. The combination of public good benefits not captured by farmers, Brazil's exceptionally high interest rates and the uncertainty inherent to farming, leaving investment decisions to farmers alone would result in far less adoption of the MultRF and SPSnuclei than is socially desirable. It is also simply unfair to have farmers shoulder the costs when the rest of society benefits. Second, while MultRF and SPSnuclei are well designed systems, improvements are always possible. If those who benefit from the ecosystem services generated do not contribute to the costs of adoption or otherwise compensate farmers for the ecosystem services they provide, future improvements are likely to focus only on benefits to farmers and may therefore fail to achieve the socially optimal balance of benefits (Schmitt Filho et al. 2013).

10.7 Participatory Action Research and Payments for Ecosystem Services

Participatory action research seeks to develop action-oriented research projects in close collaboration with stakeholders. Agroecology practices designed to restore and rehabilitate rural landscapes and forests, enhancing ecosystem services have numerous beneficiaries at multiple spatial and temporal scales. While farmers are arguably the most important participants in agroecology projects, the broad distribution of the benefits provided means it is necessary to involve other key stakeholders that share the benefits and should therefore help shoulder the costs and help with design. Coordinating research and cost sharing with all the individual beneficiaries of the public good benefits generated by agroecology would confront excessive transaction costs. Working with the public sector (governments), civil sector (NGOs) and farming cooperatives that represent all the beneficiaries can significantly reduce transaction costs (Coase 1960). In the context of our project, participatory action research has involved all of these different actors.

Throughout the process of designing the MultRF and SPSnuclei, we have therefore worked closely with local, state and even federal government to help design and incentivize the project and shoulder its costs. Recognizing the benefits for the community and local ecosystems, the municipal government has been an important partner. It has hosted several workshops, training sessions and field days tied to project and participated in others. It has also contributed numerous in-kind resources and recently signed a formal commitment to continue doing so (Farley et al. 2018; Schmitt Filho et al. 2013).

One of our major goals throughout this project has been to obtain state government support for the project. EPAGRI, the public agricultural research, development and extension office, already has a mandate to provide technological and financial support for the state's farmers, and EPAGRI extension agents have played an important role in the project for years. Many EPAGRI employees worked on this project as students, providing a depth of knowledge and support that would otherwise be difficult to achieve. However, we firmly believe that the public good benefits of ecosystem services justify stronger levels of support, especially in the context of looming ecological thresholds. Brazil is a world leader in the development and adoption of payment for ecosystem services (PES) programs (on this also chapter Rosa and Börner in this book), and Santa Catarina is one of the many states that has adopted a state-wide program designed to pay for watershed management, carbon sequestration, and/or biodiversity conservation. We have worked extensively with state ministries to design a PES program for agroecology tailored to the needs of both the farmers and the state.

Most PES programs, including those in Santa Catarina and elsewhere in Brazil, compensate landowners for land use practices designed to enhance ecosystem services, such as riparian reforestation solely for ecological benefits. Participation in the programs is typically voluntary, and landowners receive an annual payment that compensates for the opportunity cost of the new land use, which is to say the money they can no longer earn from economic production on the same land (Pagiola et al. 2013). We believe there are several serious problems with such an approach. First, PES programs of this support depend on continuous payments, which in turn depend on political will and fiscal stability, both of which can be highly unstable. For example, while Brazil's previous government did much to protect the environment, the actual government, which took power in January, 2019, is threatening to end numerous forms of environmental protection, raising serious questions about the future of national PES schemes. A similar ideological shift in Santa Catarina could threaten their state-wide program. Second, farmers are only likely to participate in voluntary programs if payments offset opportunity costs, but opportunity costs for farmers increase with food prices or with new technologies that increase yields. For example, when corn prices in the US increased in 2007–2008, 30% of the land in the national Conservation Reserve Program was withdrawn and planted to corn (USDA 2014). Global food prices are expected to increase in coming years (FAO 2011), increasing the risk that farmers would withdraw their land from voluntary payments, though this is less likely for land where forest cover is mandatory. Third, there is evidence that people have intrinsic motivations to do what is best for society. However, providing extrinsic motivations such as monetary payments can crowd out intrinsic motivations (Bowles 2008; Frey and Jegen 2001; Reeson and Tisdell 2008; Rode et al. 2014). In some cases, crowding can be so severe that the desired behavior is actually reduced (Gneezy et al. 2011; Gneezy and Rustichini 2000). In contrast, the desire to reciprocate for gifts or assistance is considered an automatic 'click-whirr' response (Cialdini 1993) suggesting that if farmers receive government assistance, they may feel a moral compulsion to reciprocate by providing ecosystem services.

Our goal was to develop a PES scheme (defined as "a transfer of resources between social actors, which aims to create incentives to align individual and/or collective land use decisions with the social interest in the management of natural resources" (Muradian et al. 2010, p. 1205) acceptable to farmers and the state that would confront these problems. The State Secretariat of Agriculture has tentatively agreed to fund a pilot project to help approximately 50 farmers in three small watersheds to adopt to MultRF and SPSnuclei agroecosystems on their properties. The state will provide subsidized loans to the farmers to cover upfront costs. If the farmers implement these agroecosystems, loans will be interest free and payments deferred until the tree crops begin producing. Given current interest rates, this would represent a substantial subsidy. Since after 5 years we expect these systems will increase farmer income, it is appropriate they share in the costs and pay back part of the loans, 75%. Twenty five percent of the loans and the interest will be the actual PES from the Fundo de Desenvolvimento Rural (FDR). The repayments will be contingent on increased revenue from NTFP from both Multifunction Riparian Forests (MultRF) and High Biodiversity Silvopastoral Systems (SPSnuclei). Since farmers receive the payments up front, the system will not collapse if in the future there are no resources or political will to continue the system. Ideally, loan repayments will create a rotating fund into FDR that can be loaned out to subsequent cohorts of farmers. We hope that farmers will interpret the PES funding as a form of government assistance rather than a payment for a specific land use, and it will therefore not crowd out the intrinsic motivation of farmers to manage their land for social benefits. Since the agroecosystems will continue producing food and other commodities, increasing commodity prices will also increase farmer incomes, even if they simultaneously increase opportunity costs.

10.8 Summary and Conclusions

The need to feed the planet while restoring essential ecosystems and the services they generate presents humanity with a serious dilemma, since agriculture is the greatest threat to global ecosystems and is simultaneously dependent on the ecosystem services they provide. There is an ongoing debate about whether the best way to achieve this goal is to use chemical intensive industrial agriculture to maximize food production on some land while leaving the remainder wild (sparing), or to adopt agroecology and agroforestry practices that synergically generate both food and ecosystem services on the same land (sharing) (Phalan et al. 2011a, b; Tscharntke et al. 2012). Conventional agriculture has displaced rural labor, driven extensive rural-urban migration of the surplus labor force, and disrupted both rural and urban communities while greatly decreasing natural ecosystems and forest cover, and pushing family farmers towards the last forest remnants of the Atlantic Forest Biome. It also places control over the research process in the hands of large corporations and universities, and uses patents to extract monopoly rents.

In Southern Brazil, most forest remnants are located on private lands, with a nation-wide legal deficit of 19 million hectares that must be restored to comply with Brazilian legislation. Thus, farmers decisions upon different land uses can determine the future of ecological restoration in the country. These decisions are often influenced by social and economic factors. In such a scenario, strategies that help local farmers restore their lands and synergically generate an economic return on investment are crucial to wide spread ecological restoration. Initiatives that incorporate income-generation activities can be the drivers of ecological restoration in a small holder dominated rural landscape. Regular income and legitimacy under the current environmental legislation provide security and confidence for family farmers.

Multifunction Riparian Forest (MultRF) as an ecological restoration alternative was designed with potential to be widely adopted among family farmers. The model was developed using participatory approach (PAR) in the rural landscape of Santa Rosa de Lima, Southern Brazil. In analyses for a 10-year period, MultRF showed a 6 year payback and an Internal Rate of Return (IRR) of 27%, four times greater than the current Selic rate. The income from NFTP on the MultRF module exceeds opportunity costs of milk production. Most investments show a lower return per hectare than ecological restoration on river banks with MultFR. Economic returns and reduction of environmental liabilities associated with the provision of multiple ecosystem services make MultRF appealing to many stakeholders, especially the local decision maker – the family farmer.

High Biodiversity Silvopastoral Systems (SSPnuclei) embrace food production, ecological rehabilitation, and multiple ecosystem services in pasture areas. It brings back tree biodiversity (50 native tree species/ha), connectivity, and an improved landscape matrix, as well as shade for livestock, improved milk yield, and revenue from NTFP. SSPnuclei have been implemented in pilot farms where biophysical variables and ES have been evaluated – soil quality and carbon, biodiversity (birds, ants and dung beetles), microclimate, carbon sequestration and landscape characteristics. The system provided shade by the third year, produced banana, rose pepper and honey in the fourth year, and açai from Juçara in the eight. Farmers will recover SPSnuclei restoration costs by the seventh – eighth year only accounting revenue from NTFP.

Finally, rather than imitating market mechanisms, we propose an unconventional approach to a Program for Payment for ecosystem services (PES), a form of co-investment in stewardship designed to align farmers' land use decisions with the broader social interest. We hope our efforts to develop viable agroecosystems that synergically restore rural landscapes will prove successful in Santa Catarina and scalable across Brazil's Atlantic Forest, while providing valuable lessons to efforts elsewhere.

References

Alarcon GG, Ayanu Y, Fantini AC, Farley J, Schmitt-Filho A, Koellner T (2015) Weakening the Brazilian legislation for forest conservation has severe impacts for ecosystem services in the Atlantic Southern Forest. Land Use Policy 47:1–11. https://doi.org/10.1016/j.landusepol.2015.03.011

Alarcon GG, Fantini AC, Salvador CH (2016) Local benefits of the atlantic forest: evidences from rural communities in Southern Brazil. Ambiente Sociedade 19:87–112

Altieri MA (1995) Agroecology: the science of sustainable agriculture. Westview Press, Boulder

Altieri MA (2002) Agroecology: the science of natural resource management for poor farmers in marginal environments Agriculture. Ecosyst Environ 93:1–24

Alvez JP, Schmitt AL, Farley JC, Erickson JD, Méndez VE (2014) Transition from semi-confinement to pasture-based dairy in Brazil: farmers' view of economic and environmental performances. Agroecol Sustain Food Syst 38:995–1014. https://doi.org/10.1080/21683565.2013.859222

Amazonas I, Zanetti V, Schmitt Filho AL, Sinisgali P, Farley J, Fantini A, Cazella AA (2016) Dynamics of land use change in southern Brazil: a case study of Santa Catarina's capital of agroecology. In: 4th Convención Internacional AGRODESARROLLO 2016 & 11th International Workshop 'Trees and Shrubs in Livestock Production', Varadeiro

Amazonas I, Schmitt Filho AL, Sinisgalli P, Zanetti V (2017) The role of riparian zone restoration in achieving Brazil's commitment of reducing carbon emission. In: V Congreso Iberoamericano Y del Caribe de Restauracion Ecologica, I Conferencia Brasileira de Restauracao Ecológica, Foz do Iguacu BR, 27-1 Setembro, 2017. Anais do VII World Conference on Ecological Restoration – SER 2017. http://ser2017.org/

Andrade J (2016) Manejo Florestal no Estado do Espirito Santo: o Cultivo da Palmeira Juçara (Euterpe edulis) como Alternativa Econômica e Ambiental.

Arroyo-Rodríguez V et al (2017) Multiple successional pathways in human-modified tropical landscapes: new insights from forest succession, forest fragmentation and landscape ecology research. Biol Rev 92:326–340. https://doi.org/10.1111/brv.12231

Baccini A et al (2012) Estimated carbon dioxide emissions from tropical deforestation improved by carbon-density maps. Nat Climate Chang 2:182. https://www.nature.com/articles/nclimate1354#supplementary-information. https://doi.org/10.1038/nclimate1354

Back F, Schmitt Filho A, Alves G, Francisco F, Surdi J, Busnardo F, Farley J (2009) Programa de Fortalecimento da Agricultura Familiar através da Produção Ecológica inserida nos Processos de Recuperação Ambiental e Gestão da Paisagem. Revista Brasileira de Agroecologia 4:1926–1930

Banco Central do Brasil (2018) Taxa selic 2018. http://www.bcb.gov.br/pt-br/#!/n/SELICTAXA. Accessed 4 May 2018

Banks-Leite C et al (2014) Using ecological thresholds to evaluate the costs and benefits of set-asides in a biodiversity hotspot. Science 345:1041–1045

Barnett H, Morse C (1963) Scarcity and growth: the economics of natural resource availability. John Hopkins University Press, Baltimore

Barreto CG, Drummond JA (2017) Pre-columbian anthropogenic changes in landscapes of the Brazilian atlantic forest. Revista de Historia Iberoamericana 10:10–33

Battisti LFZ, Schmitt Filho AL, Loss A, Sinisgalli PA (2018) Soil chemical attributes in a high biodiversity silvopastoral system. Acta Agron 67:451–132

Beckerman W (1972) Economists, scientists, and environmental catastrophe. Oxford Econ Pap 24:327–344

Bowles S (2008) Policies designed for self-interested citizens may undermine "The Moral Sentiments": evidence from economic experiments. Science 320:1605–1609

Brancalion PHS, Viani RAG, Strassburg BBN, Rodrigues RR (2012) Finding the money for tropical forest restoration. Unasylva 63:41–50

Buratto T (2019) Matas Ciliares Multifuncionais: Restauração Ecológica, Serviços Ecossistêmicos e Renda no Contexto da Agricultura Familiar. Federal University of Santa Catarina PPGA/UFSC, Florianópolis

Buratto T, Schmitt Filho A, Sinisgalli P, Kuhn V, Carvalho Filho JLS (2018a) Matas Ciliares Multifuncionais (MCmult): Quando a Restauração das Matas Ciliares Gera Renda para a Agricultura Familiar. In: Anales do VII Congreso Latinoamericano de Agroecologia da Sociedade Latinoamericana de Agroecologia, Guaiaquil, Equador, De 2 a 5 de octubre de 2018.

Buratto T, Schmitt Filho AL, Sinisgalli PA, Amazonas I, Fantini A (2018b) When restoration of agroecosystems generates multiple ecosystem services and increases farmers income. In: Proceedings of ecosystem service partnership regional conference – Latin America/ESP LAC2018, Campinas SP, October 22–26, 2018. www.espconference.org/latinamerica2018/wiki/385098/book-of-abstracts#.W-GgHpNKhPY

Buratto T, Schmitt L, Schmitt Filho AL, Sinisgalli P, Farley J (2018c) Multifunction riparian forests: when ecological restoration is an economically viable practice for family farmers and traditional communities. In: Proceedings of 2018 conference of New England branch of society for ecological restoration /SER NE, Southern CT State University, New Haven, October 11–13, 2018. https://6zvjw1i9d632in9ii1izgap9-wpengine.netdna-ssl.com/newengland/files/2018/10/Abstracts_wBios_SER_10.07.18-2.pdf

Carson R (1962) Silent spring. Houghton Mifflin, Boston

Carvalho Filho JLS, Schmitt Filho AL, Fantini AC, Farley J, Battisti LFZ, Cazella AA (2016) Matas Ciliares Multifuncionais (MCmult): Quando o agricultor familiar inova na recuperação florestal das áreas ripárias. In: 4th Convención Internacional AGRODESARROLLO 2016, Varadero, Cuba, 25 de outubro, 2016. http://agrodesarrollo2016.ihatuey.cu/

Cialdini R (1993) Influence: the psychology of persuasion. William Morrow and Co., New York

Coase R (1960) The problem of social cost. J Law Econ 3:1–44

Corbin JD, Holl KD (2012) Applied nucleation as a forest restoration strategy. For Ecol Manag 265:37–46. https://doi.org/10.1016/j.foreco.2011.10.013

Craesmeyer KC, Schmitt Filho AL, Hotzel MJ, Deniz M, Farley J (2017) Utilização da Sombra por Vacas Lactantes sob Sistema Voisin Silvipastoril no Sul do Brasil. Cadernos de Agroecologia 11:1

Daily GC (ed) (1997) Nature's services: societal dependence on natural ecosystems. Island Press, Washington, D.C.

Daly H (2014) From uneconomic growth to a steady-state economy, Advances in ecological economics series. Edward Elgar, New York

Daly HE, Farley J (2011) Ecological economics: principles and applications, 2nd edn. Island Press, Washington, D.C.

De Coster G, Banks-Leite C, Metzger JP (2015) Atlantic forest bird communities provide different but not fewer functions after habitat loss. Proc R Soc B 282:20142844. https://doi.org/10.1098/rspb.2014.2844

De Schutter O, Vanloqueren G (2011) The new green revolution: how twenty-first-century science can feed the world. Solutions 2:33–44

Dean W (1997) With broadax and firebrand: the destruction of the Brazilian atlantic forest. University of California Press, Berkeley

Denevan WM (1992) The pristine myth: the landscape of the Americas in 1492. Ann Assoc Am Geogr 82:369–385. https://doi.org/10.1111/j.1467-8306.1992.tb01965.x

Deniz M, Schmitt-Filho A, Farley J, de Quadros S, Hötzel M (2018) High biodiversity silvopastoral system as an alternative to improve the thermal environment in the dairy farms. Int J Biometeorol. doi: https://doi.org/10.1007/s00484-018-1638-8 . [Epub ahead of print]

Dorsey J, Dansingburg J, Ness R (1998) Managed grazing as an alternative manuremanagement strategy USDA-ARS land stewardship project, Madson

Fadden JM (2005) A Produção de açaí a partir do processamento dos frutos do palmiteiro (Euterpe edulis Martius) na Mata Atlântica. Federal University of Santa Catarina PPGA/UFSC, Florianópolis

FAO (2011) How to feed the world in 2050. On line: http://www.scp-knowledge.eu/sites/default/files/knowledge/attachments/How%20to%20Feed%20the%20World%20in%202050.pdf,

Farley J (2012a) Ecosystem services: the economics debate. Ecosyst Serv 1:40–49

Farley J (2012b) Natural capital. In: Craig RK, Nagle JC, Pardy B, Schmitz O, Smith W (eds) Berkshire encyclopedia of sustainability: ecosystem management and sustainability, vol 5. Berkshire Publishing, Great Barrington

Farley J, Costanza R (2010) Payments for ecosystem services: from local to global. Ecol Econ 69:2060–2068

Farley J, Aquino A, Daniels A, Moulaert A, Lee D, Krause A (2010) Global mechanisms for sustaining and enhancing PES schemes. Ecol Econ 69:2075–2084

Farley J, Schmitt Filho A, Alvez J, Ribeiro de Freitas N Jr (2012) How valuing nature can transform. Agric Solutions 2:64–73

Farley J, Schmitt Filho A, Burke M, Farr M (2015) Extending market allocation to ecosystem services: moral and practical implications on a full and unequal planet. Ecol Econ 117:244–252. https://doi.org/10.1016/j.ecolecon.2014.06.021

Farley J, Schmitt Filho AL, Sinisgalli P, Fantini A (2018) PSE Santa Rosa: leveraging social change and ecological restoration in a family farmer dominated landscape. In: Proceedings of ecosystem service partnership regional conference – Latin America 2018/ESP LAC2018, Campinas SP, October 22–26, 2018. www.espconference.org/latinamerica2018/wiki/385098/book-of-abstracts#.W-GgHpNKhPY

Farnsworth E, Tidrick TH, Smathers WM, Jorda CF (1983) A synthesis of ecological and economic theory toward a more complete valuation of tropical moist forests. Int J Environ Stud 21:11–28

Fisher B et al (2008) Ecosystem services and economic theory: integration for policy-relevant research. Ecol Appl 18:2050–2067

Freitas SR, Neves CL, Chernicharo P (2006) Tijuca National Park: two pioneering restorationist initiatives in Atlantic forest in southeastern Brazil. Braz J Biol 66:975–982

Frey BS, Jegen R (2001) Motivation crowding theory. J Econ Surv 15:589–611

Georgescu-Roegen N (1971) The entropy law and the economic process. Harvard University Press, Cambridge
Geremias V (2011) Success factors and constraints of community-based ecosystem management – a case study of the Voisin rotational grazing system in a rural community in Brazil. Wageningen University and Research Center, Wageningen
Gliessman SR (2015) Agroecology: the ecology of sustainable food systems, 3rd edn. CRC Press, Boca Raton
Globo (2009) Novo Código Ambiental de Santa Catarina entra em vigor em meio a polêmica. Globo, Rio de Janeiro
Gneezy U, Rustichini A (2000) Pay enough or don't pay at all. Q J Econ 115:791–810
Gneezy U, Meier S, Rey-Biel P (2011) When and why incentives (don't) work to modify behavior. J Econ Perspect 25:191–210
Godfray HCJ (2011) Food and biodiversity. Science 333:1231–1232
Godfray HCJ et al (2010) Food security: the challenge of feeding 9 billion people. Science 327:812–818
Homma AKO, Nogueira OL, de Menezes AJEA, de Carvalho JEU, Nicoli CML, de Matos GB (2006) Açaí: Novos Desafios e Tendências Amazônia: Ciencias & Desenvolvimento. Belém 1:7–33
IBGE (2017) IBGE Cidades: Santa Rosa de Lima SC. v4.3.18.2 Instituto Brasileiro de Geografia e Estatística. cidades.ibge.gov.br/brasil/sc/santa-rosa-de-lima/panorama
INPE (2017) Desmatamento da Mata Atlântica cresce quase 60% em um ano. Instituto Nacional de Pesquisas Espaciais. http://www.inpe.br/noticias/noticia.php?Cod_Noticia=4471
Jochims V, Leopoldino da Silva AW (2016) O leite para o Oeste Catarinense. Revista Agropecuária Catarinese 29:15–18
Joly CA, Metzger JP, Tabarelli M (2014) Experiences from the Brazilian atlantic forest: ecological findings and conservation initiatives. New Phytol 204:459–473
Joseph L, Schmitt Filho AL, Zambiazi DC, Fantini AC, Cazella AC (2016) Percepção de produtores de leite com sistema Voisin em relação à implantação de Sistemas Silvipastoris. Cadernos de Agroecologia 11:31
Kohn V (2017) Análise da viabilidade econômica de um modelo de Mata Ciliar Multifuncional no município de Santa Rosa de Lima. Universidade Federal de Santa Catarina
Kretzer SG, Schmitt Filho AL, Sinisgalli PA, Deniz M, Rover CR (2018) O Sistema Silvipastoril com Núcleos (SSPnucleos) e a Minimização dos Efeitos das Variações Microclimáticas Extremas na Pastagem. Federal University of Santa Catarina PPGA/UFSC, Florianópolis
Kuussaari M et al (2009) Extinction debt: a challenge for biodiversity conservation. Trends Ecol Evol 24:564–571. https://doi.org/10.1016/j.tree.2009.04.011
Lagi M, Bar-Yam Y, Bertrand KZ, Bar-Yam Y (2011) The food crises: a quantitative model of food prices including speculators and ethanol conversion, New England Complex Systems Institute. Available at SSRN: http://ssrn.com/abstract=1932247 or https://doi.org/10.2139/ssrn.1932247
Lagi M, Bar-Yam Y, Bertrand KZ, Bar-Yam Y (2012) Update February 2012 – the food crises: predictive validation of a quantitative model of food prices including speculators and ethanol conversion. arXiv 1203.1313, 6 Mar 2012
LASSre (2011) Silvopastoral Systems and Ecological Restoration Lab LASSre/UFSC. Laboratório de Sistemas Silvipastoris e Restauração Ecológica LASSre/UFSC. https://lass.paginas.ufsc.br/
LASSre (2014) Silvopastoral Systems and Ecological Restoration Lab LASSre/UFSC. Laboratório de Sistemas Silvipastoris e Restauração Ecológica LASSre/UFSC. https://lass.paginas.ufsc.br/
LASSre (2016) Silvopastoral Systems and Ecological Restoration Lab LASSre/UFSC. Laboratório de Sistemas Silvipastoris e Restauração Ecológica LASSre/UFSC. https://lass.paginas.ufsc.br/
LASSre (2017) Silvopastoral Systems and Ecological Restoration Lab LASSre/UFSC. Laboratório de Sistemas Silvipastoris e Restauração Ecológica LASSre/UFSC. https://lass.paginas.ufsc.br/
LASSre (2018) Silvopastoral Systems and Ecological Restoration Lab LASSre/UFSC. Laboratório de Sistemas Silvipastoris e Restauração Ecológica LASSre/UFSC. https://lass.paginas.ufsc.br/

Laureto LMO, Cianciaruso MV, Samia DSM (2015) Functional diversity: an overview of its history and applicability. Natureza Conservação 13:112–116. https://doi.org/10.1016/j.ncon.2015.11.001

Lima MM, Mariano-Neto E (2014) Extinction thresholds for Sapotaceae due to forest cover in Atlantic Forest landscapes. For Ecol Manag 312:260–270. https://doi.org/10.1016/j.foreco.2013.09.003

Mattei L (2019) Novo retrato da agricultura familiar em Santa Catarina. Federal University of Santa Catarina PGE/UFSC, Florianópolis Brazil. https://necat.paginas.ufsc.br/files/2011/10/Lauro-20100.pdf

Maurer F, Schmitt A, Farley J, Alves J, Oldra A, DaRolt L, Francisco F (2009) Serviços Ambientais e a Produção de Leite sob Pastoreio Voisin na Agricultura Familiar: Ativos Ambientais que Devem ser Considerados. Revista Brasileira de Agroecologia 4:3830–3834

Mazzucato M (2018) The value of everything: making and taking in the global economy. Hatchette Book Group, New York

Meadows DH, Meadows DL, Randers J, Behrens W (1972) The limits to growth: a report for the club of Rome's project on the predicament of mankind. Universe Books, New York

Méndez EV, Caswell M, Gliessman RS, Cohen R (2017) Integrating agroecology and participatory action Rrsearch (PAR): lessons from Central America. Sustainability 9:705. https://doi.org/10.3390/su9050705

Metzger JP (2010) O Código Florestal tem base científica? Conservação e Natureza 8(preface):92

Metzger JP, Martensen AC, Dixo M, Bernacci LC, Ribeiro MC, Teixeira AMG, Pardini R (2009) Time-lag in biological responses to landscape changes in a highly dynamic Atlantic forest region. Biol Conserv 142:1166–1177

Millar CI, Stephenson NL (2015) Temperate forest health in an era of emerging megadisturbance. Science 349:823. https://doi.org/10.1126/science.aaa9933

Millennium Ecosystem Assessment (2005) Ecosystems and human well-being: synthesis. Island Press, Washington, D.C.

Mittermeier RA, Turner WR, Larsen FW, Brooks TM, Gascon C (2011) Global biodiversity conservation: the critical role of hotspots. In: Zachos FE, Habel JC (eds) Biodiversity Hotspots. Springer Publishers, London, pp 3–22

Moore M-L et al (2014) Studying the complexity of change: toward an analytical framework for understanding deliberate social-ecological transformations. Ecol Soc 19:54. https://doi.org/10.5751/ES-06966-190454

Moreno-Peñaranda R, Kallis G (2010) A coevolutionary understanding of agroenvironmental change: a case-study of a rural community in Brazil. Ecol Econ 69:770–778. https://doi.org/10.1016/j.ecolecon.2009.09.010

Muradian R, Corbera E, Pascual U, Kosoy N, May PH (2010) Reconciling theory and practice: an alternative conceptual framework for understanding payments for environmental services. Ecol Econ 69:1202–1208

Murphy B (2010) Greener pasture on your side of the fence: better farming voisin management-intensive grazing, 7th edn. Arriba Publishing, Colchester

Myers N (1997) The world's forests and their ecosystem services. In: Daily G (ed) Nature's services. Island Press, Washington, D.C., pp 215–235

Nave LE, Domke GM, Hofmeister KL, Mishra U, Perry CH, Walters BF, Swanston CW (2018) Reforestation can sequester two petagrams of carbon in US topsoils in a century. Proc National Acad Sci 115:2776. https://doi.org/10.1073/pnas.1719685115

Orwa C, Muta A, Kindt R, Jamnads R, Anthony S (n.d.) Agroforestry database: a tree reference and selection guide version 4.0. World Agroforestry Centre (ICRAF). http://www.worldagroforestry.org/treedb/AFTPDFS/Mimosa_scabrella.pdf

Pagiola S, Glehn HCV, Taffarello D (eds) (2013) Experiências de Pagamentos por Serviços Ambientais no Brasil. Secretaria do Meio Ambiente São Paulo

Pahl-Wostl C (2009) A conceptual framework for analysing adaptive capacity and multi-level learning processes in resource governance regimes. Global Environ Chang 19:354–365. https://doi.org/10.1016/j.gloenvcha.2009.06.001

Petchey OL, Gaston KJ (2006) Functional diversity: back to basics and looking forward. Ecol Lett 9:741–758. https://doi.org/10.1111/j.1461-0248.2006.00924.x
Phalan B, Balmford A, Green RE, Scharlemann JPW (2011a) Minimising the harm to biodiversity of producing more food globally. Food Policy 36(Suppl 1):S62–S71. https://doi.org/10.1016/j.foodpol.2010.11.008
Phalan B, Onial M, Balmford A, Green RE (2011b) Reconciling food production and biodiversity conservation: land sharing and land sparing compared. Science 333:1289–1291
Pinto RS et al (2014) Governing and delivering a biome-wide restoration initiative: the case of atlantic forest restoration pact in Brazil. Forests 5:2212. https://doi.org/10.3390/f5092212
Presidência da República do Brasil (2012) Código Florestal Brasileiro. Brasilia
Reeson AF, Tisdell JG (2008) Institutions, motivations and public goods: an experimental test of motivational crowding. J Econ Behav Organ 68:273–281. https://doi.org/10.1016/j.jebo.2008.04.002
Redford KH (1992) The empty Forest: many large animals are already ecologically extinct in vast areas of neotropical forest where the vegetation still appears intact. Bioscience 42:412–422. http://www.jstor.org/stable/1311860
Reyer CPO, Rammig A, Brouwers N, Langerwisch F (2015) Forest resilience, tipping points and global change processes. J Ecol 103:1–4. https://doi.org/10.1111/1365-2745.12342
Rezende CL et al (2018) From hotspot to hopespot: an opportunity for the Brazilian atlantic forest. Perspect Ecol Conserv 16:208–214. https://doi.org/10.1016/j.pecon.2018.10.002
Ribeiro MC, Metzger JP, Martensen AC, Ponzoni FJ, Hirota MM (2009) The Brazilian atlantic forest: how much is left, and how is the remaining forest distributed? Implications Conserv Biol Conserv 142:1141–1153. https://doi.org/10.1016/j.biocon.2009.02.021
Ribeiro MC, Martensen AC, Metzger JP, Tabarelli M, Scarano F, Fortin M-J (2011) The Brazilian atlantic forest: a shrinking biodiversity hotspot. In: Zachos FE, Habel JC (eds) Biodiversity hotspots: distribution and protection of conservation priority areas. Springer Berlin Heidelberg, Berlin, pp 405–434. https://doi.org/10.1007/978-3-642-20992-5_21
Rockstrom J et al (2009) A safe operating space for humanity. Nature 461:472–475
Rode J, Gómez-Baggethun E, Krause T (2014) Motivation crowding by economic incentives in conservation policy: a review of the empirical evidence. Ecol Econ. http://www.sciencedirect.com/science/article/pii/S0921800914003668. https://doi.org/10.1016/j.ecolecon.2014.09.029
Savory A (2016) Holistic management – a commonsense revolution to restore our environment. Island Press, Washington, D.C.
Scarano FR, Ceotto P (2015) Brazilian atlantic forest: impact, vulnerability, and adaptation to climate change. Biodivers Conserv 24:2319–2331. https://doi.org/10.1007/s10531-10015-10972-y
Schmitt Filho AL, Murphy W, Busnard F, Martins FC, Nascimento AL, Ros JLD, Buss C (2008) Grass based agroecologic dairy revitalizing small family farms throughout student technical support: the Brazilian pasture outreach program. In: 22st general meeting of European grassland federation 'Biodiversity and Animal Feed', Uppsala, Sweden. Blackwell Publishing, Oxford
Schmitt Filho AL, Murphy W, Farley J (2010) Grass based agroecologic dairying to revitalize small family farms through student technical support: the development of a participative methodology responsible for 622 family farm projects. Adv Anim Biosci 1:517–518
Schmitt Filho AL, Farley J, Alarcon G, Alvez J, Rebollar P (2013) Integrating agroecology with payments for ecosystem services in Santa Catarina's Atlantic forest. In: Muradian R, Rival L (eds) Governing the provision of environmental services. Springer, Dordrecht
Schmitt Filho AL, Fantini AC, Farley J, Sinisgalli P (2017) Nucleation theory inspiring the design of high biodiversity silvopastoral system in the Atlantic forest biome: ecological restoration, family farm livelihood and agroecology. In: V Congreso Iberoamericano Y del Caribe de Restauracion Ecologica, I Conferencia Brasileira de Restauracao Ecológica, Foz do Iguacu BR. Anais do VII world conference on ecological restoration – SER. http://ser2017.org/abstract-book.php

Schmitt Filho AL, Fantini A, Farley J, Sinisgalli P (2018a) High biodiversity silvopastoral system: addressing livelihood, ecosystem services and ecological restoration in Brazil's most endangered Biome. In: Proceedings of ecosystem service partnership regional conference – Latin America 2018 /ESP LAC2018, Campinas SP, Brazil, October 22–26, 2018. www.espconference.org/latinamerica2018/wiki/385098/book-of-abstracts#.W-GgHpNKhPY

Schmitt Filho AL, Fantini A, Sinisgalli P, Farley J, Schmitt L (2018b) Ecological restoration, livelihood and ecosystem services in a smallholder dominated rural landscape. In: Proceedings of 2018 conference of New England branch of society for ecological restoration/SER NE, Southern CT State University, New Haven, October 11–13, 2018. https://6zvjw1i9d632in9ii1izgap9-wpengine.netdna-ssl.com/newengland/files/2018/10/Abstracts_wBios_SER_10.07.18-2.pdf

Schröter B, Matzdorf B, Sattler C, Garcia Alarcon G (2015) Intermediaries to foster the implementation of innovative land management practice for ecosystem service provision – a new role for researchers. Ecosyst Serv 16:192–200. https://doi.org/10.1016/j.ecoser.2015.10.007

Secretaria do Meio Ambiente do Estado-SP (2011) Estudo de Viabilidade de Plantio Florestal com Espécies Nativas Comerciais no Estado de São Paulo. Relatório sobre a Caracterização do Mercado, Seleção de Espécies e Macrolocalização Potencial em São Paulo. Ed. Da STCP Engenharia de Projetos, São Paulo

Silva Filho JLV (2005) Análise Econômica Da Produção e Transformação Em ARPP, dos Frutos de Euterpe Edulis Mart. em Açaí no Município de Garuva Estado de Santa Catarina. Universidade Federal De Santa Catarina

Silva JAA et al (2011) O Código Florestal e a Ciência. Sociedade Brasileira para o Progresso da Ciencia, SPBC. Academia Brasileira de Ciencias, Sao Paulo

Silva AA, Schmitt Filho AL, Fantini AC, Zambiazi DC, Sinisgalli PA (2018) Estimativas de biomassa e carbono em sistema silvipastoril com núcleos arbóreos (SSPnúcleos). Cadernos de Agroecologia 13. http://cadernos.aba-agroecologia.org.br/index.php/cadernos/article/view/1742

Simioni GF, Schmitt Filho AL, Joner F, Fantini AC, Farley J, Moreira A (2019) Variação da assembleia de aves em áreas pastoris e remanescentes florestais adjacentes. *Rev Ciênc Agrárias* 42:884–895. https://doi.org/10.19084/rca.17601

Simpson RD, Toman MA, Ayres RU (eds) (2005) Scarcity and growth revisited: natural resources and the environment in the new millenium. Resources for the Future, Washington, D.C.

Soares-Filho B et al (2014) Cracking Brazil's forest code. Science 344:363–364

Solow RM (1972) Notes on "Doomsday Models". Proc Natl Acad Sci USA 69:3832–3833. https://doi.org/10.2307/62096

SOS Mata Atlantica (2009) Divulgação do novo Atlas Mata Atlântica. Online at: http://www.sosmatatlantica.org.br/index.phpsection=content&action=contentDetails& idContent=392

Sovell LA, Vondracek B Fau, Frost JA, Frost Ja Fau, Mumford KG, (2000) Impacts of rotational grazing and riparian buffers on physicochemical and biological characteristics of southeastern Minnesota, USA, streams. Environ Manage. 26(6):629–41. https://doi.org/10.1007/s002670010121

Sparovek G, Berndes G, Barretto AGOP, Klug ILF (2012) The revision of the Brazilian Forest Act: increased deforestation or a historic step towards balancing agricultural development and nature conservation? Environ Sci Policy 16:65–72. https://doi.org/10.1016/j.envsci.2011.10.008

Surdi J, Schmitt Filho A, Farley J, Alvez JP, Satschumi H (2011) O fluxo de serviços ecossistêmicos na agricultura familiar da Encosta da Serra Catarinense (The flow of ecosystem services in family farming of the Encosta da Serra Catarinense). Cadernos de Agroecologia Resumos do I Encontro Pan-Americano sobre Manejo Agroecológico de Pastagens 6:1–6

Teague WR, Dowhower SL, Baker SA, Haile N, DeLaune PB, Conover DM (2011) Grazing management impacts on vegetation, soil biota and soil chemical, physical and hydrological properties in tall grass prairie. Agric Ecosyst Environ 141:310–322. https://doi.org/10.1016/j.agee.2011.03.009

TEEB (2008) The Economics of Ecosystems and BIodiversity. European Communities, Brussels

The President's Materials Policy Commission (1952) Resources for freedom: a report to the president, Foundations for growth and security, vol Vol 1. United States Government Printing Office, Washington, D.C.
Trevisan ACD, Fantini AC, Schmitt-Filho AL, Farley J (2015) Market for Amazonian Açaí (Euterpe oleraceae) Stimulates Pulp Production from Atlantic Forest Juçara Berries (Euterpe edulis). Agroecol Sustain Food Syst 39:762–781
Trevisani P (2018) Brazil's sky-high lending rates hurt consumers—and economic growth. WSJ Dow Jones & company, Inc. New York. https://www.wsj.com/articles/brazils-sky-high-lending-rates-hurt-consumersand-economic-growth-1535707800
Tscharntke T et al (2012) Global food security, biodiversity conservation and the future of agricultural intensification. Biol Conserv 151:53–59. https://doi.org/10.1016/j.biocon.2012.01.068
Uliana MP, Fronza M, da Silva AG, Vargas TS, de Andrade TU, Scherer R (2016) Composition and biological activity of Brazilian rose pepper (Schinus terebinthifolius Raddi) leaves. Ind Crops Prod 83:235–240. https://doi.org/10.1016/j.indcrop.2015.11.077
USDA (2014) Conservation progam statistics. http://www.fsa.usda.gov/FSA/webapp?area=home&subject=copr&topic=crp-st
Virapongse A, Brooks S, Metcalf EC, Zedalis M, Gosz J, Kliskey A, Alessa L (2016) A social-ecological systems approach for environmental management. J Environ Manag 178:83–91. https://doi.org/10.1016/j.jenvman.2016.02.028
Weisse M, Goldman L (2018) 2017 was the second worst year on record for tropical tree cover loss. World Resources Insitute, Washington, D.C.
Wilkie DS, Bennett EL, Peres CA, Cunningham AA (2011) The empty forest revisited. Annals NYAS 1223:120–128. https://doi.org/10.1111/j.1749-6632.2010.05908.x
Young KJ (2017) Mimicking nature: a review of successional agroforestry systems as an analogue to natural regeneration of secondary forest stands. In: Montagnini F (ed) Integrating landscapes: agroforestry for biodiversity conservation and food sovereignty, Advances in agroforestry, vol 12. Springer, Cham

Chapter 11
Forest Governance in Brazil and Chile: Institutions and Practices in the Implementation of Sustainable Management of Native Forests

Liviam Elizabeth Cordeiro-Beduschi

11.1 Introduction

Forest governance must address the challenge of sustainably managing native forests in order to ensure the best possible relationship between society and forests (Agrawal et al. 2008; Moran and Ostrom 2009; Arts et al. 2013). Forests provide a wide range of environmental products and services, as well as social, economic and cultural opportunities to build a sustainable development process. According to the FAO (2016), millions of people depend on forests to ensure their needs for food, energy and housing.

On the other hand, the degradation of native forests and deforestation are frequent in the Latin America and Caribbean countries (IPCC 2014; Cepal 2017) due to several factors that result in the substitution of native forests for other land uses, and which are reflected in the crisis of availability of water, food and environmental services (Kaimowitz and Angelsen 1998; Geist and Lambin 2001). The recent panorama presents a strong link between deforestation dynamics and national and global macroeconomic scenarios. In this scenario we can verify the progression of timber extraction and overgrazing derived from the expansion of agribusiness, especially soy crops and livestock farming, as it has been in Brazil (Nepstad et al. 2006; Tucker et al. 2016). In some countries the loss of native forest cover is also related to the change in land use to tree plantations with exotic species to supply the forestry industry, as in Chile (Miranda et al. 2015; Little et al. 2009).

Exploitation of native forests without management and planning considerations can contribute to the degradation of native forests and changes in land-use and forest cover (Geist and Lambin 2001; Sabogal 2008). The difficulties to ensure the con-

L. E. Cordeiro-Beduschi, Ph.D. (✉)
Forest Engineer, Graduate Program in Environmental Science, Institute of Energy and Environment (PROCAM/IEE – USP), University of São Paulo – USP, São Paulo, SP, Brazil

F. Fuders and P. J. Donoso (eds.), *Ecological Economic and Socio Ecological Strategies for Forest Conservation*, https://doi.org/10.1007/978-3-030-35379-7_11

servation and sustainability of native forests are also related to the precarious public policy instruments and institutions for the promotion of a diversified economy that values the knowledge of local populations and users of native forests (Miranda et al. 2015; Romero and Poblete 2016; Cepal 2017).

The challenge is to find ways to promote a new forest economy that values local knowledge and new socio-biodiversity-based trade arrangements as a driving force for sustainable forest management (Abramovay 2002; Tacon and Palma 2006; ISA 2017). In this sense, sustainable management of native forests is seen as a forestry practice that requires agreements and commitments by different social actors involved in forest management.

The United Nations Conference on Environment and Development (UNCED, Rio 92), resulted in the "Forest Principles" as one of the outstanding international agreements, aimed at the conservation and sustainable management of native forests, especially emphasizing the importance of using sustainable forest management practices and improving forums for the participation of multiple actors in the decisions for implementing forest programs and policies (Giddens 2010; Sachs 2014; Singer and Giessen 2017).

Recent studies call attention to the 'new models of forest governance' that are defined by decentralizing forest management, engaging communities in the management of forest resources, and promoting multi-stakeholder participation forums to support sustainable forest management (Ros-tonen 2007; Moram and Ostrom 2009; Arts et al. 2013; Arts 2014). However, the way forest governance is established in some countries is still a topic under construction in the literature, which demonstrates the need to improve understanding of the challenges and barriers to the practical implementation of sustainable management of native forests. In this regard, the present study focused on forest policies in two regions of South America. In Brazil, the Public Forest Management Law (Law No. 11.284) that was promulgated in 2006, and in Chile, forest governance in reference to the Native Forest Law (Law No. 20.283) that was promulgated in 2008.

Brazil and Chile participate in the main international treaties on forests (FAO 2015a, b; CEPAL 2017). These are countries with important forest ecosystems with high rates of endemism and biodiversity (Salas et al. 2016). These characteristics are associated with the social, cultural and economic elements that configure the use of resources and the forest landscape (Adams et al. 2006; Brondizio 2009; Miranda et al. 2015; Reyes 2017). In addition, there are many factors that lead to deforestation and forest degradation, resulting in changes in land-use and forest cover (Batistella et al. 2008; Geist and Lambin 2001).

In the last three decades, the implementation of forestry policies based on the global debate on forests demonstrates the limitations that exist for promoting the sustainable use of forest resources, especially at the national and local levels (Veiga 2013; Singer and Giessen 2017; Faggin and Behagel 2017). This chapter proposes to analyze the challenges and barriers to forest governance, based on the study of forest institutions and the perceptions of social actors imbricated in the institutional logic and the logic of practice for the sustainable management of native forests in Brazil and Chile (Ostrom 2011; Arts et al. 2013).

11.2 Analytical Framework: Institutions and Practices in the Forest Context

Since the 1990s, forest governance has experimented with various ways of conducting forest policies and initiatives, in which the state is no longer the sole actor in forest decisions, but begins to work together with private actors in forest design and the implementation of forest policies (Arts et al. 2013). In this sense, the concept of new forest governance has the greater challenge of implementing the sustainable forest management approach, recognizing the multiple functions of forests, which brings conservation and production needs closer together. A forest managed with sustainability principles has the capacity to adapt to long-term environmental changes (Sarre and Sabogal 2013).

Two main theoretical currents were identified to guide this research on the governance of native forests in Brazil and Chile. They stand out because they offer theoretical frameworks developed around forest governance that can help us to understand two main elements: institutions and practices (Cordeiro-Beduschi 2018). Institutions, defined as rules, norms and agreements (formal or informal), are essential elements in the elaboration and implementation of public policies (Tucker and Ostrom 2009). The approach based on the analysis of institutions for social and ecological issues has been highlighted by several studies developed by Ostrom and colleagues at the Center for the Analysis of Social-Ecological Landscape, Indiana University in Bloomington (United States). Under this approach, the debate about governance arrangements in relation to the use and maintenance of forests requires understanding about the use of so-called "Common-Pool Resources" (CPR). Common pool or common access resources are defined as resources that can be subject to exploitation and reduction, and for which the control and exclusion of users constitutes a complex and difficult process (Ostrom 1990; Ostrom et al. 2002). Common access resources include forests, watersheds, oceans, fishery resources and the stratosphere. The sustainability or maintenance of these resources depends greatly on the ability of human beings to build new institutions or to improve those in existence for the sustainable management of resources (Mckean 2000; Ostrom et al. 2002).

Considering that forests are important common resources (Tucker and Ostrom 2009), the challenge in governing these ecosystems is to consider their various elements, such as: (i) the rules governing their resource management, (ii) their ecological and biophysical characteristics, and (iii) the social, economic and political attributes of the system in which they are inserted. In this sense, the study was based on the Institutional Analysis and Development (IAD) framework, developed at the Bloomington school as an institutional analysis methodology that presents the "arena of action" as a platform where individuals interact, negotiate and construct the framework of public policy and institutions (Ostrom 2011).

To complement the institutional analysis, the Forest and Nature Conservation Policy Group of the University of Wageningen (The Netherlands) proposes the so-called 'Practice-Based Approach' (PBA), which provides a reflection on the

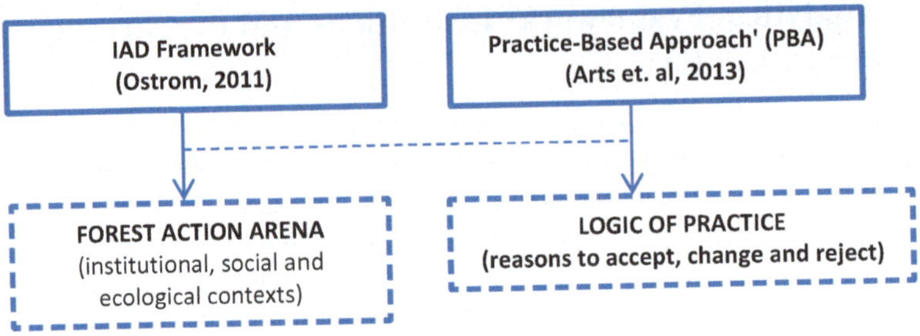

Fig. 11.1 Analytical framework for the analysis of the governance of native forests in Brazil and Chile. (Source: own elaboration)

behavior and decisions of social actors related to the management of natural resources (Arts et al. 2013; Arts 2014). The practice-based approach helps us to understand and recognize the perceptions of the various social actors in more depth, including the reasons not only for accepting established rules and institutions but also for modifying or rejecting them according to their interests. This means observing the 'forest practitioner', the actor who is involved in some way in forestry management or production activities, in offices, in multinational organizations, in the design of management projects and plans, as well as those who attend meetings, represent associations and lead community decisions (Arts et al. 2013; Behagel et al. 2017). The analytical framework developed for the present study is represented by Fig. 11.1.

Along with the logic of practice, there is also the 'institutional bricolage' (Cleaver 2012; Koning and Benneker 2013), through which forest practitioners and users reject, adapt or integrate new institutions that are introduced with those that already exist in their socio-ecological context (Faggin and Behagel 2017). In other words, forest practitioners are creative and improvise, interpret and redefine the rules (Cleaver 2012; Konning and Benneker 2013), which often results in a change in the course of forest policies in Chile and Brazil.

11.3 Methods

This study performed analyses to answer the following question: has forest governance, in fact, promoted the sustainable management of native forests in two regions of South America, Brazil and Chile? The period analyzed focused on the years between 1990 and 2015, a period in which a large number of conferences and initiatives on sustainable development and the environment took place, generating several initiatives and the evolution of the concept of sustainable natural resource management (Giddens 2010; Singer and Giessen 2017).

In almost 30 years of implementation of forestry management in these countries, the evidence points to the superficiality of public policies that aim to promote the sustainable use of forest resources, especially at the national and local scale (Veiga 2013; Singer Geissen 2017).

In order to understand the difficulties encountered in the implementation of forest management in Brazil and Chile we sought to develop interdisciplinary research, taking into consideration the contributions of the disciplines of sociology and economics in dialogue with forestry science. In this sense, the analytical framework was elaborated considering both the Institutional Analysis (Ostrom 1990, 2011) and the Practice-Based Approach (Arts et al. 2013), detailed in Sect. 2.

The study was based on a historical review of the context of the international debate on forests and the sustainable management of native forests. We identified the social actors that have promoted the main agreements and treaties, highlighting the initiatives and main products of global forest governance (Arts and Balili 2013; Secco et al. 2014). Analyses were then carried out at national and subnational scales focusing on the current policies for the management of native forests: in Brazil the Public Forest Management Law (Law No. 11.284/2006), and in Chile the Native Forest Law (Law No. 20.283/2008).

The data were collected based on qualitative research methods (IPEA 2010), which allow an exploration of the complex nature of social organization and a set of variables, both at the institutional level (legislation, agreements, rules) and in terms of the values, perceptions and practices shared by individuals and groups that implement policies or are affected by them (IPEA 2010: 662).

Information was obtained through research on secondary documents (institutional reports, gray literature and academic publications) as well as from primary information (interviews and public testimonies). The scale of analysis was concentrated on the actor involved with the management and implementation of public forest policies. Actors were categorized by their organizations and functions: governmental organization; non-governmental organization; forest users; private sector and entrepreneurs; researchers and academics; international organizations. In this way, it was possible to define the *action arena of the forest policy* of each country, which is shaped by the dynamics of the interactions between forest practitioners.

Semi-structured interviews were conducted during August 2014 and October 2018 with actors grouped into categories. There were 53 interviewees in total, 26 in Brazil, 20 in Chile and 7 informants representing multilateral organizations. Testimonials were also obtained from public speeches in seminars and institutional meetings.

11.4 The Governance of Native Forests in Brazil

Brazil is a country with approximately 494 million hectares covered by forests (59.04% of its territory), distributed in 456 million hectares of native forest and 7.7 million hectares of planted forest (Table 11.1). Of the territory covered by native forests, 71% are public forests that are registered in the Ministry of the

Table 11.1 Brazil forest cover

Forest cover	Hectares (×1000)	%
Territory	835.814	100
Total forest cover	493.538	59.04
Natural forests	485.802	58.12
Forest plantations	7.736	0.10

Source own elaboration. (Adapted from SFB 2013; FAO 2015a, b; CEPAL 2017)

Environment's National Register of Public Forests (SFB 2013; CEPAL 2017). The Amazon biome accounts for 38.30% of the national territory and is the largest tropical rainforest in the world, with a marked presence of fauna and flora species adapted to high temperatures and rainfall rates throughout the year (SFB 2013).

In addition to the intrinsic characteristics of the biome, the Amazon region is embedded in a complex context, not only in relation to its physical distribution, characterized by its water resources and biogeography, but also because it presents a great social, cultural and economic diversity. These actors coexist with the dynamics of forestry and the exploitation of natural resources due to the massive territorial occupation driven by development policies. These development policies have also contributed to the deforestation and uncontrolled exploitation of forest resources and the confusing land system in place up until now (Adams et al. 2006; Fearnside 2014).

The Public Forest Management Law (LGFP, Law No. 11.284/2006) is an important forestry institution with the objective of promoting forest development with the precepts of sustainable management of native forests and community development in the Brazilian public forests concentrated in the Amazon region (Brasil 2006; Silva 2010; Alves 2016).

From the institutional analysis, it was possible to verify that the fundamentals and concepts of sustainable forest management are present in the law. These concepts and foundations of forest sustainability were introduced in Brazil by several plans, programs and scientific research, especially since 1990, which have helped to build a new millennium public policy framework on forest issues (Nepstad et al. 2006; Moran and Ostrom 2009; Cordeiro-Beduschi 2015; Alves 2016; ISA 2017).

The LGFP brought with it new challenges for the structure of forest governance through the Ministry of the Environment with the creation of the Brazilian Forest Service (SFB) and the Public Forest Management Council, with representation from public, private and civil society stakeholders. The National Forestry Development Fund was created to direct resources to plans and programs, such as the National Environmental Policy; the National Biodiversity Plan, the Sustainable Amazon Plan; the National Forest Program, the Action Plan for Prevention and Control of Deforestation in the Legal Amazon (PPCDAm) and the National Plan for the Sustainable Development of Traditional Peoples and Communities (Brasil 2006; SFB 2017; Alves 2016). All the plans and projects indicate the existence of an action arena for native forest governance, with important socio-environmental capital and the ability to formulate proposals and launch public policies at the national level. However, there are frequent criticisms when impasses are revealed in the implementation of this policy that hinder the effective participation of native forest communities in the new forest management (Drigo 2010; ISA 2017).

Based on the analysis of official annual documents (reports of the SFB) and personal information (interviews and testimonies), it was possible to verify a change of focus in the implementation of the LGFP, mainly influenced by two events that occurred in 2012. The first was the approval of the new Forest Code (Law 12.651/2012; Brasil 2012), and the second was the insertion of resources from the Climate Fund in the agenda of the National Forest Development Fund for forest restoration in Brazil (Cordeiro-Beduschi 2015). Both events influenced changes in the structure of the SFB, mainly towards activities for the registration, regularization and restoration of private properties (SFB 2017; Sparovek et al. 2017).

The main impact of this change of focus in the implementation of the LGFP was the creation of new responsibilities for the Brazilian Forest Service by initiating the implementation of the Rural Environmental Registry (CAR – in Portuguese), an instrument created specially to regulate the country's private rural properties, where forests are also found in areas denominated Legal Reserves (RL). Most rural owners have their RL degraded or deforested, and to be "forgiven" they must register their forest areas with the new monitoring system, according to the new Forest Code (Paulino 2012; Brancalion et al. 2016).

It seems that this change in course is not a lost cause for the management of public forests in Brazil, since the decision to support the CAR and remedy private property debt due to deforestation may in the future establish connections with the sustainable management of public forests through new opportunities that the forest can offer within a vision of its multiple uses and the valuation and exploitation of environmental services.

However, the new role of registering private rural properties that was allocated to the SFB, reflected the change in the LGFP's priorities, which had previously been focused on the management of public forests. It also reflects the asymmetries of power in the debate between ruralists and socio-environmentalists in the action arena of forest policy in Brazil. The interviews and speeches confirm the evidence of the institutional bricolage (Cleaver 2012; Arts et al. 2013) where social actors with more resources were able to "change the rules of the game" and modify the course of the prevailing discourse on public policy in the action arena of forest management. In this case, the agribusiness discourse prevailed in decisions.

Recently, other changes in the course of Brazilian forestry management were announced with the transfer of the SFB to the Ministry of Agriculture (MAPA) and the abolition of national committees and commissions by the Bolsonaro government (January 2019). These changes may generate new confrontations with those social actors who participated in the construction of a forest development model focused on the best relationships between society and forests.

11.5 The Governance of Native Forests in Chile

The temperate native forests of Chile are of great ecological and human importance due to their biogeographical and sociocultural context. Their ecological value is linked to the characteristics of the humid temperate forest formation, known as Evergreen Forest and Valdivian Temperate Forest. Jointly with Argentina, and including the North Patagonian and Magallanic Temperate Rainforests to the south, this is the only temperate rainforest in South America (Donoso and Lara 1999; Miranda et al. 2015) and the second largest remaining area of this type in the world (Salas et al. 2016). Due to the geographic isolation of Chile, this forest has a high rate of endemic species of fauna and flora (Donoso 1981, 1993; Veblen et al. 1996; Luebert and Pliscoff 2005; Salas et al. 2016).

Approximately 23.9% of the Chilean territory is covered by forests (planted, natural and mixed). Of these forest territories, 19.8% are native forests, 4.1% are planted, mainly with *Pinus sp.* and *Eucalyptus sp.* And nearly 1% are mixed (planted and native) forests (Table 11.2). Currently, of the native forests, only 29% are included in the National System of Protected Wild Areas (SNASPE by its Spanish acronym) (INFOR 2016; AIFBN 2017).

Since the 1970s, and outside of the SNASPE, the property regime has been almost entirely privately owned, with 70% of the forests (planted and native) being in private properties (Reyes 2017). In recent decades, this situation has led to frequent social conflicts in the south-central region of the country, especially in the Araucania region, where family farmer Mapuche people live alongside extensive forestry monocultures cultivated mainly for the production of cellulose and paper (AIFBN 2011; Donoso et al. 2016).

The Native Forests Law (LBN by its Spanish acronym) expressed the desire of stakeholders to create a new institutionalism that could break with the hegemony of the private actors who cultivate the extensive monocultures of exotic forest species (Manuschevich 2016; Chile Política Forestal Chilena 2016). The law, which was clearly guided by a framework more in line with contemporary international debates on forest governance models, was designed to establish a new state structure – the National Forestry Service – and to respond to the demand of multiple actors who, until then, did not feel included in the country's forest development model (see also Chap. 5 by "Manuschevich").

Table 11.2 Chile forest cover

Forest cover	Hectares (thousands of hectares)	%
Territory	74.353	100
Total forest cover	17.735	23,85
Natural forest	14.691	19,75
Forest plantation	3.044	4,09

Source: own elaboration. (Adapted from INFOR 2016; FAO 2015a, b; CEPAL 2017)

For Cruz et al. (2012), the promulgation of the LBN signified an important change in the intention contained in Chilean public policies, from regulatory purposes to initiatives to promote the sustainable use of the forest. Soon after its publication, the Forest Management Incentive Fund and a Research Fund for Native Forests were created under the administration of CONAF, and topics such as forest preservation, non-timber forest products (NTFPs) and sustainable management of native forests were included in the sector's discussions.

However, the scale of difficulties for the implementation of the LBN was evident even at the moment of its elaboration. The LBN was finally promulgated in 2008 in a provisional version, but only after 16 years in the Chilean National Congress. (Manuschevich 2016, Chap. 5 in this book). In the meantime, Chile's forestry policy was guided by the well-known Forestry Law of 1931 and by Decree 701 of 1974, with the aim of promoting forest development, mainly for the production of planted forests used in the forest industry (Donoso and Reyes 2015).

The Decree Law 701 (DL 701) was the most important public policy instrument for forestry development for the consolidation of the Chilean forestry industry. It lasted approximately 38 years (1974–2012) (ODEPA 2016). Although this decree was modified to meet society's current demands in 1998, such changes were not sufficient to reverse the economic advantages for large companies in relation to small and medium-sized producers related to forest development and management in the country (De la Fuente et al. 2013; Manuschevich 2016).

The interviews carried out for the present study indicated that after 10 years of LBN implementation results were insignificant in encouraging and promoting the sustainable management of native forests, especially at the local level (rural and peasant owners). The difficulties are mainly in the precarious reach of the financial incentives and in the absence of the state in creating the necessary law enforcement and forest policy agencies. This is evidenced by the postponement of the creation of the National Forest Service, which would replace the current public-private model, embodied since the 1970s by the National Forestry Corporation (CONAF by its Spanish acronym).

The official documents available also show that between 2009 and 2012, only 16% of Forest Management projects received funding, which represents less than 4% of the total resources of the Conservation Fund (De la Fuente et al. 2013). Smallholders and indigenous community declared difficulties in drawing up Management Plans and accessing the LBN funds (Cruz et al. 2012; De La Fuente et al. 2013). As a result, native forests are replaced by plantations of exotic species supplying pulp-exporting forest enterprises.

The Chilean case demonstrates the occurrence of 'institutional bricolage' (Cleaver 2012; Arts et al. 2013) from the elaboration of LBN to its implementation, where the vision of an industrial forest sector prevails over the economy of the native forest. The structure of the National Forest Service has never been consolidated even with the launch of the Chilean Forestry Policy 2015–2035 (AIFBN 2011; ODEPA 2016).

In this way, the difficulties faced by small and medium forest producers and indigenous people to carry out forestry activities in the precepts of sustainability and legality in native forests are significant. Local populations still struggle with the scarce incentives for cooperation and associativism that could consolidate a local forest economy from the multiple use of forests, including timber, non-timber forest products (NTFPs) and the provision of ecosystem services (Burschel and Rojas 2006; INFOR 2010; Reyes et al. 2016).

11.6 Conclusions

Forest management in Brazil and Chile occurs in different ecological, political and social contexts, but both countries have in common the incorporation of the foundations of Sustainable Forest Management (SFM). This incorporation takes as reference the international debate that began in the 1970s and assumed a relevant position in the 1990s through international conventions and treaties, especially when introducing the idea of Sustainable Development.

However, these global fundamentals seem to require some time to be understood, translated and elaborated at national and local scale. Even so, they are not always clear; they are changeable and require a process of internalization by those who practice forest management. In this case, institutional structure is a key element of good governance. But it is the skills and interpretations of the social actors that guide and define how the sustainable management of native forests occurs through governance, actors, resources and capabilities.

The previous analysis suggests three pillars that support good governance for the sustainable management of native forests: (i) an institutional framework for the management of native forests; (ii) sustainability fundamentals in forest management, (iii) promotion of local development in the socio-ecological context.

It is necessary that the social actors that have participated historically in the formation of governance for the sustainable management of native forests be attentive to the governmental changes that are announced in both countries, as well as in recognizing the risks and threats of the authoritarian decisions that are emerging in Latin America. It is in this sense that this study defends the importance of the construction of participatory processes involving multiple actors.

Brazil and Chile need to learn from past experiences and value the potential of native forests. Without a doubt, the challenge for the next few years will be to strengthen the arenas of action to prevent a process of deconstruction of forest governance and to promote profound changes in the forest development model.

References

Abramovay R (2002) Muito além da economia verde. Editora Abril, São Paulo. 248 p

Adams C, Murrieta R, Neves W (2006) As Sociedades Caboclas Amazônicas: Modernidade e Invisibilidade. Annablume/FAPESP: Sociedade Caboclas Amazônicas: Modernidade e Invisibilidade, São Paulo, pp 15–32

Agrawal A, Chhatre A Hardin R (2008) Changing governance of the world's forests. Science 320 (5882), 1460–1462. Available online: https://doi.org/10.1126/science.1155369. Accessed on Oct. 2015

Agrupación De Ingenieros Forestales Por El Bosque Nativo (AIFBN) (2011) Hacia un Nuevo Modelo Forestal – Propuesta para el desarrollo sustentable del bosque nativo y el setor foorestal de Chile. Valdivia

Agrupación De Ingenieros Forestales Por El Bosque Nativo (AIFBN) (2017) Biodiversidad y áreas protegidas de Chile: El desafío de dejar atrás una precaria administración y contar con una institucionalidad fortalecida. Revista Bosque Nativo Valdivia 58:10–14

Alves CG (2016) Políticas públicas setoriais, nível de atividade econômica e fatores de rentabilidade da atividade agropecuária: vetores de pressão sobre os recursos florestais na Amazônia brasileira. Brasília,. Tese de Doutorado defendida na Universidade de Brasília, Faculdade de Tecnologia, Departamento de Engenharia Florestal. 237p

Arts B (2014) Assessing forest governance from a 'triple G' perspective: government, governance, governmentality. Forest Policy Econ 49:17–22

Arts B, Babili I (2013) Global forest governance: multiple practices of policy performance. In: Arts B, Behagel J, van Bommel S et al (eds) Forest and nature governance: a practice-based approach. Springer, Dordrecht, pp 111–130, 260 p

Arts B, Bhagel J, van Bommel S et al (eds) (2013) Forest and nature governance: a practice-based approach. Springer, Dordrecht. 260p

Batistella M, Moran E Alves D (2008) Prefácio. Abordagens Interdisciplinares na Ciência Amazônica: A contribuição do LBA e outras perspectivas. In: Batistella, M, Moran, E and Aalves, D (Org.). Amazônia: Natureza e Sociedade em Transformação. São Paulo, EDUSP,. , 303p

Behagel JH, Arts B, Turnhout E (2017) Beyond argumentation: a practice-based approach to environmental policy. J Environ Policy Plan. https://doi.org/10.1080/1523908X.2017.1295841.14p. March 2017. Available online: https://doi.org/10.1080/1523908X.2017.1295841. Accessed on 27 Aug 2017

Brancalion PHS et al (2016) A critical analysis of the native vegetation protection law of Brazil (2012): updates and ongoing initiatives. Nat Conserv (Impr) 14(Supplement):1–15

Brasil (2006) Lei de Gestão de Florestas Publicas. Lei N° 11.284, de 2 de março de 2006. Available online: http://www.planalto.gov.br/ccivil_03/_Ato2004-2006/2006/Lei/L11284.htm. Accessed on April 2018

Brasil (2012) Lei N° 12.651, de 25 de Maio de 2012. Dispõem da promulgação do Código Florestal Brasileiro. Brasília: Maio. Available online: http://www.planalto.gov.br/ccivil_03/_ato2011-2014/2012/lei/l12651.htm. Accessed on 5 Oct 2018

Brondizio E (2009) Analises inter-regional de mudança de uso da terra na Amazônia. In: Moran, E and Ostrom, E (Org.) Ecossistemas Florestais: Interação Homem-Ambiente. São Paulo: Editora SENAC.. EDUSP, São Paulo. Cap IV. 289–326

Burschel H, Rojas A (2006) Doce años acompañando los campesinos forestales. In: Catalán R et al (eds) Bosques y Comunidades del Sur de Chile. Editorial Universitaria Bosque Nativo, Santiago de Chile, pp 119–136

Chile Política Forestal Chilena 2015–2035 (2016) Ministerio de Agricultura; Corporacion Nacional Forestal, Santiago. Março, 76p

Cleaver F (2012) Development through bricolage: rethinking institutions for natural resource management. Routledge, London

Comision Economica Para America Latina Y Caribe (CEPAL) (2017) Cambio Climatico y Politicas publicas forestales em America Latina: Uma visión preliminar. Available online: http://repositorio.cepal.org/bitstream/handle/11362/40922/S1601346_es.pdf?sequence=4&isAllowed=y. Acessed on 15 Sept 2017

Cordeiro-Beduschi LE (2015) Gestão de Florestas e Mudanças Climáticas: Elaboração das Idéias na Formulação de Políticas Públicas. Brasília: Sétimo Encontro Nacional da Associação Nacional de Pós-Graduação e Pesquisa em Ambiente e Sociedade, GT 11, Mudanças Climáticas: Políticas e Governança para a Adaptação e Redução das Vulnerabilidades, 10p

Cordeiro-Beduschi LE (2018) A governança para a gestão sustentável das florestas nativas em duas regiões da América do Sul.. São Paulo: Tese (Doutorado em Ciência Ambiental) – Programa de Pós-Graduação em Ciência Ambiental, Instituto de Energia e Ambiente, Universidade de São Paulo, São Paulo, 300p

Cruz P, Cid F, Rivas et al (2012) Evaluación de la Ley No 20.283 sobre recuperación del bosque nativo y fomento forestal. Subsecretaría de Agricultura. Informe Final. Agrupación de Ingenieros Forestales por el Bosque Nativo. 240p

da Silva RC (2010) Politicas públicas, atores sociais e conhecimento: gestão sustentável das florestas públicas no Brasil. Tese de Doutorado apresentada à Universidade Estadual de Campinas, Campinas. 138p

De la fuente J, Calderón C, Torres J (2013) Informe final programa Ley de Bosque Nativo. Ministerio de Agricultura, Corporación Nacional Forestal. 171p

Donoso C (1981) Tipos forestales de los bosques nativos de Chile. Investigacíon y Desarrollo Forestal. CONAF/PNUD/FAO., Santiago. Documento de Trabajo no. 38, 70p

Donoso C (1993) Bosques templados de Chile y Argentina. Variacion, estructura y dinamica, 2nd edn. Santiago Chile, Editorial Universitaria. 484p

Donoso C, Lara A (1999) Silvicultura de los Bosques Nativos de Chile. Editorial Universitaria, Santiago, Chile

Donoso S, Reyes R (2015) La Industria de la celulosa em Chile, outra "anomalia del mercado". Revista Bosque Nativo 45:19–21

Donoso S, Reyes R, Gangas R (2016) La Industria de la celulosa en Chile, otra "anomalia de mercado". In: Revista Bosque Nativo: Nativo, Bosques y su Gente – Cinco años de labor politica y social para los bosques nativos de Chile e Argentina. N. 56, Septiembre, pp 36–40

Drigo IG (2010) As barreiras para implantação de concessões florestais na América do Sul: os casos da Bolívia e Brasil. Tese de Doutorado. São Paulo: Programa de Pós-Graduação em Ciência Ambiental, Universidade de São Paulo. L'Institut dês Sciences et Industries Du Vivant et de l'Environnement (AgroParis'Tech), Paris. 287p

Faggin JM Behagel JH (2017) Translating sustainable forest management from the global to the domestic sphere: the case of Brazil. In: Forest policy and economics. Volume 85, Part 1, December 2017. Elsevier, pp 22–31. Available online: https://doi.org/10.1016/j.forpol.2017.08.012. Accessed on Jan 2017

Fearnside PM (2014) Conservation research in Brazilian Amazonia and its contribution to biodiversity maintenance and sustainable use of tropical forests. In: 1st conference on biodiversity in the Congo Basin, 6–10 June 2014, Kisangani, Democratic Republic of Congo. pp 12–27

Food and Agriculture Organization of the United Nations (FAO) (2015a) Global forest resources assessment 2015. Available online: http://www.fao.org/forest-resources-assessment/explore-data/flude/en/. Accessed on Dec 2016

Food and Agriculture Organization of the United Nations (FAO) (2015b) Taller Regional Latinoamericano sobre Criterios e Indicadores para el Manejo Forestal Sostenible. Organización de las Naciones Unidas para la Alimentación y la Agricultura (FAO), Lima. Informe Final. 17 a 19 de Junio

Food and Agriculture Organization of the United Nations (FAO) (2016) Casos ejemplares de Manejo Forestal Sostenible en Chile, Costa Rica, Guatemala y Uruguay. 282p

Geiste, Lambin (2001) What drives tropical deforestation? A meta-analysis of proximate and underlying causes of deforestation based on subnational case study evidence. LUCC Report Series, 4

Giddens A (2010) A Política da Mudança Climática. Editora Zahar, Rio de Janeiro. 385p

Instituto de Pesquisa e Economia Aplicada (IPEA) (2010) Métodos Qualitativos de Avaliação e suas Contribuições para o Aprimoramento de Políticas Píblicas Brasília. In: IPEA, Brasil em Desenvolvimento: Estado, planejamento e políticas püblicas, V.3, cap 25, pp 661–688, 270 p

Instituto Forestal De Chile (INFOR) (2010) Influencias del Proceso de Montreal y los procesos de certificacion en la cosecha forestal. Ciência e Investigação Forestal 16(2):243–254

Instituto Forestal De Chile (INFOR) (2016) El sector forestal chileno 2016. Santiago: Área de Información y Economía Forestal de la Sede Metropolitana. Instituto Forestal, (Libreto), 48p

Instituto Socioambiental (ISA) (2017) Xingu: Historias dos produtos da floresta, 1st edn. ISA, São Paulo. 392p

Intergovernmental Panel on Climate Change (IPCC) (2014) Summary for policymakers||, climate change 2014: impacts, adaptation, and vulnerability. Part A: global and sectoral aspects.. Contribution of Working Group II to the Fifth Assessment Report of the Intergovernmental Panel on Climate Change. Cambridge University Press, Nueva York

Kaimowitz D, Angelsen A (1998) Economic models of tropical deforestation: a review. Bogor: Centro de Investigación Forestal Internacional (CIFOR)

Koning J, Benneker C (2013) Bricolage practices in local forestry. In: Arts, B, Behagel, J, Bommel, S et al. Forest and nature governance: a practice based approach. Springer Netherlands, World Forest, vol 14. pp 49–67

Little C, Lara A, McPhee J, Urrutia R (2009) Revealing the impact of forest exotic plantations on water yield in large scale watersheds in South-Central Chile. J Hydrol: 162–170

Luebert F, Pliscoff L (2005) Sobre los límites del bosque valdiviano. Chloris Chilensis Año 8

Manuschevich D (2016) Neoliberalization of forestry discourses in Chile. Forest Policy Econ 69:21–30

McKean M (2000) Common property: what is it, what is it good for, and what makes it work? In: Gibson C, Mckean M, Ostrom E (eds) People and forests: communities, institutions, and governance. MIT Press, Cambridge, MA

Miranda A, Altamirano A, Cayuela I et al (2015) Different times, same story: Native forest loss and landscape homogenization in three physiographical areas of south-central of Chile. Appl Geogr: 60, 20–28. Available online: https://doi.org/10.1016/j.apgeog.2015.02.016. Accessed on Mar 2017

Moran EF, Ostrom E (Org.) (2009) Ecossistemas Florestais: Interação Homem-Ambiente. São Paulo: Editora SENAC. São Paulo: EDUSP

Nepstad DC, Azevedo-Ramos C, Moutinho P et al (2006) Passos para uma política de gestão socio-ambiental das florestas Amazônicas. Ciência & Engenharia 32:45–54

Oficina de Estudios y Políticas Agrarias. Ministerio de Agricultura de Chile (2016) El Desarrollo Forestal. In: ODEPA. Agricultura Chilena. Reflexiones y Desafíos al 2030, pp 91–102. 298 p. Available online: https://www.odepa.gob.cl/wp-content/uploads/2018/01/ReflexDesaf_2030-1.pdf. Accessed 5 Oct 2018

Ostrom E (1990) Governing the commons: the evolution of institutions for collective action. Cambridge University Press, New York

Ostrom E (2011) Background on the institutional analysis and development framework. Policy Stud J 39:1. Wiley Periodicals, Inc., Oxford

Ostrom E, Dietz T, Dolsak N et al (eds) (2002) The drama of the commons. National Academy Press, Washington

Paulino ET (2012) A mudança do código florestal brasileiro: em jogo a função social da propriedade. Campo-Território: Revista de Geografia Agrária 7(13):40–64

Reyes R (2017) The influence of markets and culture on the use of native forests in the south of Chile. Thesis of Doctor of philosophy presented in the faculty of graduate and postdoctoral studies (forestry). The University of British Columbia, Vancouver. July, 106p

Reyes R, Blanco G, Lagarrigue A, Rojas F (2016) Ley de Bosque Nativo: Desafíos Socioculturales para su Implementación. Instituto Forestal y Universidad Austral de Chile, Informe Interno, 82p

Romero J, Poblete C (2016) El sector forestal en la institucionalidad publica y en la agenda politica. In: Revista Bosque Nativo: Nativo, Bosques y su Gente – Cinco años de labor politica y social para los bosques nativos de Chile e Argentina. N. 56. p 25–35, September 2016

Ros-Tonen M (2007) Novas perspectivas para a gestão sustentável da floresta amazônica: Explorando novos caminhos. Campinas: Revista Ambiente & Sociedade X(1):11–25

Sabogal, C, Jong, W de, Pokorny, B and Louman, B (eds) (2008). Manejo forestal comunitario en América Latina: Experiencias, lecciones aprendidas y retos para el futuro. Belem: CIFOR/CATIE,. 274p

Sachs J (2014) The age of sustainable development. Columbia University Press, New York. 606p

Salas C, Donoso PJ, Vargas R et al (2016) The forest sector in Chile: an overview and current challenges. J For 114(5):562–571
Sarre E, Sabogal C (2013) La OFS es un sueño imposible? Unasylva 64(240):26–34
Secco L et al (2014) Why and how to measure forest governance at local level: a set of indicators. Forest Policy Econ 49:57–71
Serviço Florestal Brasileiro (SFB) (2013) Florestas do Brasil em Resumo – 2013. SFB, Brasília. 188p
Serviço Florestal Brasileiro (SFB) (2017) Gestão de Florestas Públicas. Relatório 2016. Brasilia, DF: SFB, Março de 2017, p 86. Available online: http://www.florestal.gov.br/relatorios-de-gestao. Accessed on Mar 2018
Singer B, Giessen L (2017) Towards a donut regime? Domestic actors, climatization, and the hollowing-out of the international forests regime in the Anthropocene. Forest Policy Econ. 79, 69–79. Available online: https://doi.org/10.1016/j.forpol.2016.11.006. Accessed on Oct 2017
Sparovek G, Freitas FM de, Guidotti V (2017) O Código Florestal e o Portal do Jano. In: Gesisky J (Org.) Código Florestal: Haverá Futuro? Brasília: WWF Brasil, 52–59,. 104p
Tacon AC, Palma J (2006) La comercialización de los productos no madereros: una oportunidad para el manejo comunitario y la valorizacion del bosque nativo. In: Catalan R et al (eds) Bosques y Comunidades del Sur de Chile. Editorial Universitaria Bosque Nativo, Santiago de Chile, pp 253–263
Tucker Lima JM, Valle D, Moretto EM, Pulice SMP, Zuca NL, Roquetti DR, Cordeiro-Beduschi LE, Praia AS, Okamoto CPF, da Silva Carvalhaes VL, Branco EA, Barbezani B, Labandera E, Timpe K, Kaplan D (2016) A social-ecological database to advance research on infrastructure development impacts in the Brazilian Amazon. Sci Data 3:160071. Published online 2016 Aug 30. Available online: https://www.nature.com/articles/sdata201671. Accessed on: Oct 2017
Tucker CM, Ostrom E (2009) Pesquisa Interdisciplinar relacionando instituições e transformações florestais. In: Moran, E. Ostrom, E. (Org) Ecossistemas Florestais: Interação Homem-Ambiente. São Paulo: Editora SENAC.. São Paulo: EDUSP. 109–138
Veblen TT, Kitzberger T, Burns BR, Rebertus AJ (1996) Perturbaciones y regeneración en bosques andinos del sur de Chile y Argentina. In: Armesto JJ, Arroyo MK, Villagrán C (eds) Ecología del Bosque Nativo de Chile. Universidad de Chile Press, p 169–198
Veiga JE (2013) A Desgovernança Mundial da Sustentabilidade. São Paulo, nn 34 152 p

Chapter 12
Municipal Private Natural Heritage Reserves: Uses and Attributions of Natural Protected Areas in the City of Curitiba (PR)

Isabel Jurema Grimm, João Henrique Tomaselli Piva, and Carlos Alberto Cioce Sampaio

12.1 Introduction

Environmental degradation has been characterized by economic and industrial development "affecting deep interferences in the support capacity of the planet's ecosystems" (Jacobi et al. 2015: 410). Listed as a generator of large-scale change, industrial expansion has created a framework for capitalist dynamics, triggering a growing risk of scarcity of natural resources (Dowbor 1993; Sachs 1993, 2004), which is aggravated by the phenomenon of urbanization, causing degradation of living and health conditions of the populations that were concentrated in metropolis.

Given this circumstance, the establishment of protected natural areas has been one of the main global strategies for biodiversity conservation and welfare guarantee in cities and, consequently, their population. Most of the world's nations concerned with the conservation and management of their ecosystems and trying to reverse the environmental imbalance, are seeking to establish legal measures to protect or regulate land use in their territories. Among the main regulatory instruments are the

I. J. Grimm (✉)
Mestrado Profissional em Governança e Sustentabilidade, Instituto Superior de Administração e Economia – ISAE, Curitiba, Paraná, Brasil

J. H. T. Piva
Programa de Pós Graduação em Meio Ambiente e Desenvolvimento, Universidade Federal do Paraná, Curitiba, Paraná, Brasil

C. A. C. Sampaio
Environmental Management Graduate Program, Universidade Positivo, Curitiba, Paraná, Brasil

F. Fuders and P. J. Donoso (eds.), *Ecological Economic and Socio Ecological Strategies for Forest Conservation*, https://doi.org/10.1007/978-3-030-35379-7_12

Conservation Units (CU) or protected areas.[1] Conservation Units or protected areas can be understood as areas that have rules of use and management, with the specific purpose of preserving and protecting plant or animal species, possessing landscape beauties; or places of interest for scientific research, depending on the category in which they fall (Schenini et al. 2004).

The establishment of protected natural areas, as environmental policy strategy of each country, depends on its biodiversity and particularities, and determines the conservation objectives. In Brazil, Conservation Units (CU) are defined by the National System of Conservation Units (SNUC[2]), through Law No. 9.985 of 2000, which establishes the criteria and norms for the creation, implementation and management of these areas, and divides them into two groups: (i) the units of sustainable use, which include the Private Natural Heritage Reserves (RPPN[3]), which aim to restore nature conservation with the sustainable use of natural resources and where activities involving the collection and use of natural resources are allowed; (ii) the integral protection units, as Natural Park, which aim to preserve nature allowing only the indirect use of natural resources to be found there, except for the cases provided for in the Law itself (Brasil 2000).

According to Art. 21 of the SNUC Law, the RPPN are private areas, recorded with perpetuity, with the objective of biological conservation, being allowed to conduct scientific research in the area, environmental education activities and visits for tourism and recreational purposes (Brasil 2000), and it can only be implemented by the voluntary act of the owner. In addition, the property where this protection system is created becomes indivisible and its creation cannot be revoked (Curitiba 2013).

The system of creation of conservation areas according to Rocktaeschel (2006: 51) "is the main governmental mechanism of protection of the exuberant biological diversity and the scenic beauties of the country", whose evolution of the concept, according to the author, is related to the development of science, destruction of the environment, and human need for contact with nature. This scenario has stimulated processes of creation and implementation of Conservation Units constituting a new vision that presupposes the need for protection of natural resources to ensure well-being, especially in the cities.

In Brazil, according to the National Register of Conservation Units (MMA 2019), the total number of federal, state and municipal Conservation Units is 2.201, representing 250 million hectares. There are 698 units of full protection, such as Ecological Stations, Natural Monuments, National, State or Municipal Parks, Wildlife Refuges and Biological Reserves, and another 1.503 units are of sustainable use such as Forests, Extractive Reserves, Sustainable Development, Fauna, Areas

[1] The term conservation unit is restricted to Brazil and refers to certain specific types of protected areas as defined by the National System of Conservation Units (SNUC). However, the term protected areas is used internationally.

[2] Brazilian acronym.

[3] Idem.

of Environmental Protection and Relevant Ecological Interest. After the creation of SNUC in 2000, 151 federal conservation units were created, adding 134 million hectares. The result of the last 18 years represents more than three times what was done in the 65 years before. However, Drummond et al. (2010: 36) point out that there is currently the risk of discontinuities in environmental politics because of the different interests among members of the government, market and social society when it comes development.

This strategy is observed in the city of Curitiba, where public policies and legal and fiscal mechanisms have been implemented in order to preserve the private green areas. Despite being a large metropolis with 1.75 million in habitants, Curitiba, according to the Biodiversity Condominium (2014), has forest areas in good condition, with 64.5 m^2 of arboreal vegetation per inhabitant, distributed in 42 Public Municipal Conservation Units (21 parks, 17 woods, two Environmental Protection Areas, a Botanical Garden and an Ecological Station, in addition to the 24 Municipal Private Natural Heritage Reserves – RPPNMs). The city has about 20% of its area covered by forest, equivalent to more than 78 million m^2 and about 75% of this area is located in private properties and 25% is in public areas (Curitiba 2013).

In this context, in 2006 Law No. 12080 was approved, which encourages the creation and maintenance of RPPNM in Curitiba. Likewise, the possibility of a financial return for the owners was created through the sale of "the construction potential", in addition to the activities of environmental education and tourism (Panasolo et al. 2015). It's worth mentioning that "Conceptually, the instrument of Transfer of Construction Potential is the possibility that the owner has to exercise his construction potential in another property or to sell it to another owner" (Panasolo et al. 2015: 41).

Given the potential of public policy, the Municipal Private Natural Heritage Reserves, institutionalized in Curitiba, makes it possible to expand spaces of natural protection with low public investment and even combine the generation of jobs and income with conservation. The objective of this chapter is to analyze the entrepreneurial capacity of a successful RPPNM, the Airumã in Curitiba (PR). This is an interesting approach to conservation of urban forests, and this research tries to answer the question whether or not it can serve as a model for more sustainable cities and to balance urban developments with social wellbeing.

In order to achieve this purpose, bibliographical and documentary research was conducted, as well as interviews with the owner of the RPPNM Airumã. The chapter is structured into three topics. The first one seeks to understand the concept of the creation of Conservation Units in Brazil. The second deals with the theme of the creation of the Private Natural Heritage Reserves, highlighting the reserves created in the city of Curitiba. In the last one, the empirical results of the research carried out in the Airumã Reserve is given, one that was chosen for being the most active among the RPPNM, and we discuss challenges, possibilities, and opportunities in its implementation. Such results could stimulate the creation of other RPPNM in Curitiba, as well as other municipalities to implement such public policy.

12.2 Concept of the Creation of Conservation Units in Brazil

The Brazilian strategy to contain the advance of environmental degradation and loss of biodiversity, as a result of anthropic pressures arising mainly from the current model of economic development, was the establishment of Conservation Units (CUs) as a public policy of extensive application to the protection and conservation of nature. However, Irving (2013: 62) points out that "the process of creating CUs has not been continuous and reflects different political moments in terms of defining strategic priorities and institutional capacities in the sphere of environmental action".

Data from the Ministry of the Environment (MMA) reveals that there has been a substantial increase in the number of CUs in Brazil and also in the protected territorial extension. In 1997, there were 196 Protected Areas for Integral Protection and 149 of Sustainable Use, involving 47.5 million hectares protected, corresponding to 5.7% of the continental territory of the country. In 2012, this number was expanded to 548 Integral Protection CUs and 1214 of Sustainable Use, with a total area of approximately 153 million hectares protected, representing 17% of the national territory (MMA 2013).

The National System of Conservation Units (SNUC) guides the standardization and the operation of CUs, and advances to the extent that it recognizes the diversity of ecosystems and the need for protection and integrated management through the creation of ecological corridors and mosaics, making possible the establishment of privately-owned protected areas – Private Natural Heritage Reserve (RPPN). It also establishes that the creation and administration of CUs should be a process that considers public consultations and the constitution of councils, with the representation of the different social segments to ensure a more democratic management (Brasil 2007).

With regard to management tools, the SNUC also determines the requirement of management plans, which consider the population of the interior and the surrounding area, ensuring their effective participation in the elaboration, implementation and monitoring of the agreed actions, and provides for the operation of Councils, which represent instances to discuss management decisions between the State and society (Brasil 2007).

Law No. 9985 of 2000, which establishes the SNUC, defines the management plan as a technical document, based on the general objectives of the Conservation Unit, defining its zoning and standards that should govern the use and management of natural resources. However, despite the legally established obligation, the preparation and implementation of management plans is far from being consolidated. Currently, only 17.54% of the total conservation units in the country have a properly approved and updated management plan, and among the national parks, this number does not reach 40% (MMA 2019), a fact that is partly due to the elaboration of the management plans by specific teams that are external to the units and that are dedicated specifically to this task. Once the plan is finalized, it is up to the unit managers, among them the RPPN, to take care of its implementation, which in many cases is difficult conditions are not appropriate to advance in the planned actions.

12.3 Private Natural Heritage Reserves (RPPN)

According to the Instituto Chico Mendes de Conservação da Biodiversidade (ICM-Bio (2011, the origin of the RPPN refers to the Private Refuge of Native Animals – REPAN, established by Ordinance No. 327 of 1977 of the extinct Instituto Brasileiro de Desenvolvimento Florestal (IBDF). In these refuges, private owners could ban the hunting of animals even if their land was in areas cleared for hunting. In 1988, this Ordinance was revoked by a new IBDF decree, No. 217, which established the Private Reserves of Fauna and Flora – RPFF.

In accordance with Art. 21 of the SNUC Law, Private Natural Heritage Reserves are private areas, recorded with perpetuity, with the purpose of preserving nature, in which only scientific research and visitation for tourism, recreation, and educational purposes are allowed (Brasil 2000). For Kormann et al. (2010: 17) the RPPN:

> *(...) is individualized before the others because it is not created from the interest of the public power, but rather, at the initiative of the owner, who seeks to preserve an area of environmental importance. In this way, this space will be protected with perpetuity, according to the conservation provisions foreseen in the SNUC.*

The RPPNs, according to the City Hall of Curitiba (Curitiba 2013: 9) "are currently an important instrument of conservation of nature, combined with the efforts to create conservation units in the public domain". They can only be created if its importance is recognized by the public power for the conservation of the local biodiversity, by a voluntary act of the owners who decide to constitute their property, or part of it, in a Conservation Unit in perpetuity. There is no determination of maximum or minimum size for the constitution of an RPPN.

In view of this, the creation of the RPPNs in urban areas is an important strategy for the preservation of green areas, which allow cities to have spaces for leisure and improve the life quality of the population, as they provide contact with nature, environmental quality, and in general spaces for physical activities and recreation. For Lima e Amorim (2006: 71), the green areas:

> *(...) are one of the variables that are part of the urban structure and the preservation of these areas is related to its use and its integration into the city dynamics, which are reflections of human actions and are linked to the historical processes, translating the attention of the public power to the implantation and maintenance of these spaces in the urban network.*

Green areas are considered as the set of intra-urban areas that present vegetation, tree cover (native and introduced), shrub or ground (grass) and that contribute significantly to the quality of life and environmental balance in cities (MMA 2013). These green areas are present in a wide variety of situations: in public areas; in areas of permanent preservation (APP); squares, parks, forests and urban conservation units (CU).

12.3.1 Municipal Private Natural Heritage Reserves in the City of Curitiba

With a Municipal Human Development Index (IDHM) of 0.82, the municipality has a green area of 58 m^2 per inhabitant (Curitiba 2016), being a pioneer in including in its legislation the creation of a Private Natural Heritage Municipal Reserves (RPPNM), which aim to preserve the biological diversity in urban private properties covered by native vegetation in good condition (Condomínio da Biodiversidade 2013). The creation of these units according to the Association of Protectors of Green Areas of Curitiba and Metropolitan Region (APAVE) is a great contribution in several ways, avoiding negative impacts, such as irregular occupation, real estate speculation, deforestation, waste disposal, hunting, fire or other actions that may degrade these natural areas. By taking on the commitment to care for and preserve the nature of the area forever, the owner exercises an act of citizenship and environmental responsibility.

In 2006, Law No. 12080 was passed, which encouraged the creation and maintenance of private green areas in Curitiba. In 2015, Law No. 14587 restructured the Program of Municipal Private Natural Heritage Reserves – RPPNM in the city of Curitiba, revoking Law No. 12080 of December 19, 2006, and Law No. 13899 of December 9, 2011. Law No. 14587 establishes that the RPPNMs are intended to be used for the development of scientific research and visitation for therapeutic, touristic, recreational and educational purposes, provided for in the management plan.

The city of Curitiba, according to the Municipal Department of the Environment (SMMA), has a potential of about 1000 properties to become permanent preservation areas (PPAs) through voluntary action. In addition to Curitiba, 15 other Brazilian municipalities have legislations for the creation of RPPNMs, such as São Paulo, Manaus, Itamonte (MG), Passo Fundo and Santa Maria (RS). The mountainous region in Rio de Janeiro also counts on municipalities like Miguel Pereira and Petrópolis that has its municipal laws. It is worth noting that in addition to environmental benefits, such as transforming green areas into reserves, these areas also provide social contributions (Panasolo et al. 2015), as they are incentives to tourism, recreation, and increased life quality, coming from a naturally balanced environment. Nevertheless, the economic incentives of the public sector for the creation of these areas and their maintenance seems to be decisive.

It is noteworthy that the RPPNM can only be implemented by a voluntary act of the owner, it is a private domain conservation unit, and it is created with the objective of conserving biological diversity. In addition, the property where the RPPNM was created becomes indivisible and its creation cannot be revoked (Curitiba 2015). For the property to be transformed into RPPNM, it is necessary to fulfill some requirements, for example, the property must be 70% covered by remnants of native vegetation, not built or at most have a residence family nucleus, where according to the forest typology it is not possible to remove the vegetation. In the administrative process for requesting the creation of the RPPNM, the owner must present specific documentation, such as: identification document of the owner; copy of real estate registration; consultation for construction purposes (yellow guide); planimetric survey of the area; survey of vegetation and the geo-referenced perimeter in relation to the geodesic landmarks (Curitiba 2015).

According to Art. 4 of Law No. 12080, as an incentive and compensation to this conservation work carried out by the owner, the municipality offers the following advantages (Curitiba 2015):

- Exemption of the RPPNM area in the calculation of the Urban Property and Territorial Tax (IPTU);
- The possibility of using the area for the development of scientific research (through partnerships with universities or other institutions) and for visitation with ecological, tourism, recreational or educational purposes, with the possibility of generating income;
- Certificate of recognition to the owner, signed by the Mayor;
- Transfer of construction potential, in accordance with current legislation;
- Certificate of recognition of the enterprise that uses constructive potential coming from RPPNM, signed by the Mayor.

In summary, the owners of the areas have the right to require the Municipality to transfer construction potential from these areas to other properties, with the possibility of periodical renewal when the parameters set out in the specific legislation are respected, in addition to the IPTU exemption. Article 6 of Law No. 14587 provides that the granting of the constructive potential of RPPNM may be renewed every 15 years (Curitiba 2015).

Curitiba has 21 official RPPNMs (Table 12.1), the first RPPNM in Curitiba, located in the Santa Felicidade neighborhood, was created in 2007 and is called Cascatinha (Apave 2019). In these areas, only activities related to scientific research and visitation with ecological, tourist, recreational and educational objectives can be developed (Curitiba 2015). However, for these activities to be developed the property must have a management plan. Management plans deal with the activities performed in the reserve, maintenance of the environmental attributes of the area, demarcation of its limits and the warning to third parties about the existence of the RPPN and the prohibitions imposed on it. According to Bensusan (2006), a management plan defines the attributions of the owners to the supervisory bodies. However, the elaboration of the management plan and the reports suggest support from specialists, burdening the owner. The process of creating the RPPN requires a lot of documentation from the owner and the property, which are very difficult to obtain, which makes the process difficult, time-consuming and costly (Bensusan 2006).

For Panasolo et al. (2015: 37) RPPNMs contribute to environmental preservation and collaborate to maintain green areas in environmentally fragile urban centers, and also:

- Enable the private sector to participate in national conservation efforts;
- Provide leisure, tourism and recreation services;

Table 12.1 Municipal private natural heritage reserves registered in Curitiba

RPPNM	Total area (^{ha})	Total area in relation to the municipality (%)
Cascatinha	0.85	0.0019
Ecoville	11.6	0.0037
Barigui	0.21	0.0006
Bacacheri	0.5	0.0012
Bosque da Coruja	0.54	0.0012
Cedro-rosa	0.72	0.0016
Erva-mate	0.73	0.0017
Canela	0.74	0.0017
Guabiroba	0.74	0.0017
Taboa	0.75	0.0017
Jerivá	0.74	0.0017
Airumã	3	0.0068
Jataí	0.07	0.0002
Araçá	0.07	0.0002
Umbará	0.63	0.0014
Beppe Nichele	1.34	0.0031
Name	2.34	0.0054
Caxinguelê	2.03	0.0047
Refúgio do Jacu	0.53	0.0012
Geronasso	4.7	0.0108
Vô Mantino e Amélia	1.86	0.0043
Mirante do Parque Barigui	64.35	0.14
Alfred Willer	0.6	0.0012
Total	89.96	0.1412

Source: Own elaboration using data from Apave (2019)

- Contribute to the increase of landscape connectivity in areas of biological importance, such as biodiversity corridors;
- Generate financial benefits through the economic valuation of property (economic function).

The green areas can also serve to balance the urban environment offering a place of leisure, landscape, and environmental plasticity. Another important factor is the afforestation next to public roads that "serve as filters to attenuate noise, retain dust, reoxygenate the air, besides offering shade and the sensation of freshness" (Lima e Amorin 2006: 70).

12.3.2 *Private Reserve Airumã*

Airumã means in Tupi, the indigenous language, "Morning Star or Star Guide". Located in the municipality of Curitiba, it is the the largest Municipal Private

Natural Heritage Reserves registered in Curitiba (Table 12.1), with 25,000 m^2, of green preserved areas (Panasolo et al. 2015). According to the owner, the process of creating the reserve began in 2011 and was formalized in 2013, by Decree No. 521/2013, being the 12th RPPNM created in the municipality. The Reserve is the headquarter of the Association of Protectors of Green Areas of Curitiba and Metropolitan Region (APAVE) and Airumã Environmental Station,[4] created in 2004, with the mission of "contributing to the reunion of people committed to the awakening of a new socio-environmental awareness, engaged in building a more sustainable, peaceful and just land" (Vareschi 2017).

According to Chauá and Propflor (2017) in the reserve there is a spring and three streams that are born in the neighboring properties. The vegetation formation of the property corresponds to the Mixed Ombrophilous Forest, in the middle stage of secondary succession. The floristic survey identified the presence of 159 native species belonging to 62 families, and 10 exotic species. Of the native species present in the area, on the list of endangered species we find: Paraná pinus (*Araucaria angustifolia*), Xaxim-bugio (*Dicksonia sellowiana*) and Sassafras Cinnamon (*Ocotea odorifera*) (Chauá and Propflor 2017).

Residing in the area for 32 years, the main reasons of the owner that motivated the creation of the RPPNM are related to conservation of species and ecosystems in urban areas, protection of water resources and personal satisfaction. The owner also reports that one of the main factors that made it possible to make the area an RPPNM was its strategic location, inserted in the Barigui River Basin, an important river in the municipality of Curitiba, and near the Tingui Municipal Park.

In creating the RPPNM, the owner experienced difficulties related to providing unfavorable technical advice regarding the coverage of the area occupied by native vegetation. Without the support of the institutions, the owner declares, it would not have been possible to create the reserve because the studies are expensive. Another determining factor for the creation of the reserve was the support of the Society for Wildlife Research and Environmental Education (SPVS[5]) together with the O Boticário Group Foundation (FGB[6]) in the elaboration of the management plan.

Regarding the costs since the creation of the RPPNM in 2013, the owner points out that they have invested their own resources. For maintenance, several activities are carried out with the purpose of raising funds. Airumã stands out as the only reserve that develops entrepreneurial activities in the RPPMN that are economically feasible, such as the sale of tickets for visitation, rental of space for events, and courses on environment education. The reserve is open for visitation and on-site seminars are held, as well as lectures, and workshops mostly with themes related to the environment. This is facilitated with a space for events with a structure suitable

---

[4]The RPPNM and the Environmental Station Airumã have a blog to publicize the activities developed in the place. Available at: http://airumaestacaoambiental.blogspot.com.br/

[5]Brazilian Acronym.

[6]Idem.

Fig. 12.1 Residence and space for events in the Airumã RPPNM (Piva 2017). (Photo: João Henrique Piva)

to receive 40 people (Fig. 12.1). An entrance fee is established per person, according to the activity that will be developed.

The venture also receives visitors from high schools and institutions of higher education, as a way of arousing interest in the environment. Programs on environmental education and contemplation of the nature are offered. The incomes received allow obtaining resources for the maintenance of the enterprise: cleaning, renovations and investments in new projects. The property has no fixed employees, only contracted services related to maintenance and cleaning.

Regarding the sale of the Construction Potential of the RPPNM that would make it viable economically, it is noteworthy to mention that the owner has not yet received its Construction Potential, and it is necessary to initiate a legal process to obtain this right. Even without obtaining the Construction Potential that is one of the rights and incentives by the public power in benefit to the owners that make their area an RPPNM, the property is exempt of 100% of the Urban Property and Territorial Tax (IPTU) and/or Rural Property and Territorial Tax (ITR). However, even before becoming an RPPNM, Airumã had a 70% IPTU exemption because of the high level of preservation of the native forests present in the area. In 2018 RPPNM Airumã received the Environmental Ministry's grant for the policy of payment for environmental services (PSA[7]) provided to society. The values received

[7]Brazilian Acronym.

from the PSA contribute to the maintenance and conservation of urban green areas in a Protector-receiver and polluter-pays.

In the process of creating the RPPNMs, the support of non-governmental organizations (NGOs) is essential for the preparation of the necessary studies for their official recognition. These partnerships represent the common interest of civil society and third sector organizations in social and environmental causes. Partnerships are essential to ensure the viability of the enterprise and especially for the preparation and achievement of new projects. In 2017, the Airumã RPPNM received support from SOS Mata Atlântica[8] through the first edict to support the municipal conservation units. The project, named "Forest Paths: Structuring the Airumã RPPNM for Public Use" implemented an educational trail to provide access to people with reduced mobility. The trail, besides allowing access to the interior of the property, leads the visitor to see the tree that symbolizes the excellent state of preservation of the Airumã reserve, the "Vovó Airumã" (Grandma Airumã), a tercentennial Araucaria (*Araucaria angustifolia*).

Other activities of an environmental nature are developed in the reserve, such as the partnership with the municipal government to develop the "Jardins do Mel" Program, with the aim of promoting natural pollination in the city, and the environmental education of municipal school students on beekeeping, with native stingless bees. Also, in partnership with the company Eco-Guaricana[9] an Environmental Education program is being formalized, the "Guardians of the Araucaria Forest", which seeks to affiliate municipal schools and carry out activities in network with other RPPNMs of the municipality. The focus of the project is to awaken students' interest in preserving the environment. In addition to the environmental education program, the company ECO-Guaricana also carries out activities related to cycling tours. The RPPNM Airumã is located in front of a bicycle lane, so it becomes a stopping point to bikers.

The Airumã reserve has the characteristic of symbolizing the importance of the preservation of green areas in urban centers, where its managers are activists in environmental causes with the Association of Protectors of Green Areas and the Metropolitan Region. It is worth to mention that the RPPNM Airumã has a comparative advantage because it is located in a municipality, Curitiba, recognized worldwide for sustainable urban management. In the context of the urban space, it is an alternative space where several actions are made with the intention of showing society the importance of preserving natural areas in urban environments, with a view to making the unit a model to be replicated in other properties that seek the creation of Conservation Units.

[8]The SOS Mata Atlântica Foundation has established a public notice to support the creation and implementation of municipal Conservation Units (CUs) in the country. The objective is to encourage cities to strengthen the environmental management of their territories by investing in planning and implementing measures to ensure protection and the sustainable use of the environment (SOSMA 2017). Available in: https://www.sosma.org.br/105146/conheca-os-projetos-aprovados-edital-de-apoio-ucs-municipais/.

[9]http://www.ecoguaricana.com.br/

12.4 Final Considerations

The top-down model to the implementation and management of conservation units in Brazil has been in many cases inefficient in the face of the objectives of their creation, which aims to contain the advance of environmental degradation and the loss of biodiversity, as a result of anthropic pressures, mainly due to the current model of economic development. As an alternative to this model, the RPPNM system represents a bottom-up model for conservation, in this case in urban stings. The RPPNM Airumã represents a successful example of a viable urban, socio-environmental enterprise related to environmental conservation.

With regard to conservation units called RPPNMs, the strategy of their creation allows to create, where possible, urban ecological corridors that connect other conservation units in peri-urban and non-urban perimeters, characterizing systemic and resilient spaces of biodiversity and socio-biodiversity, as well as pedagogical and leisure spaces (recreational and contemplative). The Airumã RPPNM has become an exemplary model related to the protection of green areas and sustainable development in areas of permanent preservation. Concerning their managers, they have an entrepreneurial profile, but the main purpose is related to the preservation of native forests, where ownership makes income generation compatible for maintenance of native forests and ecosystem services while contributing to the improvement of the well-being of people and sustainable development of the city.

The fact that the municipality of Curitiba has a history of public policies that promote sustainable urban development, beyond the fact that it is pioneering in the creation of a specific legislation for Conservation Units, creates a political environment conducive to the increase of RPPNMs. When comparing Curitiba with other states that have legislations for RPPNM, as e.g. Rio de Janeiro, it is possible to see that the municipality is advanced in numbers of RPPNMs: the entire state of Rio de Janeiro only adds 17 RPPNMs, while Curitiba has already 24 (RIO 2017).

However, even after 13 years of creation of the first RPPNMs in Curitiba, there is still much to be done in terms of public policies, mainly in the financial incentive tools. These tools are important for encouraging the creation of new RPPNMs and avoid the risk of not only value urban neighborhoods as financial speculation strategy. In order for this model to be replicated in other countries, it is necessary to have a federal legislation of Conservation Units that provides the guidelines, as is the case with SNUC in Brazil. However, those public policies must be formulated with meaningful civil society participation.

Finally, it should be understood that natural systems are interconnected with social systems, so it makes sense to call it socio-environmental systems that preserve the natural environment and protect life in the urban environment.

References

APAVE – Associação dos Protetores de Áreas Verdes de Curitiba e Região Metropoli tana (2019) RPPNMs Oficializadas em Curitiba. Retrieved from: http://apavecuritiba.blogspot.com/. Accessed 19 Feb 2019

Bensusan N (2006) Conservação da biodiversidade em áreas protegidas. In: 1st. FGV Editora, Rio de Janeiro

Brasil (2000) Lei n° 9.985/00. Institui o Sistema Nacional de Unidades de Conservação da Natureza e dá outras providências. Diário Oficial, Brasília-DF. Retrieved from: http://www.planalto.gov.br/ccivil_03/leis/l9985.htm. Accessed 19 Sept 2017

Brasil (2007) Decreto n. 6.040/07. Institui a PNPCT – Política Nacional de Desenvolvimento Sustentável dos Povos e Comunidades Tradicionais. Diário Oficial, Brasília-DF. Retrieved from: http://www.mds.gov.br/backup/institucional/secretarias/secretaria-de-articulacao-institucional-e-parcerias/ arq uivo-saip/povos-e-comunidades-tradicionais-1/decreto_6040_2007_pnpct .pdf/ view. Accessed 12 Dec 2017

Condomínio da Biodiversidade – ConBio (2013) Projetos. Disponível. http://www.spvs.org.br/projetos/condominio-da-biodiversidade-conbio/. Accessed 17 Jul 2018

Curitiba (2013) Reserva Particular do Patrimônio Natural Municipal (RPPNM) em Curitiba – roteiro para criação e elaboração do plano de manejo e conservação. Prefeitura Municipal de Curitiba e Sociedade de Pesquisa em Vida Selvagem e Educação Ambiental (SPVS), Curitiba

Curitiba (2015) Lei n° 14.587, de 14 de janeiro de 2015. Reestrutura o programa das Reservas Particulares do Patrimônio Natural Municipal do município de Curitiba. Retrieved from: https://leismunicipais.com.br/a/pr/c/curitiba/lei-ordinaria/2015/1458/14587/lei-ordinaria-n-14587-2015-reestrutura-o-programa-das-reservas-particulares-do-patrimonio-natural-municipal-rppnm-no-municipio-de-curitiba-revoga-as-leis-n-12080-de-19-de-dezembro-de-2006-e-lei-n-13-899-de-9-de-dezembro-de-2011. Accessed 23 June 2017

Curitiba (2016) Perfil de Curitiba. Retrieved from: http://www.curitiba.pr.gov.br/conteudo/perfil-da-cidade-de-curitiba/174. Accessed 12 Jan 2017

Dowbor L (1993) Formação do Terceiro Mundo. Brasiliense, São Paulo, p 52

Drummond JÁ, JL de Andrade Franco, Oliveira D (2010) Uma análise sobre a história e a situação das unidades de conservação no Brasil. In: Conservação da Biodiversidade, Legislação e Políticas Públicas, (SI), pp 341–385

ICMBIO – Instituto Chico Mendes de Conservação da Biodiversidade (2011) História das RPPNs. Retrieved from: http://www4.icmbio.gov.br/rppn//index.php?id_menu=149. Acesso em 09 Sept 2017

Irving MA, Rodrigues CGO, Rabinovici A, Costa HA (2013) Parques Nacionais do Rio de Janeiro: paradoxos, contextos e desafios para a gestão social da biodiversidade. Folio Digital, Letra e Imagem, Rio de Janeiro, pp 19–78

Jacobi PR, Giatti L, Ambrizzi T (2015) Interdisciplinaridade e mudanças climáticas: caminhos para a sustentabilidade. In: Philippi A Jr, Valdir F (eds) Práticas da interdisciplinaridade no ensino pesquisa. Manole, Barueri, pp 408–422

Kormann TC, Thomas BL, Nascimento DB, Foleto EM (2010) Contribuição geográfica na criação de uma Reserva Particular do Patrimônio Natural (RPPN) em Itaara – RS. Revista Geografar, Curitiba 5(2):13–31. jul./dez

Lima V, Amorim MCT (2006) A importância das áreas verdes para a qualidade ambiental das cidades. Revista Formação, n°13, pp 69–82

MMA – Ministério do Meio Ambiente (2013) Sistema Nacional de Unidades de Conservação. Retrieved from: http://www.mma.gov.br/areas-protegidas/sistema-nacional-de-ucs-snuc. Acesso em 01 Dec 2017

MMA – Ministério do Meio Ambiente (2019) Cadastro Nacional de Unidades de Conservação. Retrieved from: http://www.mma.gov.br/areas-protegidas/cadastro-nacional-de-ucs

Panasolo A, Silva JCGL, Peters EL, Santos AJ (2015) Áreas verdes urbanas privadas de Curitiba: uma proposta de valorização para conservação. Enciclopédia Biosfera, Centro Científico Conhecer, Goiânia 10(19):27–31
Piva JHT (2017) Viabilidade econômica de empreendimentos sustentáveis: Reserva Particular do Patrimônio Natural Municipal (RPPNM). Trabalho de Conclusão de Curso em Administração da Universidade Positivo, Curitiba. 90p
RIO (2017) Prefeitura de Rio de Janeiro. Estudo sobre a regulamentação de reserva particular do patrimônio natural pelo município do rio de janeiro
Rocktaeschel BMMM (2006) Terceirização em áreas protegidas: estímulo ao ecoturismo no Brasil. SENAC, São Paulo
Sachs I (1993) Estratégias de transição para o século XXI: desenvolvimento e meio ambiente. Studio Nobel/Fundap, São Paulo
Sachs I (2004) Desenvolvimento includente, sustentável, sustentado. Garamond, Rio de Janeiro
Schenini PC, Costa AM, Casarin VW (2004) Unidades de conservação: aspectos históricos e sua evolução. Congresso Brasileiro de Cadastro Técnico Multifinalitário, UFSC Florianópolis
Sociedade Chauá Propflor consultoria ambiental (2017) Plano de manejo da Reserva Particular do Patrimônio Natural Municipal Airumã – Curitiba, PR. Curitiba, 2013. Condomínio da Biodiversidade. Retrieved from: http://http://www.condominiobiodiversidade.org.br. Accessed 07 June 2017
SOS Mata Atlântica – SOSMA (2017) Relatório Anual de Atividades. Disponível. https://www.sosma.org.br/wp-content/uploads/2019/10/AF_RA_SOSMA_2017_web.pdf. Accessed 4 Jun 2018.
Vareschi T (2017) Reserva Particular do Patrimônio Natural Municipal Airumã. Entrevista concedida em março de 2017

Chapter 13
Understanding Adoption and Design of Incentive-Based Forest Conservation Policies: A Case Study of the SISA Program in Acre, Brazil

Hugo Rosa Da Conceição and Jan Börner

13.1 Introduction

Since the 1980s the state of Acre, in the northwest of Brazil, has been a pioneer in formulating and implementing public policies for forest conservation (May et al. 2016). It is also the birthplace of some of Brazil's most prominent environmental activists and a symbolic place for Brazilian environmentalism. Acre's latest institutional innovation is the System of Incentives for Environmental Services (SISA). Led by Acre's state government SISA promotes a governance structure for the jurisdictional implementation of incentive-based conservation policies.

Due to the state's prominence as an incubator of conservation policy innovations, Acre has been closely scrutinized by scholars and practitioners alike. Building on this body of knowledge, this chapter scrutinizes the design process of SISA, which, as we will argue, reflects a historically grown common concept of conservation among the involved stakeholders. "The everyday, even mundane, tasks involved in institutional construction and maintenance" (Jespersen and Gallemore 2018, p. 508), can generate important insights to the study of conservation programs. We thus aim to understand how Acre's actors, institutions, and historical idiosyncrasies led to the specific design features of SISA. To frame our analysis from the perspective of

H. Rosa Da Conceição (✉)
Center for Development Research, Rheinische Friedrich-Wilhelms-Universität Bonn, Bonn, Germany

J. Börner
Center for Development Research/Institute for Food and Resource Economics, Rheinische Friedrich-Wilhelms-Universität Bonn, Bonn, Germany

F. Fuders and P. J. Donoso (eds.), *Ecological Economic and Socio Ecological Strategies for Forest Conservation*, https://doi.org/10.1007/978-3-030-35379-7_13

the actors involved in the design of SISA, we will use the established concept of "institutional governance",[1] which is introduced in detail below.

The research for this chapter was based on the analysis of policy documents, available drafts, meeting minutes, and other written materials, as well as interviews with policy-makers directly and indirectly involved in program design, and with current and former technical staff in Acre's government institutions. The sampling of initial respondents was based on previous knowledge by the researcher of key decision makers in Acre (reputational criteria) and on the selection of actors in core institutional positions (positional criteria), to be complemented with further actors suggested in the initial interviews (snowball/chain referral approach). Such strategy follows the methodology suggested by Tansey (2007) for interviewing elites.

In Sect. 13.2, the main aspects of institutional governance and its suitability as a framework for the analysis are presented. Section 13.3 describes the main institutional aspects of SISA. Section 13.4 narrates the most relevant aspects of SISA's design process and identifies the instances of institutional governance performed by the actors involved in that process. Section 13.5 discusses our findings and presents conclusions.

13.2 Institutional Governance

Institutional governance is defined by its main proponents (Lawrence and Suddaby 2006, p. 215) as "the purposive action of individuals and organizations aimed at creating, maintaining, and disrupting institutions". It highlights "the intentional actions taken in relation to institutions, some highly visible and dramatic, as often illustrated in research on institutional entrepreneurship, but much of it nearly invisible and often mundane, as in the day-to-day adjustments, adaptations, and compromises of actors" (Lawrence et al. 2009, p. 1). It is, therefore, especially suitable for our analysis of SISA's design process, which is basically a process of creating and maintaining institutions.

In a literature review on Payments for Environmental Services (PES),[2] Jespersen and Gallemore (2018) found no explicit use of institutional governance concepts, but identified examples of institutional governance especially in terms of creating institutions. Most institutional analyses of PES, however, emphasize "constituted institutions and their behavioral effects rather than the process by which institutions are created or maintained" (Jespersen and Gallemore 2018, p. 507).

[1]The analytical framework is referred to as "institutional work" in the original references. It will be named "institutional governance" throughout the chapter to facilitate understanding.

[2]According to a widely used definition, PES are "voluntary transactions between service users and service providers that are conditional on agreed rules of natural resource management for generating offsite services" (Wunder 2015, p. 241).

Institutional governance can take several forms. Table 13.1 presents the typology proposed by Lawrence and Suddaby (2006), encompassing forms of institutional governance for creating and for maintaining institutions. It also provides some examples from the PES literature as identified by Jespersen and Gallemore (2018). Evidently not all of Lawrence and Suddaby's categories will apply in all cases, but they serve as a useful framework for acquiring analytical insight into SISA's policy design process. Since our study does not deal with the disruption of institutions, we ignore this particular category here. Since PES is a relatively new 'kid on the block' of conservation policy arrangements, it may not surprise that the literature so far mainly focused on institutional governance in the "creating institutions" category.

13.3 Acre's State System of Incentives to Environmental Services (SISA)

SISA was created in 2010 by the state of Acre in Brazil. Instead of being solely a payments-providing program, with specific activities to be implemented, SISA is an institutional umbrella that organizes the financing and implementation of ecosystem service-related programs and projects in the state of Acre. SISA builds on previous state policy experiences, such as the Ecological-Economic Zoning (ZEE) and the Valuation of Forest and Environmental Assets Policy (Duchelle et al. 2014). The program's overarching objective is the conservation and enhancement of environmental services in the state (Acre 2010a). The system has the status of a State law, promulgated on October 22nd, 2010.

The core institution of SISA is the Institute for Climate Change and Regulation of Environmental Services (*Instituto de Mudanças Climáticas* – IMC). The Institute is responsible for the overall coordination of the program, refining norms, defining strategies, overseeing monitoring, evaluation activities, and the accomplishment of policy objectives. With the support of SISA's Scientific Committee, it is also responsible for carrying out inventories and estimates of environmental services provision and conservation.

Another core institution established by SISA is the Commission for Validation and Oversight (*Comissão Estadual de Validação e Acompanhamento* – CEVA). CEVA is SISA's main link to the civil society, aiming to ensure transparency and social control of projects and activities carried out within the SISA framework, by analyzing and approving norms and other documents presented by the IMC. CEVA is composed of eight members, four stemming from the civil society and four from the government.

The third core institution of Acre, and arguably the most innovative in SISA's structure, is the Company for the Development of Environmental Services (*Companhia de Desenvolvimento de Servicos Ambientais* – CDSA). CDSA's main functions are to raise funds for activities within the SISA and to head the interaction with potential private project implementers (Interview 8). Such function entails

Table 13.1 Types of institutional governance, definitions and examples

Type	Definition Lawrence and Suddaby (2006)	Examples in the PES literature, as identified by Jespersen and Gallemore (2018 p. 510–511)
Creating institutions		
Advocacy	The mobilization of political and regulatory support through direct and deliberate techniques of social suasion	In the PSA-CABSA PES program in Mexico, a consortium of environmental and agricultural groups lobbied to set the program's sectoral scope and funding priorities
Defining	The construction of rule systems that confer status or identity, define boundaries of membership or create status or hierarchies within a field	Definitional work tends to involve adopting understandings of PES developed in the academic and policy literature and then attempting to implement them on the ground, as well as taking place while creating offset or certification standards
Vesting	The creation of rule structures that confer property rights	Clarifying the types of rights bestowed upon the owners of relevant resources and whether or not owners possess agency over the property they inhabit
Constructing identities	Defining the relationship between an actor and the field in which that actor operates	N.a.
Changing normative associations	Re-making the connections between sets of practices and the moral and cultural foundations for those practices	N.a.
Constructing normative networks	Constructing of inter-organizational connections through which practices become normatively sanctioned and which form the relevant peer group with respect to compliance, monitoring, and evaluation	Socially inclusive governance systems deemed as important for successful PES projects
Mimicry	Associating new practices with existing sets of taken-for-granted practices, technologies, and rules in order to ease adoption	Adopting conditional cash transfer approaches is a mimicry approach to ecosystem protection
Theorizing	The development and specification of abstract categories and the elaboration of chains of cause and effect	N.a.
Educating	The educating of actors in the skills and knowledge necessary to support the new institutions	Stakeholder education as a central component of empowering communities to effectively engage in negotiations about project expectations and outcomes

(continued)

Table 13.1 (continued)

Type	Definition Lawrence and Suddaby (2006)	Examples in the PES literature, as identified by Jespersen and Gallemore (2018 p. 510–511)
Maintaining institutions		
Enabling work	The creation of rules that facilitate, supplement and support institutions, such as the creation of authorizing agents or diverting resources	N.a.
Policing	Ensuring compliance through enforcement, auditing, and monitoring	N.a.
Deterring	Establishing coercive barriers to institutional change	N.a.
Valorizing and demonizing	Providing for public consumption positive and negative examples that illustrate the normative foundations of an institution	N.a.
Mythologizing	Preserving the normative underpinnings of an institution by creating and sustaining myths regarding its history	N.a.
Embedding and routinizing	Actively infusing the normative foundations of an institutions into the participants' day to day routines and organizational practices	N.a.

Source: Own elaboration based on data provided by Jespersen and Gallemore (2018), Lawrence and Suddaby (2006)

more than going after potential donors, but also being the executive agency of projects and placing SISA's activities within potential markets for environmental services, managing and marketing potential credits generated by SISA's initiatives. CDSA is defined as a public company (*empresa pública*), regulated by private law, having more legal flexibility to do business than public agencies. Legal advice on the SISA is provided by the State Attorney General's Office (*Procuradoria Geral do Estado* – PGE), which also participates in CEVA and hosts SISA's ombudsman. In addition to the structure described above, SISA entered a multitude of agreements with partners ranging from local NGOs, to federal government agencies to international organizations to support SISA's activities.

As stated above, SISA is a general institutional framework for the support and execution of activities on the conservation and generation of environmental services. Since its enactment, it has supported already existing activities carried out by the government (Acre 2015). It will also provide support for activities implemented by non-governmental and private institutions in the state (Interview 8). SISA will be ultimately composed of a series of thematic programs, focused on priority areas, namely water resources conservation, socially beneficial biodiversity conservation, carbon and climate regulation, the valorization of traditional knowledge, and soil conservation and improvement.

The program related to reducing greenhouse gasses (GHG) emissions from deforestation and degradation within the planned SISA structure, named *ISA-Carbono*, was the first to be created and is currently the one undergoing the initial stages of implementation. SISA intends to generate tradable carbon credits as one of the means to finance its execution. Acre intends to validate their emission reductions through the application of independent auditors such as the Voluntary Carbon Standard (VCS) and the Climate, Community & Biodiversity Alliance (CCBA) (WWF 2013).

Monitoring activities within the SISA framework will be carried out by the Central Geoprocessing and Remote Sensing Unit (*Unidade Central de Geoprocessamento* – UCEGEO), which, with the support of SISA, intends to "monitor deforestation occurring at a smaller scale than the Brazilian national monitoring institution, the National Institute for Space Research (INPE), can detect [. . .] particularly important in Acre, where most deforestation occurs at a small scale." (Duchelle et al. 2014, p. 36).

Acre is among the smallest and poorest states in Brazil, so raising funds for conservation activities in the state has been one of the motivations and is one of the main concerns of SISA. The first funding source achieved within the context of SISA came from the German Development Bank (Kreditanstalt für Wiederaufbau – KfW), within the context of their REDD+[3] for Early Movers (REM) program. Of the resources provided by the first tranche of REM resources, 70% were directly provided to program beneficiaries and that the remaining 30% may be used for project's structuring activities (Acre 2015). The program has decided to use the beneficiary-related resources to further finance the rubber-tapper subsidy program (for more on this program see Sills and Saha (2010)) and to the productive restructuring of properties activities. The remaining resources will be applied to provide technical assistance for the institutions within SISA's structure, studies to improve SISA's monitoring and program development, and technical-scientific advice (Acre 2015).

13.4 Institutional Governance for Incentive-Based Forest Conservation: The Case of SISA-Acre

We conceptualize the design of SISA as a dual process of institutional creation and maintenance. The program has been described as "a pioneer initiative globally", having "faced the formidable challenge of having to innovate at every step of its development" (Duchelle et al. 2014, p. 47), but it clearly also perpetuates

[3]Reducing emissions from deforestation and forest degradation and the role of conservation, sustainable management of forests and enhancement of forest carbon stocks.

long-standing policy processes in the state's environmental sector. Additionally, several elements of the program were set up specifically to ensure long-standing institutional maintenance.

The history of environmental discourse formation and policymaking in the state yielded several institutional practices relevant to the design of SISA. Hence, Subsect. 13.4.1 deals with the historical background of SISA and Sect. 13.4.2 focuses on the details of SISA's own design. In both sections, we will identify instances of institutional governance carried out by the actors involved. Table 13.2 provides a summary of the section.

13.4.1 Environmental Policymaking in Acre: Activism and Innovation

Development plans for the Amazon region promoted by the Brazilian government in the 1970s attracted many investors from the south of Brazil, whose main activity was cattle ranching, which sometimes was used as a façade for land speculation (Bakx 1988). The arrival of the ranches caused clashes with Acre's rubber tappers, who had settled in the state since the rubber boom at the end of the nineteenth century. From 1974, the rubber tappers started to organize to protect their lands and livelihoods, facing increased violence (Keck 1995). During the 1980s, environmentalists' concern with negative impacts of large development programs in the forest reached out to partners of the rubber tappers, and both groups agreed that framing the rubber tappers' struggles within an appeal to save the forest would be beneficial to both (Keck 1995). The second half of the 1980s witnessed a rise in the profile of the rubber tappers, and they were viewed as guardians of the forest, personified in the rubber tappers' leader Chico Mendes (Keck 1995). Therefore, at the root of Acre's environmental policy history, we can observe *changing normative associations* and *identity construction* when the rubber-tapper movement started framing their struggle as not only one for land and labor but also for the conservation of the Amazon. Indeed, as stated by Hochstetler and Keck (2007, p. 109), "by the beginning of the 1990s most Brazilian environmental organizations had adopted a social environmental discourse, one that made poverty and environmental degradation part of the same causal story".

During this time, the proposal for the creation of 'extractive reserves' (*Reservas Extrativistas* – Resex) took shape, intended to be a protected forest area which would also "make rubber production economically viable, diversify economic survival strategies, and provide basic social infrastructure for those living on the reserve" (Keck 1995, p. 416). Mendes's assassination in 1988 and the subsequent national and international outcry strengthened the position of the rubber tappers and their supporters (Hochstetler and Keck 2007). The first concrete conquest of the *advocacy* work done by Acre's social movement was the creation by the federal government of two Resex in the state in 1990 (Hochstetler and Keck 2007). Acre's activists and

Table 13.2 Instances of institutional governance identified in the history of Acre's environmental policy and in SISA's design process

Creation	
Instances in SISA's history and design	**Institutional governance category**
The rubber-tapper movement re-framed their struggle as not only for land and labor but also for forest protection	Changing normative associations.
Creation of Resex	Constructing identities
	Advocacy
Acre's activists and their associates have consistently brokered alliances, especially with local and transnational actors, to highlight the importance of the rubber tappers' struggles	Constructing normative networks
Several rule systems were created in the 1990s by Acre's government to implement their intended policies	Defining
Creation of the governors' climate and forests task force	Constructing normative networks
SISA's switch from a program-based approach to a jurisdictional rules and norms system	Defining
Sustainable productive activities mimic the central tenets of ICDPs, as they are of interest to most stakeholders	Mimicry
Possible change from a holistic view of conservation towards conservation aimed at climate change mitigation	Changing normative associations
Maintenance	
Subsidies to rubber-tapping and other extractive activities to sustain an occupation that has a central symbolic role in the state's history	Mythologizing
SISA's legal status as a state law, instead of a weaker governmental decree	Defining
Diversion of resources to finance newly created conservation programs	Enabling work
Participation, a normative underpinning of Acre's grassroots environmentalism, was actively infused in SISA's multi-stakeholder governance structure	Embedding and routinizing
Welfare and conservation are seen as inextricably linked in Acre's government's discourse, and SISA's design features reflect that view	Embedding and routinizing
Great attention to monitoring and enforcement	Policing

Source: Own elaboration

their associates have also consistently brokered alliances, especially with local and transnational actors, to bring attention to the rubber tappers' struggles, successfully *constructing normative networks* that would be highly influential in the state's politics in the coming decades (Hochstetler and Keck 2007).

One result of that network construction was the emergence in the 1990s of what Lemos and Roberts (2008, p. 1899) call "donor-led socio-environmental management", meaning the implementation of conservation projects involving a "new coalition for sustainable development in the Amazon [that] included both international and domestic NGOs, supportive officials within the World Bank, western developed countries and Brazilian governments, epistemic communities of scientists, and grass-roots groups, such as rubber tappers and indigenous people". The establishment of the G7's Pilot Program to conserve the Brazilian rainforest (PPG7), for example, made significant amount of funding available for the implementation of conservation programs in the Amazon. The PPG-7, and other programs like Planafloro and Prodeagro, experienced implementation setbacks but helped to strengthen networks and the professionalization of environmental policies at national and subnational levels.

In addition to the nation-wide developments, during the 1990s the position of Acre's environmental movement continued to be strengthened, as "Mendes's murder had the effect of a centrifugal force, catapulting onto a national stage, and into political spaces that had been suddenly opened up, many of the people who had been associated with him in the struggles of rubber tappers in Acre" (Hochstetler and Keck 2007, p. 155). Among Mendes's associates who took up political spaces was Jorge Viana, a forest engineer with long-standing ties to the rubber tapper movement. In 1998 he was elected on the Worker's Party (Partido dos Trabalhadores – PT) ticket as the state's governor, and his party remained in power in the state until 2018.

Since 1999, Viana's government implemented policies that would set the tone for later developments in the state. Despite the establishment of extractive reserves, traditional populations were increasingly turning to other activities such as cattle-raising, as rubber tapping was not financially viable without governmental support (Sills and Saha 2010). For that reason, the government established the provision of subsidies for rubber production in the state, through the so-called "Chico Mendes law". While the subsidy was primarily aimed at improving the rubber-tappers living standards, it was expected to generate secondary benefits, such as forest conservation and reduced rural exodus (Sills and Saha 2010).

Also, in 1999, the state instituted the first Phase of its ZEE, further refined in 2006, which identified areas for conservation and others suitable for forest management, among others, and became the planning basis for most policies further implemented by the state (Acre 2010b). 2008 saw the creation of two policies that would become very relevant in the framing of SISA. The Valorization of the Environmental Assets (VEA) program aims to "support both the development of sustainable product chains and protection of the standing forest" (Schmink et al. 2014, p. 34). The Certification of the Sustainable Rural Properties law (CSRP) aims at incentivizing smallholders to gradually adopt sustainable land-use practices through the provision of financial incentives, rural technical assistance and preferential credit lines (WWF 2013).

Defining has been the key type of institutional creation work realized in Acre during this period, as several rule systems were created to implement the policies as intended by the government. As the new programs were building upon previous instances of institutionalization in the state (such as the presence of traditional activists in positions of power and the centrality of socio-environmentalism in the government's discourse), we can also identify elements of institutional maintenance. *Enabling work*, such as the diversion of resources to finance the VEA and CSRP programs, has also been consistently adopted in the period. It can also be argued that the implementation of VEA is an instance of *mythologizing*, as by keeping rubber-tapping and other extractive activities economically possible, the government has artificially sustained an occupation that has a central symbolic role in the state's history.

In 2003, the Partido dos Trabalhadores (PT) took the Brazilian presidency, politically aligning Acre and the federal government. In the same year, the government established the Bolsa Família conditional cash transfer (CCT) program. It consists of a monetary payment to families conditioned to school enrollment and vaccination of the family's children. Bolsa Família would later be influential in the adoption and acceptance of incentive-based environmental policies in the Amazon such as the Bolsa Floresta program in the Amazonas state (Bakkegaard and Wunder 2014).

While Acre was experimenting with those policies in the 2000s, new ideas regarding forest conservation were gaining traction in the international debate. The concept of ecosystem services became more common in environmental policy debates (Gomez-Baggethun et al. 2010). Incentive-based conservation policies, such as PES, became increasingly popular during that decade, especially due to their increasingly contested promise of a win-win outcomes for conservation and poverty alleviation (Muradian et al. 2013). The most influential of those conservation ideas was to integrate forest conservation the international climate change regime, aiming at generating financial resources through REDD+.

Government-led REDD+ activities in Brazil have been pioneered by the states, in the absence of a national REDD+ strategy (Gebara et al. 2014), which was only concluded in 2015. As such, the national policy can be considered a REDD+ implementation laggard in Brazil. SISA is not an isolated action at the state level, as exemplified by the participation of six Amazonian states in the Governors' Climate and Forests Task Force (Wunder and Duchelle 2014), which represent a concrete result of the *construction of a normative network*. The emergence of PES and REDD+ was an additional motivation for the Acre government to try tapping into the potential resources available for both initiatives by setting up a state incentive-based project. Therefore, in 2009, the government started working towards that goal, a process that will be described in the following section.

13.4.2 The Design of the SISA Program

In 2009, the government of Acre initiated the preparation of a proposal for a REDD+ project for the state, to be implemented in seven to eight priority areas of high deforestation, as identified by the ZEE (Duchelle et al. 2014). The initial proposal was subjected to an extensive consultation process (described in detail below), which sharply reshaped the initial proposal. The implementation in priority areas was deemed "insufficient and politically untenable, since areas with a low risk of deforestation also needed to be included for two main reasons: (i) people living in these areas should be rewarded for having conserved forests; and (ii) areas with a low deforestation risk could quickly become areas of high risk." (Duchelle et al. 2014, p. 39). The focus from a program-based approach to be carried out in priority areas was, thus replaced by a statewide jurisdictional approach, the first of its kind in Brazil, SISA serving as an umbrella for forest conservation activities implemented in the state. The change from a program-based approach to a system based on jurisdictional rules and norms is a clear instance of the *defining* type of institutional governance.

SISA performed a long participatory consultation process with the state's stakeholders. An initial version of the project prepared by the government was made available to scientists, local, national and international NGOs, leaders of potential beneficiary groups, fundraising and carbon market specialists, mayors, state and national assembly representatives, as well as other governmental agencies' representatives. The initial consultation process occurred between August 2009 and April 2010 and also involved meetings, workshops, and a technical seminar, generating 357 recommendations from stakeholders (Acre 2012). After 2 months spent in the analysis of the recommendations, reformulation of the program's initial draft, and drafting of the law project, a second consultation process was performed, lasting from July to October 2010 (Acre 2012). Figure 13.1 provides a breakdown of the issues addressed by the consultation process.

Stakeholder participation was not only a core feature of the design process, but it is also one of the pillars of SISA's governance, as evidenced by the centrality

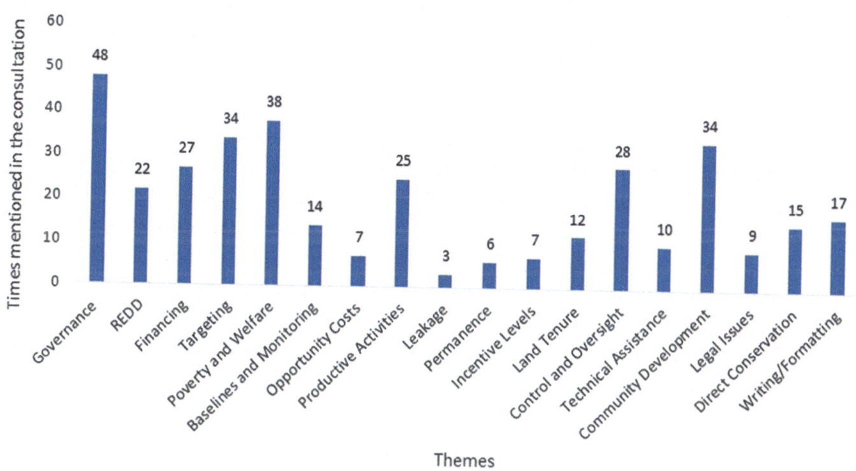

Fig. 13.1 Breakdown of themes addressed by SISA's consultation process. (Source: own elaboration based on Acre 2012)

of CEVA in decision making. The multi-stakeholder governance structure actively promoted by SISA can be seen as an instance of *embedding and routinizing*, as participation is central to the normative underpinnings of Acre's government's discourse. As described, the state's central political leaders constructed their political careers together with the emergence of the Acrean grassroots movement. Acre's governmental group has a historical association with the state's social movements (Hochstetler and Keck 2007), and much of the governing group's political discourse was built upon their connections with forest peoples, which have historically vied for being considered in the state's decision-making processes (Keck 1995).

One relevant remark related to the multi-stakeholder governance of SISA is that leaders of social movements in the state were given work positions within the government. That, according to a respondent (Interview 6) and one remark made during the consultation process (Acre 2012) might have generated some questioning about the allegiances of members of social movements, as they became part of the government. That situation led some to fear a weakening of the role of the civil society in the state, as leaders that before pushed the civil society's agenda ahead could now be discouraged to criticize the government in case of disagreement. That point was also raised by other organizations working in the state, which identified a "change in the nature of the relations of the social movements and the government", and a "strong cooptation of leaders, injection of public money in the civil society through agreements and salaries, generating a political and partisan binding" (IUCN et al. 2008, p. 29, author's translation).

Another instance of *embedding and routinizing* institutional governance is SISA's dual aim to conserve forests and improve the welfare of Acre's population. The association between social equity and environmental politics in Acre has been long identified by scholars (Keck 1995). The decades-long social struggle of rubber tappers in Acre was initially related more to the Brazilian labor movement than to environmental concerns (Keck 1995). Their struggle to ensure land-use rights (on the problem of land use rights see also Chapter Von der Mühlen et al. in this book) and fight encroachment by farmers eventually led, as stated above, to the creation of extractive reserves (Vadjunec et al. 2011). The strengthening of Acre's grassroots social movements, "whose participants had a strong dependence on Acre's abundant forest resources for their livelihoods, and a well-defined common identity as rubber tappers" (Kainer et al. 2003, p. 875), both in local politics and in its international image further solidified the normative importance of linking welfare and conservation in the government's discourse. Several elements of SISA's design reproduce this stance, and it was constantly embedded and routinized in the state's policy and discourse.

Poverty reduction and welfare improvement concerns permeate the whole design of SISA. Benefit-sharing is stated as one of the guiding principles of SISA and is one its objectives (Acre 2010a). Figure 13.1 above also shows that poverty and welfare issues were the second most frequent concerns of the process' participants, indicating its importance to stakeholders. Various respondents also emphasized that reconciling environmental conservation with welfare concerns is one of the core philosophies of state policies in Acre. Hence, policies that seek to conserve forests are considered incomplete if they lack a component to ensure that their benefits also reach forest-dependent people (Interviews 1, 2, 8).

Sustainable production activities, similar to integrated conservation and development programs (ICDPs), with the aim of generating income while conserving forests, are central to the state's conservation strategy, despite their questionable effectiveness (Ferraro and Kiss 2002; Wunder 2005). Recommendations from the Brazilian Agricultural Research Company (*Empresa Brasileira de Pesquisa Agropecuária* - EMBRAPA), for example, state that conservation incentives may originate "from the adoption of sustainable practices" (Acre 2012, p. 57, author's translation). The SISA team implicitly agreed by arguing that "the term 'payments for environmental services' does not reveal the innovative dimension of the program" (Acre 2012, p. 57, author's translation) and that it was important to make it clear that the project supports the adoption of production techniques that seek to have a lower environmental impact than traditional forms of agricultural production, while not compromising the income-generation purpose of the rural properties (Acre 2012). Also, potential local beneficiaries favored that type of intervention: the National Confederation of Rubber-Tappers (*Confederação Nacional dos Seringueiros* - CNS) worried that "the problem with incentives, for example, the rubber subsidy, is that the rubber tappers get complacent and, when the incentive ends, they want to move to the city". In addition, local population representatives claimed that "what the people want are incentives to generate income and not deforest, we do not want just payments for not deforesting" (Acre 2012, p. 58, author's translation). The quotes above from the consultation process show a clear interest to associate the new (the intended innovative nature of SISA) with the old (a taken for granted practice such the implementation of ICDP-like activities) to "make the new structure understandable and accessible" (Lawrence and Suddaby 2006, p. 226).

The idea that the programs can tackle deforestation and poverty at the same time is also very attractive to external actors, especially international donors. The international debate on the establishment of a global REDD+ regime gained prominence concomitantly with SISA's design period. The design team and the stakeholders consulted in the participation process took notice of the emergence of REDD+ as a potential tool for financing conservation activities in the state. That represented an additional incentive for the initial proposal of a government-led PES project in the state to morph into a system with a strong intention to relate to the REDD+ international debate. SISA was intentionally designed to insert itself in a future REDD+ regime, incorporating much of the terminology used in REDD+ (i.e. safeguards, register, carbon credits) and showing commitment to generate certifiable emissions that can later be negotiated in carbon markets or used as the basis for the receipt of grants by international organizations.

It is likely that this possibility of drawing additional financial resources to the state is the most attractive aspect of integrating activities with a budding REDD+ regime for a relatively poor, isolated jurisdiction like Acre, with fewer resources to implement an ambitious and likely expensive large-scale conservation program. In addition, as stated in the consultation report, the design team understood that the term REDD+ could bring communication benefits (Acre 2012), furthering the state's attractiveness for potential donors.

SISA's broader idea is to address several environmental services, but the prominence of REDD+ in the international conservation policy arena made it clear that climate change mitigation would be the priority. The prioritization of carbon might indicate a *change in normative associations*, from long-standing local discourses and practices founded on a more holistic view of conservation towards conservation actions connected with climate change as the current issue at the top of the international environmental agenda. Some authors have indeed pointed out the possibility that incentive-based conservation may represent the commodification of nature (Kosoy and Corbera 2010; Norgaard 2010) or crowd-out intrinsic motivations for forest protection (Muradian et al. 2013; Rode et al. 2015). It may, on the other hand, be a more pragmatic decision since, as put by Lemos and Roberts (2008, p. 1901) "the networks forming around the issue of climate/forests are constantly repositioning themselves in terms of political opportunity, international pressure, and organizational resources".

Acre's decision to attract long-term financing for SISA through the generation of certified emissions reductions required strict methodologies for forest conservation. Acre's already existing monitoring activities form a solid base for deforestation monitoring of SISA's activities. Most of the monitoring infrastructure in the state had been created by previous policy efforts in tandem with the federal government's monitoring structure, but SISA intends to further improve the state's capacity with the creation of UCEGEO. Monitoring and the participation of PGE in SISA's executive committee represent instances of *policing* as they are necessary to ensure compliance of the conservation objectives.

Despite the lack of guaranteed long-term financing, Acre's government was committed to making sure that SISA should be as strongly institutionalized as possible. For that reason, SISA was established through a state law, a clear instance of *defining* institutional governance. Being legally defined as a law, instead of a weaker governmental decree, as it was the case in other contemporary incentive-based policies in the Amazon (Rosa da Conceição et al. 2015), may increase the prospects of SISA remaining a state policy in the long-term. The law proposal was discussed in the state's legislative assembly and passed without much opposition (Acre 2012).

13.5 Discussion and Conclusions

This chapter analysed the creation of an institution (SISA) as a process of turning ideas and interests into concrete rules and norms. Making such processes transparent is a precondition for understanding the making of policies including the constraints that limit the transformation of scientific knowledge into action. Through the lens of institutional governance, our findings support the conjecture that new formal institutions are seldom created from scratch. Rather new ideas must be moulded to accommodate previously existing informal institutions and to attend to the need of ensuring that newly created structures can be maintained in the future.

Acre is seen not only as a symbol of Brazilian environmentalism but as an example of forward-thinking, innovative and effective forest governance. Indeed, in a recent evaluation of 39 subnational jurisdictions, Stickler et al. (2018) rated Acre as the only one in an advanced stage in the development of successful low carbon emission development strategies. May et al. (2016) also state that "today, Acre is considered by some REDD+ investors at national and international levels as being able to provide compliance-grade emission reductions". Positive signs can also be identified in the structure of SISA in the attempt to integrate welfare concerns in cost-efficient conservation policies. The embedded structure designed in Acre may, for example, mean that a subprogram or project will target areas with high potential to achieve additional emission reductions, while respecting welfare-related safeguards established by the government. With SISA being a core component of Acre's low carbon emission development strategy, our study case thus stands out as a good practice example, well positioned to produce cost-effective and socially acceptable climate change mitigation outcomes.

Whether Acre's approach will be adopted by other jurisdictions remains uncertain, however. Our analysis highlighted its idiosyncratic historical development and a peculiar setting of administrative and political economy elements. The pool of policy-makers in SISA was small and closely connected, being largely part of the high-level bureaucracy of the state for years and sharing similar priorities and ideologies. The stability and cohesion of Acre's governing and administrative group have without a doubt played a role in SISA's design. Additionally, when environmental issues are higher on the governments' agendas, environmental agencies tend to have more cooperative relations with the rest of the government, and their priorities will be more in line with the work of other sectors, as in the case of Acre. Clearly, Acre boasts a less diverse system of economic and political actors than many other Brazilian jurisdictions. This relative uniformity of interests reduces pressure on sectoral agencies and perhaps explains the perception of high policy coordination levels in the state. It is, however, unclear if SISA's design process translates into effective and stable institutional mechanisms in the long run.

In fact, Acre's deforestation rates have been largely following those of other Northern Brazilian states, despite the state's policies and discourse. Additionally, SISA is experiencing an extremely slow starting phase, showing that a more thorough design may indeed hinder swifter implementation. It remains to be seen in the medium term how much Acre's design model will be able to fulfill its own requirements, and if they will generate the desired conservation gains.

The recent electoral defeat in the October 2018 election ended the Worker Party's 20-year lasting government in the Acre, and this may affect the future of SISA. As we have seen, SISA's creation is tightly connected with PT's government and the history of its members. The new government will likely be completely different, both in terms of personnel and ideology, so the fate of SISA might be in check. Will the new government carry on with strengthening the system? If not, will it slowly disenfranchise the system, or quickly revoke the law that created

SISA? Will the measures undertaken by the government to ensure the stability of SISA survive the new administration? Who will be the actors involved in any of these scenarios? Considering the short-term view of much public policy in Brazil, observing the fate of SISA may provide valuable insights on the reasons why a public policy 'sticks' across opposed administrations or if it gets eliminated as soon as its creators step out. PT was in power for almost 20 years in Acre, providing a political continuity that is unusual for Amazonian states in Brazil, hinting at a strong electoral support for Acre's "socio-environmental" discourse and policies. The recent changes indicate erosion in that support, and the following years will be key in determining whether future scholars of institutional governance will select SISA as a case of institutional maintenance or disruption.

Appendix – List of Interviews

Acre

1. High-level decision maker
2. High-level decision maker
3. SISA staff
4. SISA staff
5. SISA staff
6. Foreign technical cooperation staff
7. NGO staff
8. SISA staff
9. Environment secretariat staff
10. SISA staff
11. Family production secretariat staff
12. Family production secretariat staff

References

Acre, Governo do Estado do (2010a) Lei 2308 – Cria o Sistema Estadual de Incentivos a Serviços Ambientais – SISA, o Programa de Incentivos por Serviços Ambientais – ISA Carbono e demais Programas de Serviços Ambientais e Produtos Ecossistêmicos do Estado do Acre e dá outras providências. Rio Branco.

Acre, Governo do Estado do (2010b) Zoneamento Ecológico-Econômico do Acre – Fase II – Documento Síntese

Acre, Governo do Estado do (2012) Construção Participativa da Lei do Sistema de Incentivos a Serviços Ambientais – SISA do Estado do Acre. IMC Acre, Rio Branco

Acre, Governo do Estado do (2015) Relatório de Gestão 2015. Instituto de Mudanças Climáticas e Regulação de Serviços Ambientais, Rio Branco

Bakkegaard RK, Wunder S (2014) Bolsa Floresta, Brazil. In: Sills E (ed) REDD+ on the ground. A case book of subnational initiatives across the globe. Centre for International Forest Research (CIFOR), Bogor Barat, pp 51–67

Bakx K (1988) From proletarian to peasant: rural transformation in the state of Acre, 1870–1986. J Dev Stud 24(2):141–160. https://doi.org/10.1080/00220388808422060.

Duchelle A, Greenleaf M, Mello D et al (2014) Acre's state system of incentives for environmental services (SISA), Brazil. In: Sills E (ed) REDD+ on the ground. A case book of subnational initiatives across the globe. Centre for International Forest Research (CIFOR), Bogor Barat

Ferraro P, Kiss A (2002) Direct payments to conserve biodiversity. Science 298(5599):1718–1719

Gebara MF, Fatorelli L, May P, Zhang S (2014) REDD+ policy networks in Brazil. Constraints and opportunities for successful policy making. Ecol Soc 19(3):53. https://doi.org/10.5751/ES-06744-190353

Gomez-Baggethun E, de Groot R, Lomas P, Montes C (2010) The history of ecosystem services in economic theory and practice. From early notions to markets and payment schemes. Ecol Econ 69:1209–1218. https://doi.org/10.1016/j.ecolecon.2009.11.007.

Hochstetler K, Keck M (2007) Greening Brazil. Environmental activism in state and society. Duke University Press, Durham

IUCN, GTA, WWF (2008) Análise da Participação da Sociedade Civil e da Governança de Cinco Espaços de Definição de Políticas Públicas do Estado do Acre

Jespersen K, Gallemore C (2018) The institutional work of payments for ecosystem services: why the mundane should matter. Ecol Econ 146:507–519. https://doi.org/10.1016/j.ecolecon.2017.12.013

Kainer K, Schmink M, Pinheiro AC et al (2003) Experiments in forest-based development in Western Amazonia. Soc Nat Resour 16:869. https://doi.org/10.1080/08941920390231306.

Keck M (1995) Social equity and environmental politics in Brazil. Lessons from the rubber tappers of acre. Comp Polit 27(4):409–424

Kosoy N, Corbera E (2010) Payments for ecosystem services as commodity fetishism. Ecol Econ 69(6):1228–1236. https://doi.org/10.1016/j.ecolecon.2009.11.002

Lawrence T, Suddaby R (2006) Institutions and institutional work. In: Clegg S, Hardy C, Lawrence T, Nord W (eds) Sage handbook of organization studies, 2nd edn. Sage, London

Lawrence T, Suddaby R, Leca B (2009) Introduction: theorizing and studying institutional work. In: Lawrence T, Suddaby R, Leca B (eds) Institutional work: actors and agency in institutional studies of organizations. Cambridge University Press, Cambridge, pp 1–26

Lemos MC, Roberts JT (2008) Environmental policy-making networks and the future of the Amazon. Philos Trans Royal Soc Lond Ser B Biol Sci 363(1498):1897–1902. https://doi.org/10.1098/rstb.2007.0038

May P, Gebara MF, Barcellos LM, Rizek MB, Millikan B (2016) The context of REDD+ in Brazil. Drivers, agents, and institutions, 3rd edition

Muradian R, Arsel M, Pellegrini L et al (2013) Payments for ecosystem services and the fatal attraction of win-win solutions. Conserv Lett 6(4):274–279

Norgaard RB (2010) Ecosystem services. From eye-opening metaphor to complexity blinder. Ecol Econ 69(6):1219–1227. https://doi.org/10.1016/j.ecolecon.2009.11.009

Reid E (2000) Understanding the word "advocacy": context and use. In: Reid E (ed) Structuring the inquiry into advocacy. The Urban Institute, Washington

Rode J, Gómez-Baggethun E, Krause T (2015) Motivation crowding by economic incentives in conservation policy. A review of the empirical evidence. Ecol Econ 117:270–282. https://doi.org/10.1016/j.ecolecon.2014.09.029.

Rosa da Conceição H, Börner J, Wunder S (2015) Why were upscaled incentive programs for forest conservation adopted? Comparing policy choices in Brazil, Ecuador, and Peru. Ecosyst Serv 16:243–252

Schmink M, Duchelle A, Hoelle J et al (2014) Forest citizenship in Acre, Brazil. In: Katila P, Galloway G, de Jong W, Pacheco P, Mery G (eds) Forests under pressure: local responses to global issues, vol 32. IUFRO (IUFRO World Series, Vienna, pp 31–48

Sills E, Saha S (2010) Subsidies for rubber. Conserving rainforests while sustaining livelihoods in the Amazon? J Sustain For 29(2–4):152–173. https://doi.org/10.1080/10549810903543907
Stickler C, Duchelle A, Ardila JP et al (2018) The state of jurisdictional sustainability: a summary for practitioners and policymakers. EII, CIFOR, GCF, San Francisco
Tansey O (2007) Process tracing and elite interviewing. A case for non-probability sampling. Political Sci Politics 40(4):765–772
Vadjunec J, Schmink M, Gomes CV (2011) Rubber tapper citizens. Emerging places, policies, and shifting rural-urban identities in Acre, Brazil. J Cult Geogr 28(1):73–98. https://doi.org/10.1080/08873631.2011.548481
Wunder S (2005) Payments for environmental services – some nuts and bolts. CIFOR, Bogor. Occasional Paper
Wunder S (2015) Revisiting the concept of payments for environmental services. Ecol Econ 117(1):234–243
Wunder S, Duchelle A (2014) REDD+ in Brazil: the national context. In: Sills E (ed) REDD+ on the ground. A case book of subnational initiatives across the globe. Centre for International Forest Research (CIFOR), Bogor Barat, pp 31–32
World Wildlife Fund (WWF) (2013) Environmental service incentive system in the state of acre, Brazil – lessons for policies, programmes and strategies for jurisdiction-wide REDD+. WWF

Conclusion

Pablo J. Donoso and Felix Fuders

The current crisis of natural resources conservation and management is at the heart of global change. While we need to keep oil under the ground if we expect to keep future temperature raise below the 2.0 °C threshold (Paris accord; Overpeck and Conde 2019), we also need to mitigate climate change effects by diverse means, and one of major relevance and impact could be through carbon sequestration in forest ecosystems. However, these forest ecosystems are also exposed to changes and uncertain future disturbances, such as those triggered by climate change, and to adapt to these new scenarios ecosystems need to be resilient, which depends upon many factors. Olsson et al. (2004) suggest that institutional and organizational landscapes should be given the same importance as ecological matters in order to clarify features that contribute to the resilience of social-ecological systems, such as forests. These features include vision, leadership, trust, collaboration, appropriate legislation for ecosystem management, funding, monitoring, information flow through social networks, and knowledge. This book has precisely conceived that forest conservation problems have profound roots that relate to a lack of transdisciplinary visions towards natural resource management, and to inappropriate economic and institutional foundations that inhibit adaptation.

Chapters in this book have aimed to illustrate some of the forest conservation problems and challenges in two very different countries in South America, Brazil and Chile. We have intended to provide a perspective to understanding the problems of forest conservation in these countries mainly from the point of view of transdisciplinarity, ecological economics, and governance. We think these perspectives

P. J. Donoso
Institute of Forests and Society, Universidad Austral de Chile, Valdivia, Chile
e-mail: pdonoso@uach.cl

F. Fuders
Economics Institute, Faculty of Economics and Administration, Universidad Austral de Chile, Valdivia, Chile

F. Fuders and P. J. Donoso (eds.), *Ecological Economic and Socio Ecological Strategies for Forest Conservation*, https://doi.org/10.1007/978-3-030-35379-7

help to start viewing natural resource management as a challenge that requires collaboration, generosity, passion, and new perspectives and models. As clearly stated in one chapter, historical examples of social downfalls, or even the collapse of entire civilizations, which are likely to be triggered during the current century, have had the common cause of a crisis of perception, i.e. a cognitive disconnection of a specific culture with the socio-environmental context from which it arose. This disconnection is revealed today by a dominant economic model based on fossil fuels and historical high levels of consumption which are seen as the drivers that are damaging the planet and, especially, increasing its temperatures. We already perceive unprecedented climate disturbances that in the near future could unleash huge damages to human populations, biodiversity and entire ecosystems.

To understand the principal drivers of the overexploitation of our natural resources in general, and of our forests in particular, it is crucial to understand that the overexploitation is strongly driven by the obligation to grow the real economy. The ever-growing levels of consumption and energy throughput can be regarded as symptoms of this growth imperative. It was explained that this obligation to grow is significantly triggered by the money interest rate as an opportunity cost of any productive investment, including investments in the exploitation of natural resources. This pressure to grow applies to private goods as well as to freely accessible ones. In other words, regardless of whether or not natural resources are freely accessible goods, they will be overexploited in the long term because of this logic of growth. The obligation to grow can be seen as the primary and transversal failure of our economic model, though it is not recognized as such in the conventional teaching of economic sciences.

Given this growth imperative, we might take into consideration a reform of resource property law that allows to protect natural resources and, additionally, to promote justice. While in the classic liberal view all production factors –even nature– are privately owned, and while on the contrary in Marxism all production factors are commonly owned, it is time to think of a scheme of defining property rights as it had been envisioned already by John Stuart Mill, Henry George and Silvio Gesell: only for the products of humans private property should exist, while for pure nature-made products (land, soil, water, air, mineral resources, virgin forests, oceanic fishing grounds), where humans have not added any value, private property should not exist. This would not only be fair but improve allocative efficiency and avoid monopoly rents. Possible regulation policies depending on the nature of the resource to come as close as possible to this rule are: (i) concessions and quotas, (ii) leasing of land and coastal management areas and (iii) private property but levying monopoly rents.

Contents in this book have aimed to illustrate some locally successful experiences of forest governance that have been developed in some regions (especially in Brazil) and critical views of some dominant contexts of forest governance that continue to prevail at national scales (such as in Chile), including proposals for change. This book therefore represents one of multiple efforts that many researchers have conducted everywhere to move things to eventually achieve increased levels of forest conservation along with social and economic satisfaction of forest landowners and local societies in general. In this sense, Franklin et al. (2018) mention that the five major economic, social and political drivers of forest conservation and management include ownership, markets, investment/income strategies, policy formation processes, and social acceptability. A country or region where forest management strategies are not only environmentally and ecologically sound, but also responsive to an increasingly diverse set of demands by society and local communities, is one in the ultimate stages of forest development, the Social stage (Kimmins 1997).

A bit more than a decade ago Donoso and Otero (2005) made an analysis to evaluate where Chile was located in terms of development in its forestry sector according to Kimmin's classification and determined that it can be found in between the stages "Regulation" and "Sustainable management". When tensions continue to dominate the social scape of the forestry sector, as currently is the case in Chile as well as in Brazil, this means that at least one of the social-economic-political drivers mentioned above is failing. When searching for the ultimate causes of that failure, it will be apparent that any economic system based on steady economic growth, consumption and, thus, overexploitation of natural resources will make at least one of the three pillars of sustainable development (social, economic, environmental; Purvis et al. 2018) to be the looser, and generally the winner will be the economic component, but usually benefiting a minority.

The different authors in this book have intended to go a little further in regard to understanding forestry conflicts in countries like Brazil and Chile, which are in many ways representative of Latin American and other developing countries. The different chapters have basically dealt around the needs to embrace novel economic approaches to managing forest resources along with mutual understanding and collaboration of different disciplines (transdiciplinarity) to move in the direction of achieving sustainable forestry sectors.

As a sort of overall conclusion, we could say that sustainable forest economies in Brazil and Chile (and elsewhere) will be reached when they meet solid forest governance, i.e. when organizations, people, rules, instruments and processes through which decisions are made relating to forests are in balance. However, when trying to answer why in so many regions of the world forest degradation continues at a rapid pace, an ultimate explaining factor seems to lay in the existing incoherencies between neoclassical economics and natural sciences. Nearly two decades ago, Mery et al. (2001) compared forest economies of Chile, Brazil and Mexico, and observed a common element in them in the sense of lacking coherent long-term forest policies and programs that would take into account the interests of all key stakeholders, a source of conflicts that could jeopardize the good forest development prospects of these countries. They mentioned by then that a definition of an optimal mix of markets and policies is an important pending issue in the three countries. This mix needs to develop over the foundations of an economic system that can assure lower impacts of human activities upon forest resources as a key to increasing ecosystem services from forests, including carbon sequestration in current and future forests.

Past and current learning in both regions (and elsewhere), however, set the stage to continue pushing the forestry sectors of these countries and regions towards models conducive for more social justice in landscapes where forest conservation and management should provide multiple benefits, especially ecosystem goods and services in the long term. In other words, ecological economic and socio ecological strategies for forest conservation are urgently needed especially at this juncture of planetary crisis where forests can play such important roles for mitigation of climate change and welfare of local communities and societies in general.

References

Donoso PJ, Otero L (2005) Hacia una definición de país forestal: ¿Dónde se sitúa Chile? Bosque 26(3):5–18

Franklin JF, Johnson KN, Johnson DL (2018) Ecological forest management. Waveland 552 Press, Inc, USA

Kimmins H (1997) Balancing act: environmental issues in forestry, 2nd edn. UBC Press, Vancouver

Mery G, Kengen S, Lujim C (2001) Forest-based development in Brazil, Chile and Mexico. In: Palo M, Uusivuori J, and Mery G (eds) World forests, markets and policies. Springer Science+Business Media, B.V., pp 243–262

Olsson P, Folke C, Berkes F (2004) Adaptive co-management for building resilience in social–ecological systems. Environ Manag 34(1):75–90

Overpeck JT, Conde C (2019) A call to climate action. Science 364(6443):807

Purvis B, Mao Y, Darren R (2018) Three pillars of sustainability: in search of conceptual origins. Sustain Sci. https://doi.org/10.1007/s11625-018-0627-5. (0123456789().,-volV)(0123456789().,-volV)

Index

F. Fuders and P. J. Donoso (eds.), *Ecological Economic and Socio Ecological Strategies for Forest Conservation*, https://doi.org/10.1007/978-3-030-35379-7

The manufacturer's authorised representative in the EU is Springer Nature Customer Service Centre GmbH, Europaplatz 3, 69115 Heidelberg, Germany. If you have any concerns regarding our products, please contact ProductSafety@springernature.com

Printed and bound by CPI Group (UK) Ltd, Croydon, CR0 4YY
23/09/2026
02221542-0007